AF539961

Nanomaterials And Nanostructures

NANOMATERIALS AND NANOSTRUCTURES

By

Darren Samuels

2016

SBS Publishers & Distributors Pvt. Ltd.
New Delhi

ISBN 13 : 9789380090610

First Published in 2016

Published by:

SBS PUBLISHERS & DISTRIBUTORS PVT. LTD.

2/9, Ground Floor, Ansari Road, Darya Ganj,

New Delhi - 110002,

INDIA

Tel: 0091.11.23289119 / 41563911

Email: mail@sbspublishers.com

www.sbspublishers.com

Preface

Nanomaterials are one of the main products of nanotechnologies – as nano-scale particles, tubes, rods, or fibres. The properties of nanomaterials are not always well-characterised, and they call for risk assessment of possible exposures arising during their manufacture and use. Nanomaterials research takes a materials science-based approach to nanotechnology, leveraging advances in materials metrology and synthesis which have been developed in support of microfabrication research. Materials with structure at the nanoscale often have unique optical, electronic, or mechanical properties. A nanostructure is a structure of intermediate size between microscopic and molecular structures. In describing nanostructures it is necessary to differentiate between the number of dimensions on the nanoscale. Nanotextured surfaces have one dimension on the nanoscale, i.e., only the thickness of the surface of an object is between 0.1 and 100 nm. Nanotubes have two dimensions on the nanoscale, i.e., the diameter of the tube is between 0.1 and 100 nm; its length could be much greater. Finally, spherical nanoparticles have three dimensions on the nanoscale, i.e., the particle is between 0.1 and 100 nm in each spatial dimension. The terms nanoparticles and ultrafine particles (UFP) often are used synonymously although UFP can reach into the micrometre range. The term 'nanostructure' is often used when referring to magnetic technology. Nanomaterials and Nanostructures provides balanced and comprehensive coverage of the fundamentals and processing techniques with regard to synthesis, characterization, properties, and applications of nanostructures and nanomaterials. Both chemical processing and lithographic techniques are presented in a systematic and coherent manner for the synthesis and fabrication of 0-D, 1-D, and 2-D nanostructures, as well as special nanomaterials such as carbon nanotubes and ordered mesoporous oxides.

Editor

Contents

Chapter 1

FMR MEASUREMENTS OF MAGNETIC NANOSTRUCTURES

Manish Sharma, Sachin Pathak and Monika Sharma

Indian Institute of Technology Delhi, India

INTRODUCTION

Ferromagnetic nanowires showed solitary and tunable magnetization properties due to their inherent shape anisotropy. The fabrication of such nanowires in polycarbonate track-etched and anodic alumina membranes have been widely studied during the last 15 years [1-2]. Their potential applications might be explored in spintronic devices and more specifically in magnetic random access memory (MRAM) and magnetic logic devices [3-5]. Furthermore, microwave devices, such as circulators or filters for wireless communication and automotive systems can be fabricated on ferromagnetic nanowires embedded in AAO substrates [6-9]. This chapter begins with a brief overview of the historical development of the theory of ferromagnetic resonance in magnetic nanostructures. State-of-the-art calculations for resonance frequency in ferromagnetic nanowires (solid and hollow) and multilayer nanowires are presented. In addition, experimental approach to synthesis such structures and detecting material properties using various techniques will be discussed in brief. Recently, due to the development of spintronics, there have been increasing interests in the microwave dynamics of one-dimensional structures such as nanowires and two dimensional structures like multilayer magnetic films. The most important parameters that control dynamic behaviors are the internal fields and damping constant. The ferromagnetic nanowires in anodic alumina (AAO) templates seem to be attractive substrates for microwave applications. Since they have high aspect ratio, electromagnetic waves can easily penetrate through them. They exhibit ferromagnetic resonance (FMR) even at zero bias fields and, due to their high saturation magnetization, operating frequency can be tuned with DC fields.

FMR is a useful technique in the measurement of magnetic properties of ferromagnetic materials. It has been applied to a range of materials from bulk ferromagnetic materials to nano-scale magnetic thin films and now a day's people have started research to characterise nanoparticles and nanowires systems. The dynamic properties of magnetic materials can be easily perturbed by ferromagnetic resonance (FMR), as they can excite standing spin waves due to magnetic pinning [10-11, 31, 32]. It also yields direct information about the uniform precession mode of the nanowires which can be related to the average anisotropy magnitude [12-15]. Several measurement techniques which have been used to characterise magnetization dynamics such as femtosecond spectroscopy [16-18], pulse inductive microwave magnetometer [19], FMR force microscopy [20], network analyzer FMR [21,31,32] and high-frequency electrical measurements of magneto dynamics [22]. All these techniques can be used for modern application for example in telecommunications and data storage systems. In the present chapter, we deal with magnetic nanostructures such as nanowires and multilayered nanodiscs and rings which exhibit unique FMR responses since their various anisotropy energies are strongly influenced by size and shape [33]. In addition, interactions between multilayered segments can be tailored such that the FMR response is not only angle-dependent but also influenced strongly at certain frequencies [32]. In this chapter, we shall be describing three different aspects of this topic:

1. We first develop the theory of FMR response of densely packed nanowire arrays that can be treated as two-dimensional periodic nanostructures. We then extend the theory to three-dimensional structures, which can be made using multilayered nanowires.
2. We then describe how to synthesize such nanowire arrays and also direct measurements of such structures. Several different experimental techniques are discussed.
3. We then continue the treatment to describe the use of such periodic nanowire arrays in microwave devices to exhibit nonlinear responses and also for circulators and isolators. Although the effects seen till now are weak, these are still quite promising.

FUNDAMENTAL THEORY OF FERROMAGNETIC RESONANCE

Ferromagnetic resonance (FMR) is a very powerful experimental technique in the study of ferromagnetic nanomaterials. The precessional motion of a magnetization M of ferromagnetic material about the applied external magnetic field H is known as the Ferromagnetic resonance (FMR). In the physical process of resonance, the energy is absorbed from rf transverse magnetic field hrf, which occurred when frequency matched with precessional frequency w. The precession frequency depends on the orientation of the material and the strength of the magnetic field. It allows us to measure all the most important parameters of the material: Curie temperature, total magnetic moment, relaxation mechanism, elementary excitations and others. A single domain magnetic particle with ellipsoid shape was considered as in [Kittel] to drive the resonance condition for the phenomenon of ferromagnetic resonance. A uniform, static magnetic field H is applied along the z-axis and set the sample in a microwave cavity. A resonance is observed at a frequency given by

$$\hbar\omega = g\mu_B\sqrt{[H+(N_x-N_z)M][H+(N_y-N_z)M]}. \tag{1}$$

N_x, N_y, N_z are the demagnetization factors in the x, y, z directions, and so on, and where g is the spectroscopic splitting factor (Lande factor) and μB = eħ/2me is the Bohr magneton. The demagnetizing factors affect the shape anisotropy of the magnetic material depending upon its geometry to be ellipsoid, sphere, thin film etc.

Part of the classical approach to ferromagnetism is to replace the spins by a classical microspin vector M magnetization. The time-dependence of the magnetization can be obtained directly by calculating the torque acting on M by an effective field H_{eff},

$$\frac{dM}{dt} = -\gamma M \times H, \tag{2}$$

where λ = gμb/ħ is a gyromagnetic ratio. This equation represents an undamped precession of the magnetization. From experiments actual changes of the magnetization are known to decay in a finite time. The occurrence of a damping mechanism leads to reversal of the magnetization towards the direction of H within several nanoseconds. The damping is just added as a phenomenological term to Eq. 2.

$$\frac{dM}{dt} = -\gamma M \times \left(H_{eff} - \frac{\alpha}{\gamma M_s}\frac{dM}{dt}\right) = -\gamma M \times H_{eff} + \frac{\alpha}{M_S} M \times \frac{dM}{dt} \tag{3}$$

where α is the dimensionless Gilbert damping constant, of order 10-2 in ferromagnetic thin films. Eq. 3 is known as Landau-Lifshitz-Gilbert equation after Gilbert introduces the damping term. Heff is the total effective magnetic field which is a sum of static applied magnetic field (H), dynamic magnetic field (hrf) and internal magnetic field (Hin). Internal field constitutes various magnetic anisotropies such as magneto crystalline anisotropy, shape anisotropy, and magneto elastic anisotropy etc. This equation therefore describes torque acting on M. This torque leads to a rotation of the magnetization towards the direction of the external magnetic field. The damping causes decay in precessional motion which by applying a dynamic magnetic field becomes continuous as shown in Fig. 1.

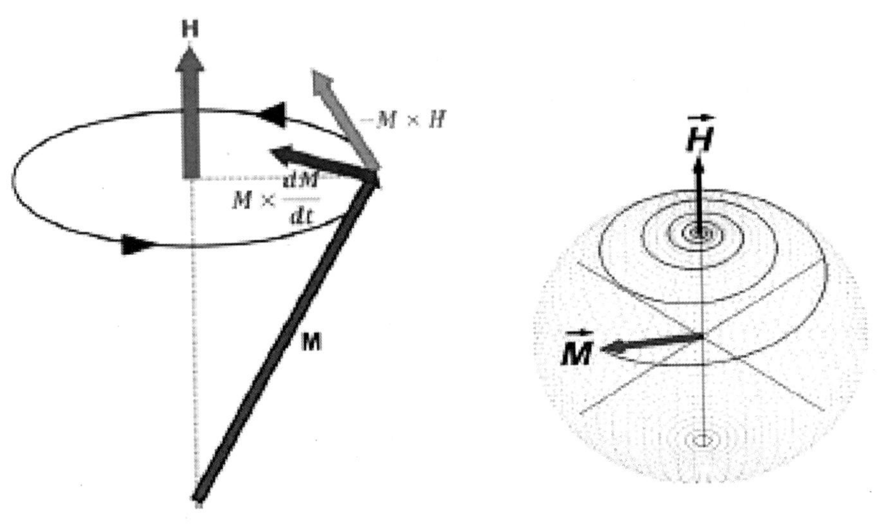

Figure 1 (a): Torque components exerted on the magnetization M by rotational field H (b) Motion of M for constant H.

FMR DISPERSION RELATION FOR NANOSTRUCTURES

Due to miniaturization from bulk to the nanoscale, material properties shows a drastic change like surface-to-volume ratio, electron transport, thermodynamic fluctuations, defects etc. The nanomagnetic material demonstrates distinct magnetic response due to various anisotropies. The one-dimensional nanostructures such as nanowires, nanotubes, nanorods and nanorings are the area of recent research for data storage applications, sensors, biomedical drugs, microwave devices. To perturb the dynamic magnetization in these structures, ferromagnetic resonance is an effective tool. In case of a system which incorporates an array of nanowires dipole-dipole interaction, shape anisotropy, crystalline anisotropy and Zeeman energy interaction plays a complex role. Therefore the total energy will be the sum of all internal energies.

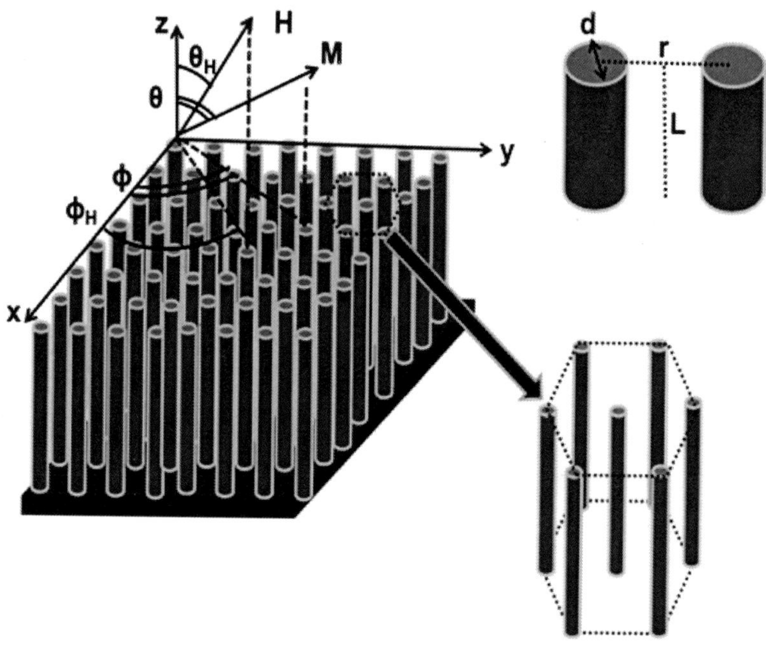

Figure 2: Coordinate system for an array of nanowires.

Fig. 2 shows the schematic of an array of nanowires with relative orientation of the magnetization M and the applied magnetic field H w.r.t nanowire axis in spherical coordinate systems. The free energy density equation for an array of magnetic nanowires in the presence of external magnetic field at angles θ_H from nanowires axis can be written as

$$E \approx -MH\left(\sin\theta \sin\theta_H \cos\left(\phi - \phi_H\right) + \cos\theta \cos\theta_H\right) + K_{eff} \sin^2\theta \tag{4}$$

where K_{eff} is the effective uniaxial anisotropy which can be written as

$$K_{eff} = \pi M^2\left(1 - 3P\right) + K_U \tag{5}$$

The first term includes the dipole-dipole interaction between the nanowires and second term represents second-order uniaxial anisotropy along the wire axis. P is the porosity which can be obtained from

$$P = \frac{\pi}{2\sqrt{3}} \frac{d^2}{r^2} \tag{6}$$

where d is the diameter of the pore and r is the centre-to-centre inter-wire distance between the pores. The equilibrium values for polar angles are obtained by minimizing the energy term

$$E_\theta = \frac{\partial E}{\partial \theta} = 0 \tag{7}$$

$$E_\theta = -MH\left(\cos\theta \sin\theta_H \cos(\phi - \phi_H) - \sin\theta \cos\theta_H\right) + K_{eff} \sin 2\theta = 0 \tag{8}$$

From eq (8) we retrieve the dispersion relation which can be written as

$$\left(\frac{\omega}{\gamma}\right)^2 = \left[H\cos(\theta - \theta_H) + H_{eff}\cos^2\theta\right]\left[H\cos(\theta - \theta_H) + H_{eff}\cos 2\theta\right] \tag{9}$$

Here $H_{eff} \approx 2K_{eff}/M_S$ is the effective anisotropy field that comes from a combination of effects including shape, magneto crystalline and magneto elastic anisotropy. Therefore from eq (5)

$$H_{eff} = 2\pi M_S(1 - 3P) + 2\frac{K_U}{M_S} \tag{10}$$

For the case of multilayer nanowires, the effective anisotropy field H_{eff} is given by

$$H_{eff} = 2\pi M_S(1-3\{1-f(1-P)\}) + 2\frac{K_U}{M_S} \qquad (10a)$$

Where, $f \approx \frac{h_m}{h_{nm}+h_m}$, and hnm represents magnetic and non-magnetic thicknesses in multilayer section. If hnm = 0 (i.e. no non-magnetic spacer layer), the above equation represents the case of a single-element nanowire. Depending upon the direction of the external magnetic field along the easy axis of the nanowires, we can determine the various cases:

Case 1: H| | to the wire and H_{eff}>0

$$\frac{\omega}{\gamma} = H + H_{eff} \qquad (11)$$

Case 2: H| | to the wire and Heff<0

$$\frac{\omega}{\gamma} = H_{eff} - H \qquad (12)$$

Case 3: H$\perp$ easy axis and H<H_{eff}

$$\left(\frac{\omega}{\gamma}\right)^2 = \left(H^2_{eff} - H^2\right) \qquad (13)$$

Case 4: H$\perp$ easy axis, H>H_{eff}>0

$$\left(\frac{\omega}{\gamma}\right)^2 = H(H - H_{eff}) \qquad (14)$$

The frequency-field characteristics can be studied from these relations for various cases of the direction of the applied field and corresponding angular variation with resonance field of the nanowires as shown in Fig. 3. The horizontal line shows the intersection of the dispersion relation and indicates where in an FMR spectrum the resonance lines would be found for a fixed frequency in Fig. 3(a).

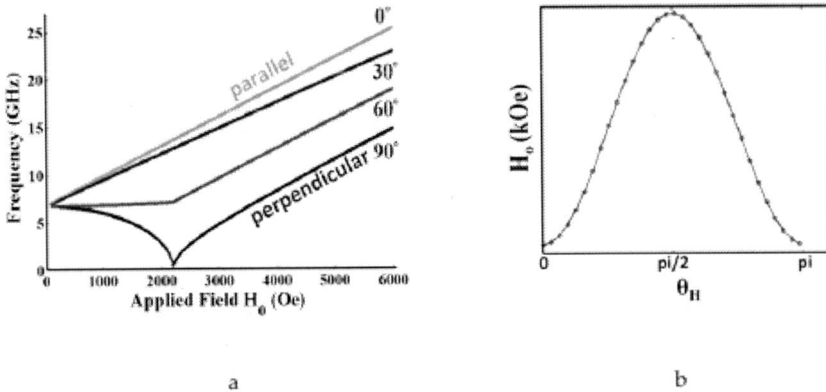

Figure 3 Simulated plots (a) dispersion relation for Ni nanowires for various angles in which the external magnetic field is applied (b) resonance field as a function of the externally applied field angle θ_H.

Sample Preparation and Characterization

During last decade magnetic nanowires have attracted enormous research attention in many areas of advanced nanotechnology, including patterned magnetic recording media, materials for optical and microwave applications. Template assisted growth of nanostructures under constant potential in several electrolytes has been carried out by researchers for over 30 years. In comparison to other deposition techniques (sputtering and MBE) electrode position is a low cost and simple technique to fabricate magnetic nanowires and multilayers. Arrays of Co nanowires were fabricated by template assisted electrochemical deposition into the nanometer-sized pores.

GROWTH OF NANOSTRUCTURES

Templates

Nanowires were grown by using two different types of templates (commercially available) polycarbonate track etched (PCTE) and aluminum oxide (AAO) from Whatman. These templates are completely different from each other due to their qualities as well as the preparation methods. The methods used for fabrication and properties of templates are given below;

Polycarbonate Track Etched (PCTE)

In PCTE template, the pore size varies from 20nm to 200nm and the thickness of ~6μm. To fabricate the PCTE template, initially high energy particle are used to bombard it to produce the path. Later these paths are etched in different chemical bath. The size of the pores is determined by the etching process. Fig.4(a) shows a SEM picture of a commercial PCTE membrane, with a reported pore size of 100 nm. Although the pores seem to have similar diameters, the pore placement is random. The pores density is quite low (in the range of 1010 to 1012 per cm^2) in case of PCTE.

Anodic Aluminum Oxide (AAO)

In 1995, Masuda and Fukuda reported the method to fabricate highly ordered nanohole arrays on aluminum foil. Double anodization of aluminium in acidic solution is adopted to fabricate porous alumina template. The beauty of these templates is their cylindrical pores of uniform diameter, arranged in hexagonal arrays with a thin oxide layer exist at the bottom. Anodic Aluminum Oxide (AAO) templates have been successfully used as templates for the growth of nanowires by electrodeposition. In order to fabricate the AAO template the aluminum is cleaned in the acidic medium to remove the surface impurities. Further the cleaned aluminium processed by two step anodization described by Masuda. The pore size and spacing between pore in the alumina templates are controlled by the anodization voltage.

Temperature of electrolyte also plays a important role in pore parameters. The pores in AAO template cylindrical and highly dense which makes it a good candidate to study the interaction effect between the nanowires on magnetic properties. Fig. 4(b)-(c) shows the SEM micrograph of AAO templates.

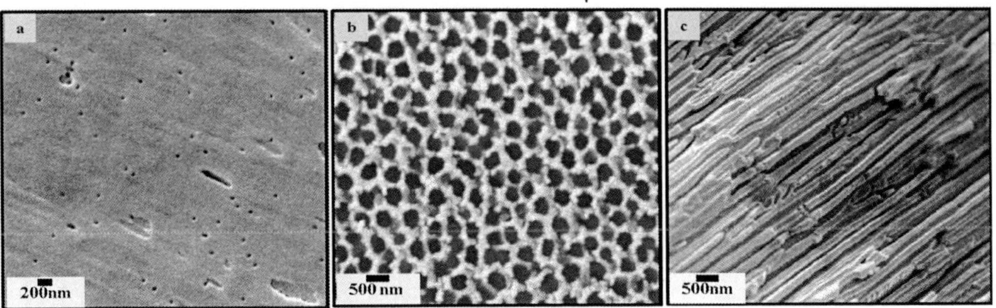

Figure 4 SEM micrograph of membranes (a) Polycarbonate (b) Anodic alumina (Top view) and (c) anodic alumina (cross-section view).

Template Assisted Electrode Position Technique

Electrode position process involves the electric current to reduce cations from electrolyte and deposited that material as a thin film onto a conducting substrate. At the cathode the metal reduction takes place and metal deposits according to:

$$Mn^{+} + n\,e^{-} = M(s)$$

In order to form the nanowires, the cations from electrolyte move through the nonconducting template (AAO/ PCTE) having nanosized pores and deposited on the conducting substrate.

The desired material properties depends upon the various process parameters like electrolyte composition, bath pH, mode of deposition (DC, pulse and AC) and deposition temperature.

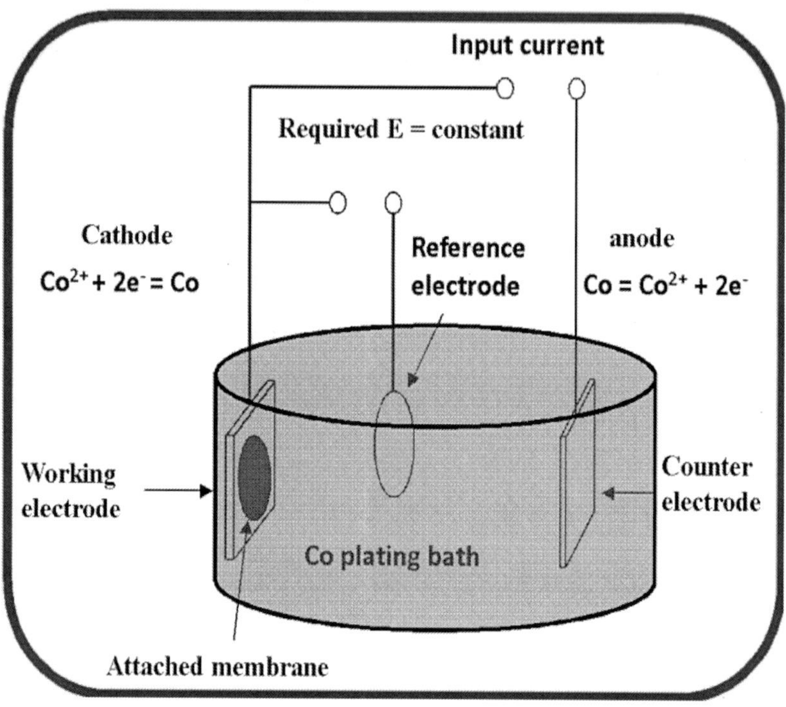

Figure 5 Schematic representation of three-electrode electrochemical cell set-up employed. AAO template mounted electrodes act as a working electrodes (WE), platinum foil counter electrode (CE) and Saturated calomel electrode (SCE), reference electrode (RE).

Fig.5 illustrates the three-electrode cell set-up used in this study. A platinum foil and a saturated calomel electrode (SCE) were used as the anode (or counter electrode) and as a reference electrode respectively. The steps of preparation of nanostructure are as followed;

Steps Involved in the Template Assisted Synthesis of Nanowires

- Step 1: The porous anodic alumina (AAO) /polycarbonate templates were taken and one side of the AAO were sputtered with Au by RF sputtering, which acted as the working electrode in a three-electrode electrochemical cell.
- Step 2: The electrodeposition solution was restrained to the

other side of the membrane so that deposition was initiated onto the Au layer within the pores. The array of Co nanowires was deposited from a solution of 25 gm/L CoSO4.7H2O, 5 gm/L H3BO3 and sodium Lauryl Sulphate (SLS), which is used to reduce the surface tension of water for proper wetting of pores.

- Step 3: The cyclic voltammetry is used to figure out the constant deposition potential of the working electrode with respect to a standard reference electrode RE to get the favourable condition for deposition.
- Step 4: The growth of nanowires is carried at the optimized potential.
- Step 5: For further investigation of freely standing nanowires, the templates is used to dissolve in an appropriate solution. Etching solution for AAO and PCTE are sodium hydroxide and dichloromethane respectively

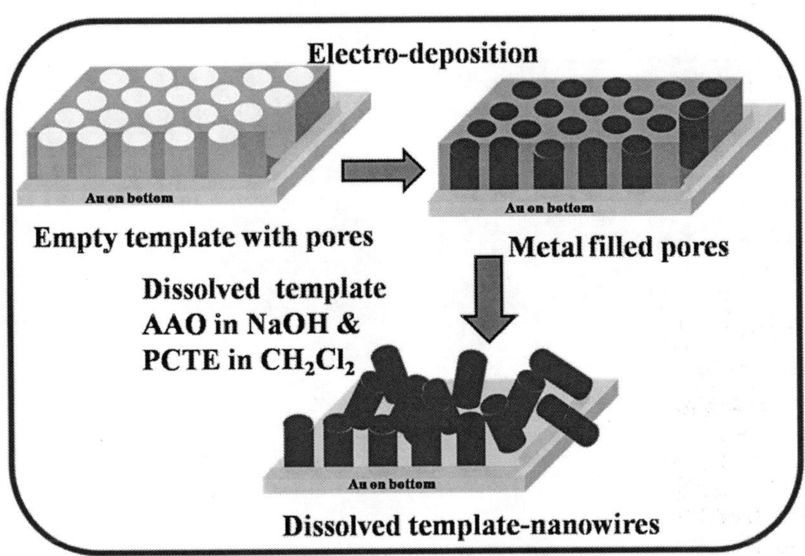

Figure 6 Schematic illustration of the growth of magnetic nanowire in alumina template by electrode position process.

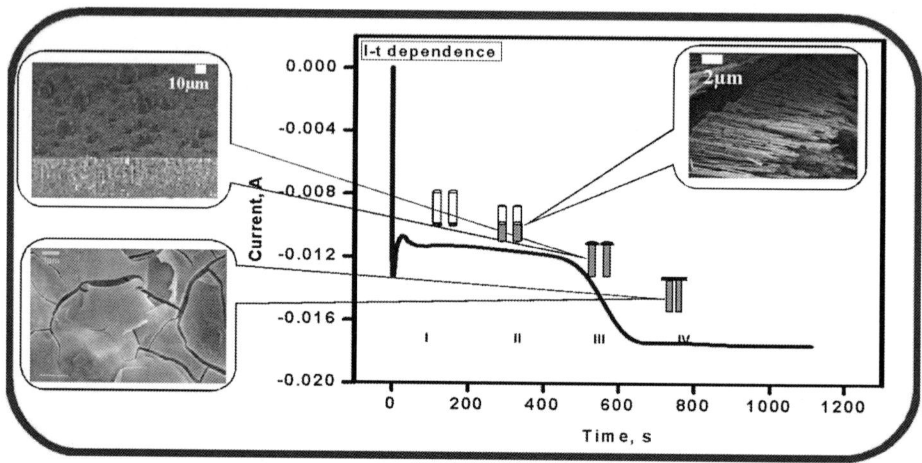

Figure 7 Typical choronoamperometery plot during potentiostatic electrode position taken during the fabrication of Co nanowires. The various stage of pore filling during deposition is shown as insets at the respective current-time positions

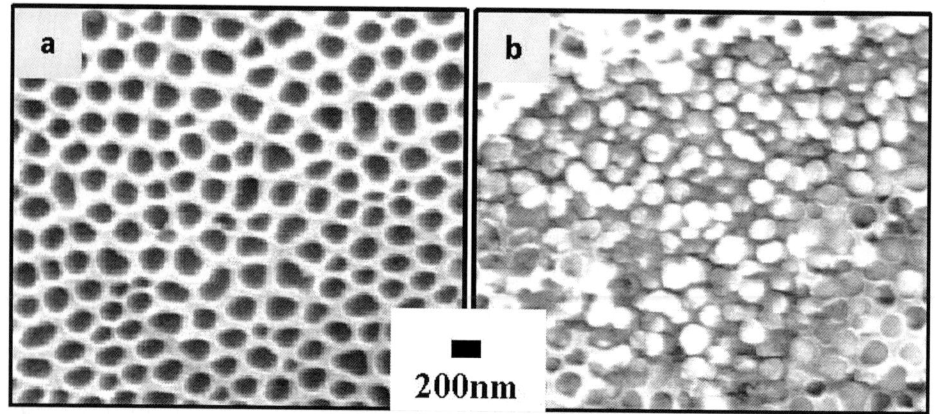

Figure 8 Scanning electron micrograph (SEM) of empty and filled surface morphology of the template.

CHARACTERIZATION OF NANOSTRUCTURES

Scanning Electron Microscopy

For scanning electron microscopy (SEM), the template containing nanowires is partially released from their template by appropriate solution. To remove the residual part of template the etched sample is cleaned by deionised water. To carry out the surface analysis secondary electron imaging in scanning electron microscope has been utilized. Surface morphology of Co nanowire was investigated by scanning electron microscope (SEM: ZEISS EVO 50) operating at 20 kV accelerating voltage by secondary electron imaging. Fig. 8 shows the SEM micrograph of empty AAO and filled with nanowires. The morphology of grown nanowires in PCTE template can be clearly seen in Fig. 9(a). Side and top view of AAO assisted nanowires are presented in Fig. 9(b & c). The growth density of nanowires in AAO template is more as compared to PCTE template.

Figure 9 Scanning electron micrograph (SEM) of electrodeposited nanostructure in (a) Polycarbonate and (b) & (c) Anodic alumina.

Transmission electron microscopy

Transmission Electron Microscope (TEM) is a widely used instrument for characterizing the interior structure of materials. For TEM, the template was completely etched and rinsed several times with deionised water to clean the residual template part. Template etching is an important step of sample preparation of nanowires for TEM characterization. Cobalt nanowires were scratched from the substrate and ultrasonicated in acetone for 15 min so that the nanowires could disperse properly. Few drops of the suspension were then transferred on to a carbon coated copper grid and the microstructures were analyzed by high resolution TEM (HRTEM: Technai G20 S-Twin model) operating at 200 kV. To get the actual diameter of the nanowire, the complete dissolving of residual template from surrounding the nanowires is very important. Typical TEM results obtained are shown in Figure 10. While nanowires grown in PCTE templates typically have a tapered cross-section due to the non-uniform diameters of the pores, the AAO template-based nanowires grow in a more uniform manner.

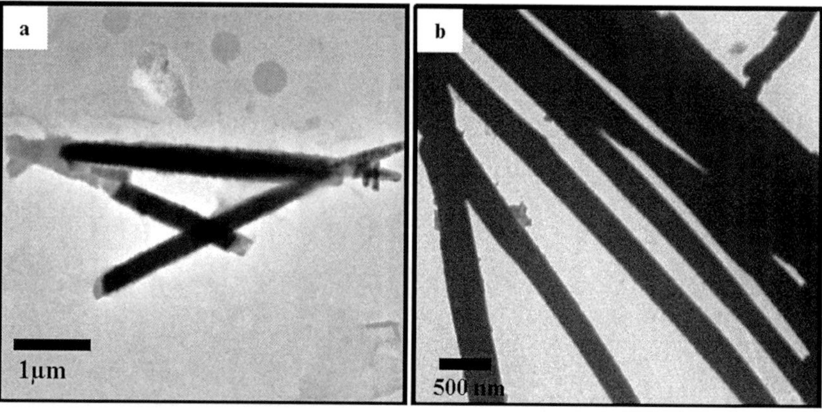

Figure 10. Transmission electron micrograph (TEM) of electrodeposited nanostructure in (a) Polycarbonate and (b) Anodic alumina.

ELECTRON PARAMAGNETIC RESONANCE

EPR/FMR measurements obtained from ferromagnetic nanowire arrays give detailed information on the size of the nanowires. The spectra can be used to calculate the interwire magnetic interactions quite accurately [31]. In Fig. 11 are shown typical spectra obtained from Co nanowire arrays. Here, both single-Co nanowire arrays and multilayered Co/Pd nanowire arrays are compared. There is clearly an angular dependence of the applied microwave field with respect to the easy axis of the nanowires, as has been studied previously [32].

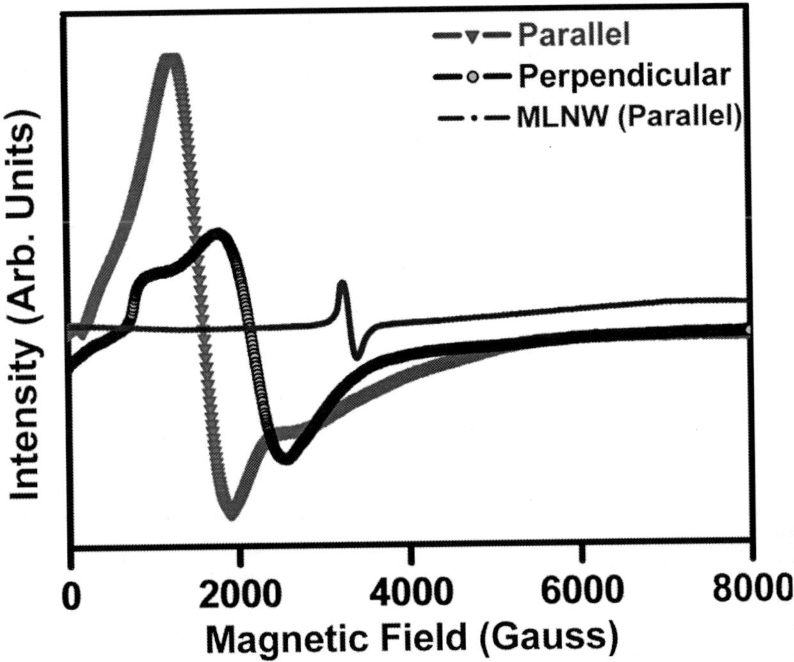

Figure 11 Electron paramagnetic resonance (EPR/FMR) spectra of single element Co nanowire arrays in parallel and perpendicular orientation at room temperature.

Fig.12 shows the FMR spectra of Co/Pd MLNW arrays for the temperature range from 90K to 290K. It is clear from the data that two different peaks in the FMR spectrum have been observed. The peak present at the ~ 3310 G is corresponding to the main peak of FMR and itis due to the quantum confinement in the nanostructures which is more in case of Co/Pd multilayered nanowires. The spin waves are confined in the magnetic nanostructure and it is found to be dominant when we reduce the third dimension (in z-direction) of the nanostructure. This main FMR peak present at all the temperature and the variation of resonance field at different temperature is very slow. As we go down to the 90K with decrease of 20K from room temperature the resonance field increases slowly. This slow variation can be explain as temperature decreases the Ms (T) increase and resultant head-totail alignment between FM segments increases, which further reduces the effective field as a result slow increase in the resonance field. Inset in Fig.12 shows the variation of resonance field with temperature.

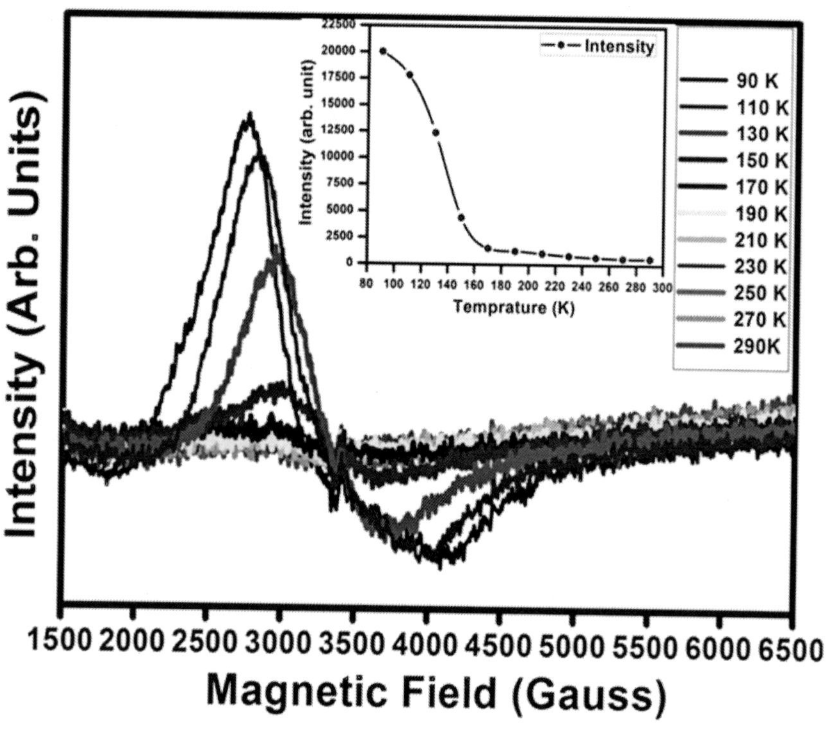

Figure 12. Electron Spin resonance (ESR/FMR) spectra of multilayered Co/Pd nanowire arrays in parallel orientation with temperature range (90K-290K). (Inset) Peak height change with temperature.

MICROWAVE DEVICES

The increasing demand for higher frequency magnetic microwave structures triggered a tremendous development in the field of magnetization dynamics over the past decade. In order to develop smaller and faster devices, a better understanding of the complex magnetization precessional dynamics, the magnetization anisotropy, and the sources of spin scattering at the nanoscale is necessary [23-25]. Magnetic data storage with its promise of non-volatility, robustness, high speed and low energy dissipation attracted long back. It has been encountered in a number of applications such as smart cards, hard drives, thin film read heads or video tapes [26-28]. Industry activities are also directed to replace semiconductor random access memories by magnet based memory devices. Logical 0 and 1 are encoded by the direction of the magnetization of a small magnetic element, such as a giant magnetoresistance (GMR) or tunneling magnetoresistance (TMR) element. For data manipulation i.e. reading and writing, the magnetization has to be switched between the two equilibrium positions. In order to push forward data transfer rates which are already in the GHz range, spin waves transportation mechanism has to be employed in place of current semiconductor based devices [29-32]. To this end an understanding of the dynamic motion and the mode spectrum is necessary.

On the road towards the size reduction of microwave devices, ferromagnetic nanowires embedded into porous templates have proven to be an interesting route to ferrite based materials. The main advantages of what we call ferromagnetic nanowired (FMNW) substrates are that they present a zero-field microwave absorption frequency that can be easily tuned over a large range of frequencies, as well as being low cost and fast to produce over a large area as compared to standard ferrite devices. Conventional ferrite circulators need to be biased by a magnetic field to operate. This biasing field is generally provided by permanent magnet and in view of the volume reduction, we need to think of FMNW substrates which work at zero fields. A rigorous theory of microwave devices is very cumbersome due to ferromagnetic resonance (FMR) and spin-wave phenomena.

VNA-FMR is an important technique for the investigation of magnetization dynamics of lowdimensional magnetic structures and patterned microwave devices. Also, broadband flip-chip technique can be used to measure the material intrinsic and extrinsic properties by applying external magnetic field. In this chapter, we will demonstrate both techniques to have better understanding of ferromagnetic resonance. We propose fabrication and measurements of various non-reciprocal microwave devices like band-stop filter, isolator and electromagnetic band-gap (EBG) structures using FMNW substrates using Ni and Co nanowires.

CONVENTIONAL FERROMAGNETIC RESONANCE – FLIP-CHIP BASED TECHNIQUE (FMR)

In the frequency domain measurements, the magnetic excitation is sinusoidal magnetic field hrf and the response of the sample is detected by vector network analyzer. The magnetic field can be applied along parallel and perpendicular direction of the nanowires satisfying the FMR condition. The main components of the experimental setup are shown in Fig. 13. The VNA is connected to a coplanar waveguide (CPW) having a characteristic impedance of 50 Ω using coaxial cables and microwave connectors. For such radio frequency connections often coaxial cables with Teflon insulation and SMA connectors are employed. They are comparably low priced and offer a bandwidth of typically 18 GHz. The used cables should not have a metallic reinforcement. Flip-chip based measurements are done to extract the material parameters, which are technologically important for data storage applications [31]. Fig. 13 shows the schematic of the set-up used for flip-chip measurements. Magnetic field is applied parallel to the nanowires and perpendicular to RF magnetic field, so that resonance condition must be satisfied.

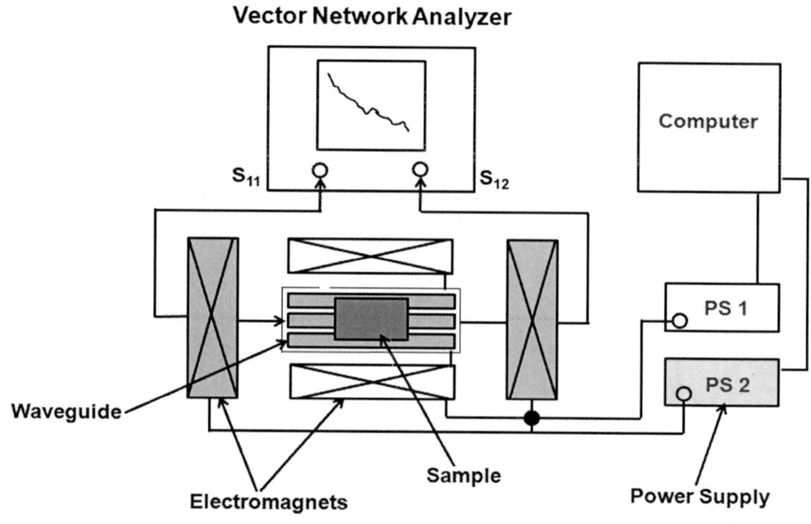

Figure 13. Schematic representation of VNA-FMR system

The transmission response of Ni NWs placed on a co-planar waveguide transmission line (Fig. 14(a)) for different static magnetic field is shown in Fig. 14(b). A small but perceptible and repeatable change was observed when the applied magnetic field was turned on/off. The change is however too small to show up in the plot of Fig. 14(b). This is partly due to masking of the shifts due to reflections from soldering and other undesirable metal deposits near the patterned transmission line and partly due to the relatively small interaction between the RF fields (largely confined to the substrate holding the CPW line) and the nanowires. It is thus expected that clearer shifts can be seen with more careful preparation and patterning of the substrate and ultimately by printing the CPW line on the nanowire template itself.

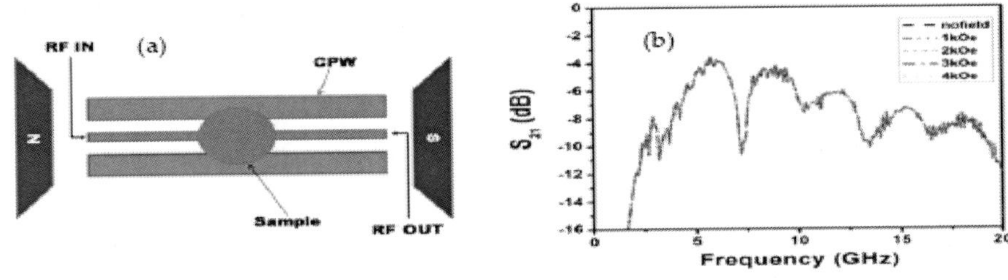

Figure 14 (a) CPW geometery for magnetic field applied along perpendicular direction (b) transmission response as a function of frequency of Ni nanowires for various applied magnetic field

SUBSTRATE INTEGRATED MAGNETIC MICROWAVE DEVICES

Newly developed techniques enable to characterize and re-arrange matter at nanometer scale. Now a day, automotive and wireless communication requires reduction in dimensions of nanodevices for yielding higher cut-off frequencies. Here, we propose integrated

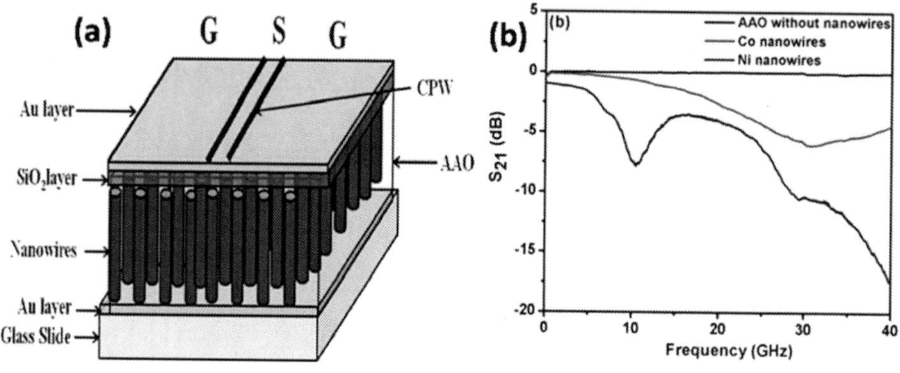

Figure 15: Schematic of band stop filter (b) S-parameter of the DUT for three different samples magnetic band-stop filter fabricated on FMNWS and studying their microwave properties like permittivity and permeability using CPW.

Fig. 15(a) shows the schematic of the device under test which is a coplanar waveguide on NW substrate. The transmission coefficient of the device for three samples having bare AAO, Ni and Co NWs are shown in Fig. 15(b) at zero biasing. It is observed that by applying the magnetic field the resonance frequency shift towards higher range, so that we can tune our operating frequency of the device. We also observe that material properties also influence the Device-Under-Test (DUT).

Non-reciprocal Devices

It is important to note that it is possible to obtain non-reciprocal structures also using nanowire-based devices. Essentially, the microwave properties like permittivity and permeability can be made to be asymmetric. This is useful to obtain non reciprocal devices such as isolators and circulators. Prototypes of such devices are presently under fabrication. Some of the proposed devices are shown in Fig. 16 below. The detailed studies of these devices are out of scope of this chapter.

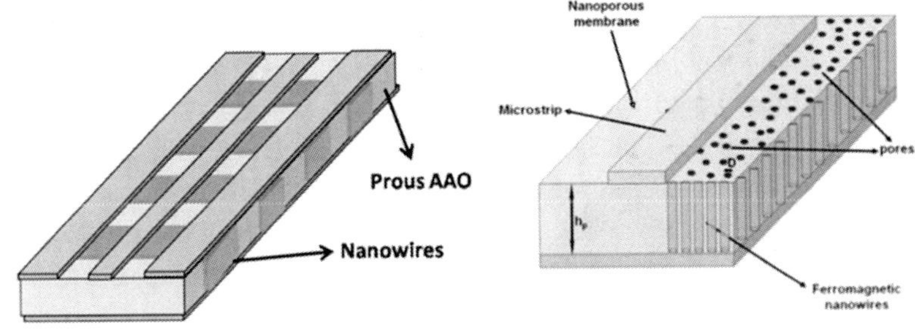

EBG structure in porous AAO template Isolator

Figure 16: Possible non-reciprocal structures using nanowires.

AUTHOR DETAILS

Manish Sharma, Sachin Pathak and Monika Sharma

Indian Institute of Technology Delhi, India

REFERENCES

1. A. Fert and L. Piraux, "Magnetic nanowires", J. Magn. Magn. Mater. 200 (1999) 338-358.
2. S. Shamaila, R. Sharif, S. Riaz, M. Khaleeq-ur-Rahman and X. F. Han, "Fabrication and magnetic characterization of CoxPt1-x nanowire arrays", Appl. Phys. A, 92 (2008) 687- 691.
3. Z. Z. Sun and J. Schliemann, "Fast domain wall propagation under an optimal field pulse in magnetic nanowires", Phys. Rev. Lett. 104 (2010) 037206.
4. M. Yan, A. Kakay, S. Gliga and R. Hertel, "Beating the walker limit with massless domain walls in cylindrical nanowires", Phys. Rev. Lett. 104 (2010) 057201.
5. C. T. Boone, J. A. Katine, M. Carey, J. R. Childress, X. Cheng and I. N. Krivorotov, "Rapid domain wall motion in permalloy nanowires excited by a spin-polarized current applied perpendicular to the nanowire", Phys. Rev. Lett. 104 (2010) 097203.
6. B. K. Kuanr, V. Veerakumar, R. Marson, S. Mishra, R. E. Camley and Z.

Celinski, "Nonreciprocal microwave devices based on magnetic nanowires", Appl. Phys. Lett. 94 (2009) 202505.

7. M. Darques, J. De La T. Medina, L. Piraux, L. Cagnon and I. Huynen, "Microwave circulator based on ferromagnetic nanowires in alumina template", Nanotechnology 21 (2010) 145208.

8. J. De La T. Medina, J. Spiegel, M. Darques, L. Piraux and I. Huynen, "Differential phase shift in nonreciprocal microstrip lines on magnetic nanowired substrates", Appl. Phys. Lett. 96 (2010) 072508.

9. C. Kittel, "On the theory of ferromagnetic resonance absorption", Phys. Rev. 73 (1948) 155.

10. C. Kittel, "Excitation of spin waves in a ferromagnet by a uniform rf field", Phys. Rev. 110 (1958) 1295-1297.

11. L. Kraus, G. Infante, Z. Frait and M. Vazquez, "Ferromagnetic resonance in microwires and nanowires", Phys. Rev. B 83 (2011) 174438.

12. J. De La T. Medina, L. Piraux, J. M. Olais Govea and A. Encinas, "Double ferromagnetic resonanace and configuration-dependent dipolar coupling in unsaturated arrays of bistable magnetic nanowires", Phys. Rev. B 81 (2010) 144411.

13. A. Encinas-Oropesa, M. Demand, L. Piraux, I. Huynen and U. Ebels, "Dipolar interactions in arrays of nickel nanowires studied by ferromagnetic resonance", Phys. Rev. B 63 (2001) 104415.

14. C. A. Ramos, M. Vazquez, K. Nielsch, K. Pirota, R. B. Rivas, R. B. Wherspohn, M. Tovar, R. D. Sanchez and U. Gosele, "FMR characterization of hexagonal arrays of Ni nanowires", J. Magn. Magn. Mater. 272-276 (2004) 1652-1653.

15. C. A. Ramos, E. Vassallo Brigneti and M. Vazquez, "Self-organized nanowires: evidence of dipolar interactions from ferromagnetic resonance measurements", Physica B 354 (2004) 195-197.

16. E. Beaurepaire, J. C. Merle, A. Daunois and J. Y. Bigot, "Ultrafast spin dynamics in ferromagnetic nickel", Phys. Rev. Lett. 76 (1996) 4250-4253.

17. M. R. Freeman, and W. K. Hiebert, "Stroboscopic microscopy of magnetic dynamics", In: B. Hillebrands, K. Ounadjela (Eds.), Spin Dynamics in Confined Magnetic Structures I. Springer, Berlin, pp. 93-126.

18. W. K. Hiebert, A. Stankiewicz, and M. R. Freeman, "Direct observation of magnetic relaxation in a small permalloy disk by time-resolved scanning Kerr microscopy", Phys. Rev. Lett. 79 (1997) 1134-1137.

19. T. J. Silva, C. S. Lee, T. M. Crawford, and C. T. Rogers, "Inductive measurement of ultrafast magnetization dynamics in thin-film permalloy", J. Appl. Phys. 85 (1999) 7849- 7862.

20. M. M. Midzor, P. E. Wigen, D. Pelekhov, W. Chen, P. C. Hammel, and M. L. Roukes, "Imaging mechanisms of force detected FMR microscopy", J. App. Phys. 87 (2000) 6493- 6495.

21. O. Mosendz, B. Kardasz, D. S. Schmool, and B. Heinrich, "Spin dynamics at low microwave frequencies in crystalline Fe ultrathin film double layers

using co-planar transmission lines", J. Magn. Magn. Mater. 300 (2006) 174-178.

22. N. Mecking, Y. S. Gui, and C. M. Hu, "Microwave photovoltage and photoresistance effects in ferromagnetic microstrips", Phys. Rev. B 76 (2007) 224430.
23. B. K. Kuanr, M. Buchmeier, D. E. Buergler, P. Gruenberg, R. E. Camley, and Z, Celinski, "Dynamic and static measurements on epitaxial Fe-Si-Fe", J. Vac. Sci. Technol. A 24 (2003) 1157.
24. J. Curiale, R. D. Sanchez, C. A. Ramos, A. G. Leyva, A. Butera, "Dynamic response of magnetic nanoparticles arranged in a tubular shape", J. Magn. Magn. Mater. 320 (2008) 218-221.
25. Can-Ming Hu, "Recent progress in spin dynamics research in Canada", La Physique Au Canada 65 (2009).
26. X. Kou, X. Fan, R. K. Dumas, Q. Lu, Y. Zhang, H. Zhu, X. Zhang, K. Liu and J. Q. Xiao, "Magnetic effects in magnetic nanowire arrays", Adv. Mater. 23 (2011) 1393-1397.
27. T. Schrefl, J. Fidler, D. Suss, and W. Scholz, "Hysteresis and switching dynamics of patterned magnetic elements", Phys. B 275 (2000) 55-58.
28. X. Kou, X. Fan, H. Zhu and J. Q. Xiao, "Ferromagnetic resonance and memory effects in magnetic composite materials".
29. K. Nagai, Y. Cao, T. Tanaka and K. Matsuyama, "Binary data coding with domain wall for spin wave based logic devices", J. Appl. Phys. 111 (2012) 07D1301-07D1304.
30. M. Sharma, S. Pathak, S. Singh, M. Sharma and A. Basu, "Highly ordered magnetic nickel nanowires: structural properties and ferromagnetic resonance measurements", AIP Conf. Proc. 1347 (2011).
31. O. Yalçın, et al, Ferromagnetic resonance studies of Co nanowire arrays, Journal of Magnetism and Magnetic Materials 272–276 (2004) 1684–1685.
32. G. Kartopu, O. Yalçın, et al. Size effects and origin of easy-axis in nickel nanowire arrays, Journal of Applied Physics 109, 033909 (2011).
33. G. Kartopu, O. Yalçın, M. Es-Souni, and A. C. Başaran, Magnetization behavior of ordered and high density Co nanowire arrays with varying aspect ratio, Journal of Applied Physics 103, 093915 (2008).

Chapter 2

FMR STUDIES OF [SNO_2/CU-ZN FERRITE] MULTILAYERS

R. Singh and S. Saipriya[1]

[1]School of Physics, University of Hyderabad, Central University P.O., Hyderabad, India

INTRODUCTION

The artificially structured multilayers (ML) have opened a new field of interface magnetism. It is possible to fabricate a multilayer sample with a specific design according to the requirement consisting of magnetic, non-magnetic, metallic or non-metallic components. Various types of interfaces may be synthesized in ML by combining a magnetic component with a non-magnetic one or from two different magnetic components. The interface between two non-magnetic components is particularly interesting if a magnetic anomaly happens to be induced at the interface atom layer due to some interface effect.

The oscillations of magnetic parameters as a function of the number of interfaces, spacer layer or magnetic layer thickness in the ML have been gaining attention in the recent years. Bruno et al [1] employed RKKY model and Edwards et al [2] used spin dependent confinement of electrons in the quantum well provided by the spacer layer (quantum interference) to explain the oscillations initially observed in metallic ML. Exchange coupling in ML is also affected by direct dipolar coupling like the correlation of spins at rough interfaces (orange peel coupling), which is magnetostatic in origin and occurs due to the interaction of the dipoles which appear due to the roughness of the material [3]. This coupling favors parallel or anti-parallel alignment of spins depending upon the interplay between the magnetostatic exchange and anisotropy energy. The oscillations in exchange coupling are also observed in oxide ML apart from the metallic ML [4]. Oxide materials are more stable and their ML can form magnetic structures which do not exist in bulk form. There are reports on ML of ferrites used along with NiO [5,6], MgO [7] , CoO [8] and SnO_2 [9-12].

The mixed spinel structure of Cu-Zn ferrites with composition $Cu_{0.6}Zn_{0.4}Fe_3O_4$ (CZF) gives high magnetization when deposited in Ar environment [13]. The properties of non-magnetic

(NM) SnO2 are sensitive to oxygen vacancies [14]. The interfacial region between these two materials may give rise to interesting magnetic phenomena. The objective of the present work is to study the magnetic properties of SnO2 and Cu-Zn Ferrite multilayers and interpret them in terms of suitable magnetization mechanism.

This chapter is about the Ferromagnetic Resonance (FMR) studies of (SnO_2/Cu-Zn Ferrrite) ML as a function of CZF layer thickness and SnO2 layer thickness.

EXPERIMENTAL

Alternate layers of SnO_2 and $Cu_{0.6}Zn_{0.4}Fe_3O_4$ (CZF) were deposited on quartz substrates at room temperature (RT) using rf-magnetron sputtering in Ar environment from SnO2 and CZF targets at a power of 50W and 70W respectively. The rate of deposition was estimated by depositing thin films of SnO_2 and CZF separately. The synthesis method of targets and the rf sputtering system are described in our earlier work [15]. The field emission scanning electron microscopy (FESEM) studies were carried out to observe the multilayer structure. The FMR studies were carried out in the temperature range 100 – 450 K using JEOL x-band spectrometer. FMR studies were carried out at different temperatures and by varying the angle θH between the thin film normal and the direction of the applied field. The peak-to-peak FMR signal intensity (Ipp) which is proportional to the power absorbed was measured between the positive and negative peak points along the y-axis. The peak-to-peak line width (ΔH) was measured as a function of temperature.

FERROMAGNETIC RESONANCE (FMR) STUDIES

Effect of CZF Layer Thickness

The ML samples [SnO_2 (46 nm)/CFZ(x nm)]$_5$ where x = 42 nm, 83 nm and 249 nm werem synthesized. A final capping layer of 46 nm of SnO_2 was deposited on all the ML samples. The error in the thickness is ± 5 nm. The FESEM images (Figure 1) portray a stack of alternate dark and light layers corresponding to CZF and SnO_2 respectively as confirmed by the EDAX data. The CZF layer exhibits a columnar growth as observed in many of the sputtered films [16]. The protrusions of CZF column into SnO_2 layers visible from the FESEM images indicate that the interfaces are diffused.

FMR studies were carried out at different temperatures and by varying the angle Θh between the thin film normal and the direction of the applied field. Figure 2 shows the room temperature FMR spectra of the ML in θH = 90° (parallel) and 0° (perpendicular) configuration. In the parallel geometry, the FMR signal for the ML with x = 249 nm is highly asymmetric. The steep increase in the low field side and a slow increase in the high field side indicate negative anisotropy in the ML [17]. The line shape of FMR signal is sensitive to magnetic interactions in the material and hence a change in line shape is an indication of change in the magnetic interactions in the ML. Many a times the observed single FMR signal may arise from overlapping of 2 or more FMR signals.

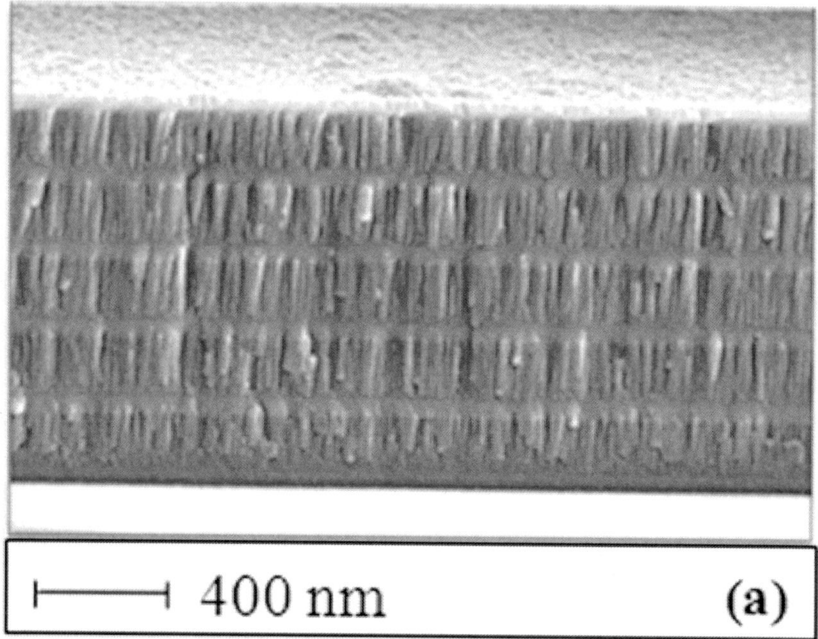
400 nm
(a)

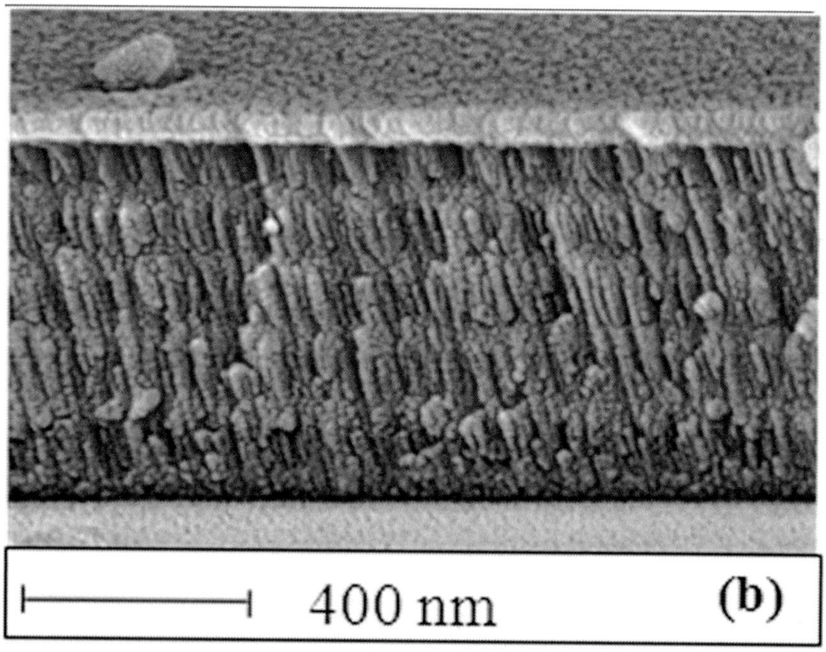
400 nm
(b)

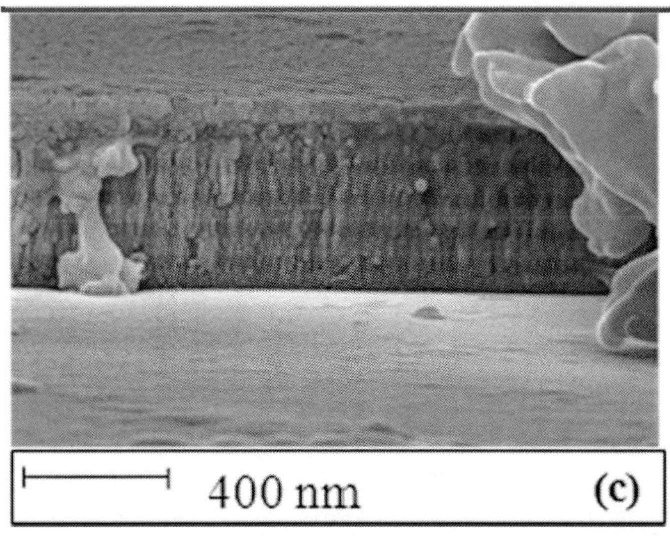

Figure 1 FESEM images of [SnO_2 (46nm) CZF (x nm)]$_5$ for x = 249 (a), 83(b) and 42 nm (c).

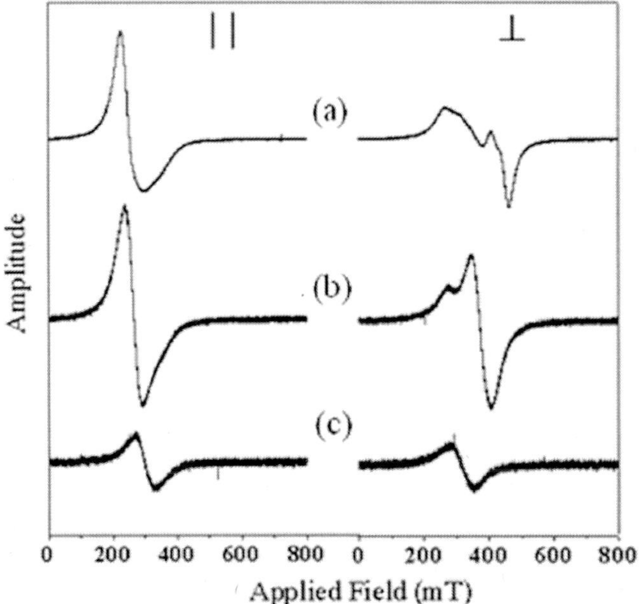

Figure 2 RT FMR spectra of [SnO_2 (46 nm) CZF(x nm)]$_5$ in the parallel and perpendicular configuration for x = 249 (a), 83(b) and 42 nm (c).

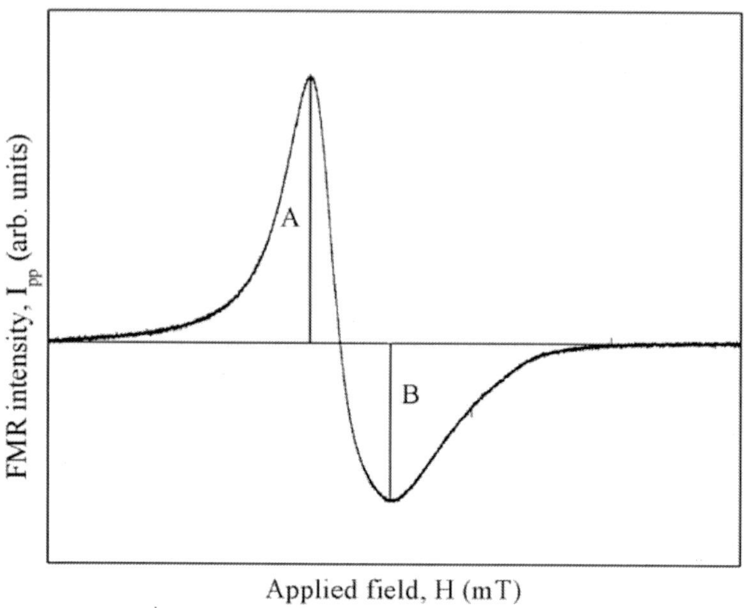

Figure 3 A representative FMR plot depicting the A/B ratio.

The asymmetry parameter, A/B ratio, was estimated by dividing the intensity of the upper part of the FMR signal by that of the lower part (Figure 3). A/B > 1 indicates that the intensity of the upper half of the FMR signal is higher than the lower one and vice versa. A/B = 1 indicates a signal with both its upper and lower half with equal intensity. The A/B ratio at room temperature in the parallel configuration is 1.27, 1.12 and 0.69 for x = 249, 83 and 42 nm respectively. A/B > 1 gives Dysonian line shape to the spectrum which results from lower skin depth and higher conductivity of the sample [18]. Hence the possibility of occurrence of dispersive component due to diffusion of electrons into and out of the skin region cannot be ruled out for the ML with x = 249 and 83 nm.

The asymmetric nature of the FMR spectra of x = 249 nm may also be attributed to the interlayer coupling in the ML. The asymmetry decreases with decrease in CFZ layer thickness presumably due to the decrease in the interlayer coupling strength between the layers. As a consequence of interlayer coupling, the coupled spins process out of phase with the uncoupled ones. Hence the FMR spectra of spins which precess at different frequencies overlap to give an asymmetric signal [19]. The peak to peak FMR signal intensity (Ipp) is measured between the positive and negative peak points along the y-axis and is proportional to the power absorbed. In the parallel configuration Ipp values, normalized to the volume of the CZF layers, increases as the magnetic layer (CZF) thickness increases from x = 42 to 83 nm followed by a decrease for x = 249 nm. The resonance field (H_{res}) decreases with increase in the CFZ layer thickness in this geometry. This is attributed to the increase in internal field with increasing CZF layer (magnetic layer) thickness because of which the magnetization also increases. The FMR parameters for both parallel and perpendicular configuration are listed in Table 1.

The intrinsic line shape and linewidth of an FMR signal is sensitive to variation of magnetization at the interfaces, surface pits, grain boundaries etc. The peak to peak linewidth (ΔH) is sensitive to inhomogeneity, surface roughness, internal field etc. ΔH values for the ML reported in this study initially decreases from 631 to 541 Oe as x changes from 42 to 83 nm followed by an increase to 704 Oe for x = 249 nm. The ratio of the active layer thickness to dead layer thickness is low for ML with x = 42 nm. As x changes from 42 to 83 nm, the contribution from the dead layer becomes relatively negligible due to large active layer thickness. The increase in ΔH for x = 249 nm may be due to the presence of interlayer coupling [20].

The ML with x = 249 nm exhibits multiple resonances in θH = 0° configuration at RT (Figure 2). The resonances occur at 363 and 418 mT for this ML. The occurrence of multiple resonances may be attributed to the excitation of spin waves due to the inter layer coupling of CZF layers mediated by the non-magnetic SnO_2 layers [21]. This implies that the entire ML is coupled by the interlayer exchange coupling as a single magnetic entity. The spin waves propagate through the SnO_2 layers and are sustained by the entire multilayer film. This confirms that for x = 249 nm the CZF layer thickness is ample enough to polarize SnO_2 layer. With decreasing x it loses its capacity to polarize SnO_2 layer. The occurrence of multiple resonances may also be attributed to the existence of two different magnetic phases pertaining to the interfacial region and the bulk of the magnetic layer.

The number of resonances decreases with decreasing CZF thickness in the perpendicular geometry. As the thickness of CZF layer increases, the ML provides sufficient length for the spin wave modes to be formed.

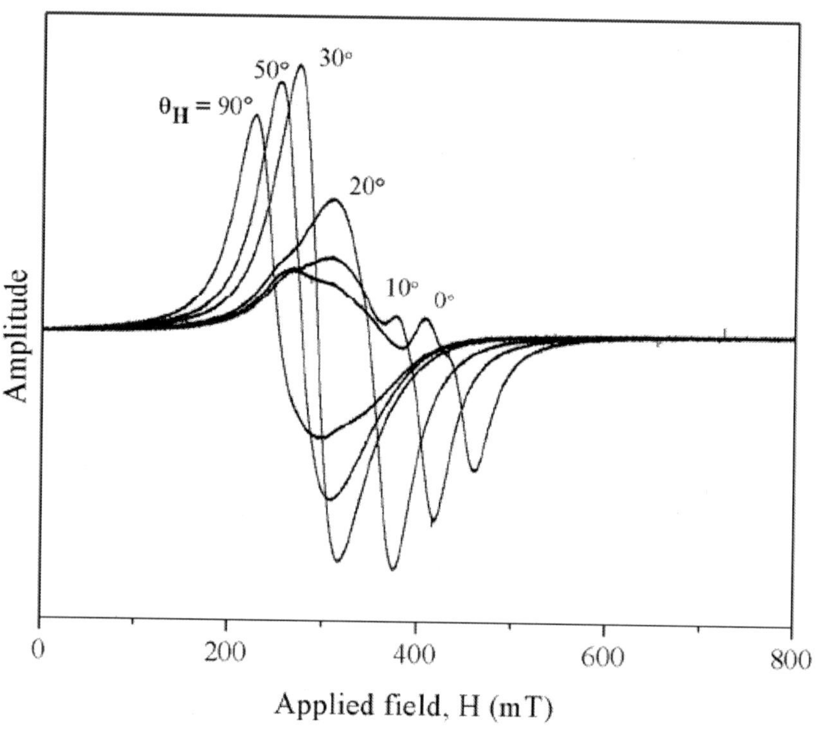

Figure 4 Room temperature FMR spectra of [SnO_2(46 nm)CFZ(249 nm)]5 at different orientations.

The FMR spectra of [SnO_2(46 nm)/CZF(249 nm)]$_5$ recorded at different orientations is shown in figure 4. For θH > 20° there is only single resonance in the FMR spectra. Small shoulder appears at θH = 20°. The multiple resonances appear at θH < 20°. The relative intensity of the FMR signals is also angle dependent. As θH approaches 0°, the amplitude of the shoulder peaks increases at the cost of that of the main peak. The separation between the resonances also increases and the intensity of the FMR signal decreases as θH approaches 0°. This may be due to the prominent demagnetization effects in the perpendicular configuration. A similar behavior is exhibited by the ML with x = 86 nm. There are no multiple resonances for the ML with x = 42 nm.

Figure 5 is a representative plot of the angular dependence of A/B ratio. A/B is close to 2.1 for the perpendicular configuration and decreases as the ML approaches parallel configuration. A/B = 1 for θH ~ 30° showing that the spectra is symmetric at an orientation of 30° of the sample normal with the applied field. Below 30° A/B is < 1 and above 30° A/B is > 1 evidencing the variation in the intensity of the lower and upper part of the FMR spectra with θH.

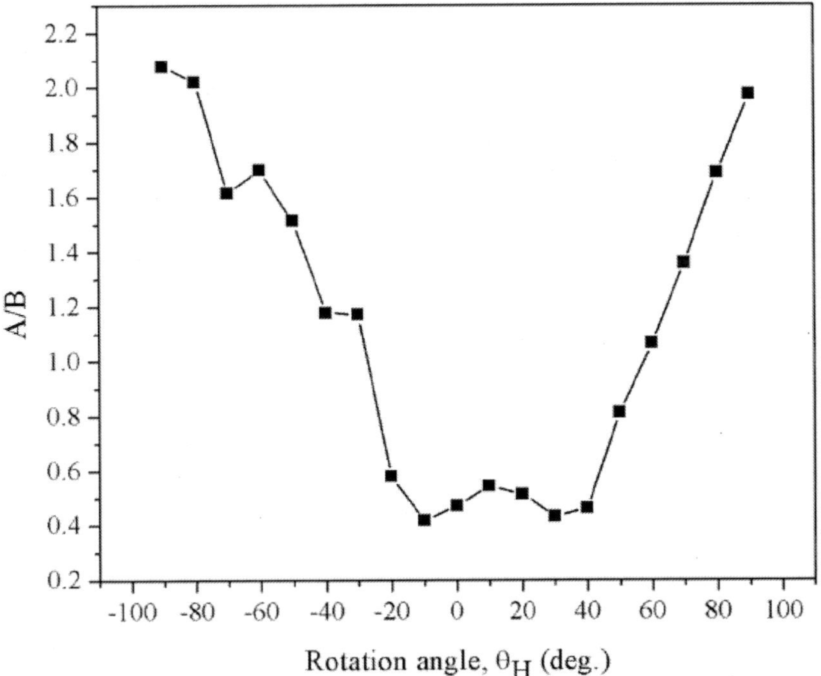

Figure 5 A representative plot of angular dependence of A/B ratio of [SnO_2(46 nm)CFZ(249 nm)$]_5$ ML.

Figure 6 depicts the angular variation of H_{res}, ΔH and Ipp. For all the ML there is no significant change in H_{res} and ΔH the FMR spectra for θH ≥ 60°. Below 60°, H_{res} and ΔH increase continuously. This shows that the ML is anisotropic in nature. The Ipp value initially increases from 0° till 40° followed by a decrease with further increase in θH. Thus the ML is homogeneous when placed at an orientation of 40°. When an applied field is oriented along the film plane, H_{res} increases with decreasing thickness of the CZF layers. But H_{res} decreases with decreasing thickness of the CZF layers when the external magnetic field is oriented perpendicular to the film plane. This behavior is attributed to the decrease of the perpendicular anisotropy energy with decreasing CZF thickness [22]. The difference between the resonance fields in the parallel and perpendicular configuration decreases with decreasing CZF thickness evidencing the fact that the ML with x = 42 nm is relatively less anisotropic compared to the ML with x = 249 and 83 nm. In the perpendicular configuration, the ML with x = 249 nm exhibits multiple resonances presumably due to the presence of spin waves. The spin wave resonance spectrum at RT and HT for ML with x = 249 nm were initially analyzed using the dispersion relation developed by Kittel as follows [23].

$$H_n = H_0 - \Lambda M_0 \left(\frac{\pi}{L}\right)^2 n^2 \quad (1)$$

where $\Lambda = \frac{2A}{{M_0}^2}$

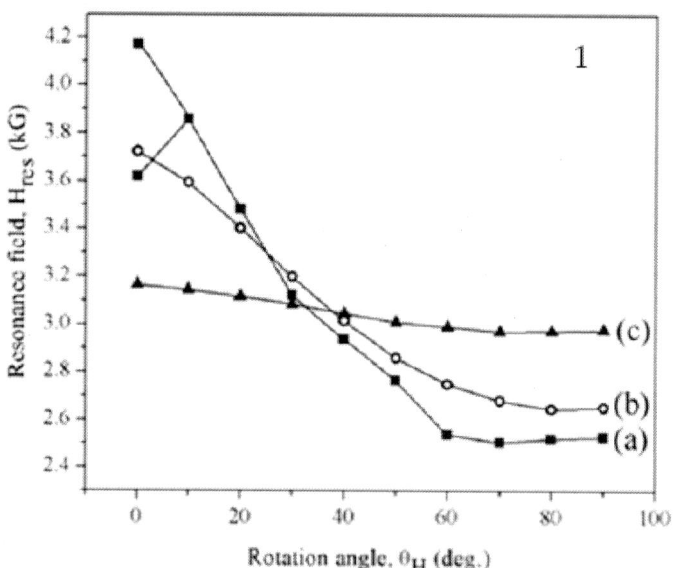

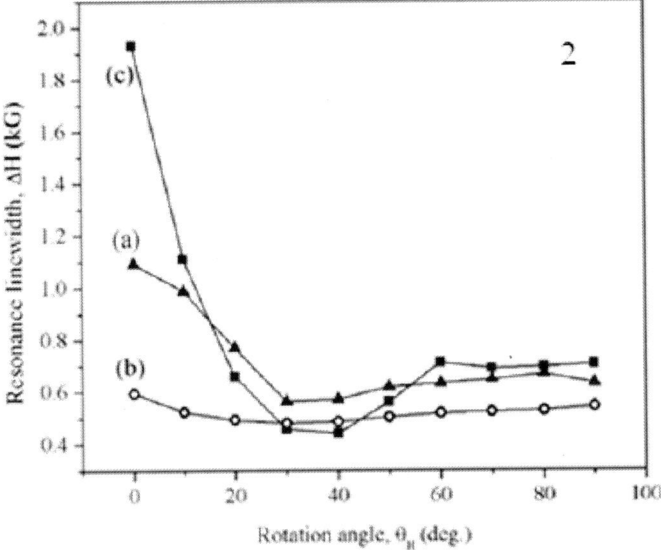

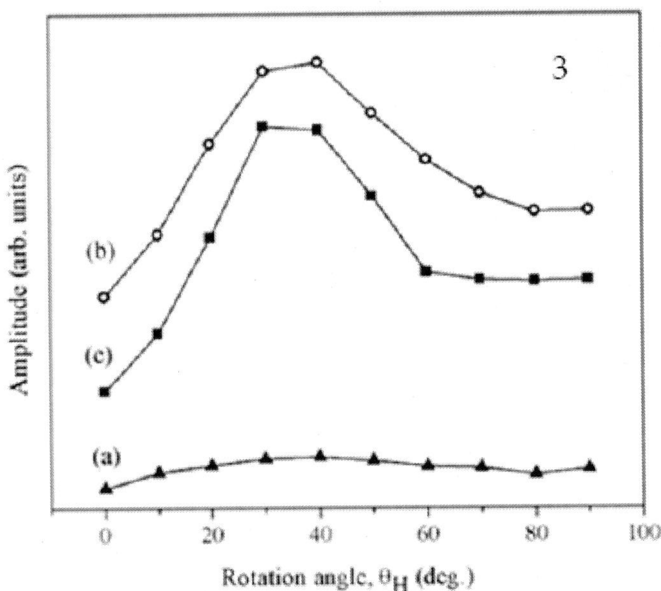

Figure 6. Angle dependence of resonance field (1), resonance linewidth (2) and amplitude (3) of [SnO_2(46 nm)CFZ(x nm)]5 for x = 249 (a), 83(b) and 42 nm (c).

Here Hn is the position of the nth mode of the spin wave resonance, H0 is the position of the FMR resonance, L is the thickness of the sample, M0 is the magnetization of the film, n is an odd integer and A is the exchange constant. This relation is valid for uniformly magnetized film. The results obtained do not fit well into the Kittel model where Hn is plotted against n_2. They rather exhibit a linear behavior with n. According to Portis model [24] for films with non-uniform magnetic profile, the magnetization shows a parabolic drop away from the center of the film. The spin wave modes are given by the following relation

$$H_n = H_0 - \frac{4M_0}{L}(4\pi\epsilon)^{1/2}\left(n + \frac{1}{2}\right) \qquad (2)$$

Here ε is the distortion parameter in the film.

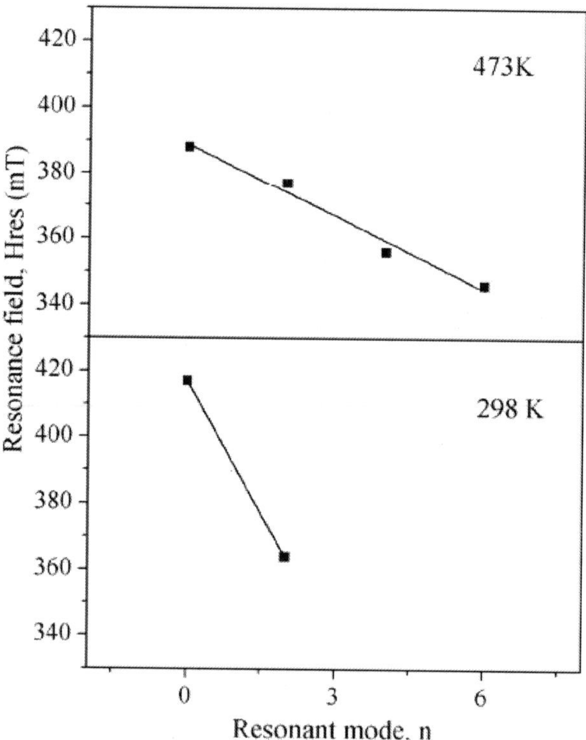

Figure 7. Variation of resonance field with the mode number according to Portis model.

Portis considered that the pinning in the film arises due to non-uniform magnetization in the film. This model proposes a linear relation between Hn and n. Also from the above equation, the difference between the resonance modes decreases with decrease in magnetization.

Figure 7 shows the linear fits corresponding to Portis model. The exchange constant is proportional to the slope of the line. The slope of the line decreases with increasing temperature. Hence the exchange constant decreases with increasing temperature leading to weak coupling and hence lower magnetization. Also, the separation between the resonance modes decreases at higher temperatures, indicating a decreasing magnetization.

The effective magnetization (4пMeff) and effective g value (geff) calculated using the Kittel's relations are listed in table1. At room temperature 4пMeff increases and geff decreases with increasing CZF thickness. This may be due to increasing FM layer thickness [25]. The various factors which can account for deviation of geff from that of the free electron are spin orbit coupling, coupling at the interfaces etc [26].

Table 1. FMR parameters for [SnO2(46 nm)/ CZF (x)]5 ML for various values of x.

x (nm)	H_{res} (kG)		ΔH_{res} (kG)	g (± 0.01)		$4\pi M_{eff}$ (Oe)	g_{eff}	ΔH (G)	
	$\theta_H = 90°$	$\theta_H = 0°$		$\theta_H = 90°$	$\theta_H = 0°$			$\theta_H = 90°$	$\theta_H = 0°$
42	2.977	3.165	0.188	2.21	2.08	125.76	2.16	631	750
86	2.649	3.720	1.071	2.48	1.77	728.39	2.19	541	590
249	2.526	3.633, 4.173	1.107, 1.647	2.60	1.81, 1.58	754.41, 1132.88	2.28, 2.14	703	1930

The perpendicular anisotropy field HK is obtained from the following equation [27]

$$H_K = 4\pi M_S - 4\pi M_{eff} \tag{3}$$

The values of K are listed in Table 2. In multilayers, there are two factors which contribute to the effective anisotropy - anisotropy due to the interfacial region (KS) and anisotropy due to the bulk volume of the magnetic layer (KV). The relation is given as follows [28].

$$K = K_V + 2\frac{K_S}{x} \tag{4}$$

Where x is the thickness of the magnetic layer. The factor of 2 in the second term on the (r.h.s) of the equation is due to an assumption that the magnetic layer is bound by identical interfaces on either side. When (K.x) is plotted against t (figure 8), the slope gives the value of KV and the intercept gives the value of KS. The negative slope indicates a negative volume anisotropy which tends to confine the magnetic moment in the plane of the film [29]. The intercept at x = 0 gives positive value of KS which tends to align the magnetic moment perpendicular to the surface of the film. The intercept on the x axis gives the thickness at which the contribution from KV outweighs that from KS. From the figure, the intercept on x-axis is found to be 56 nm. Thus below 56nm, major contribution comes from interfaces. The KS and KV values obtained from the plot are 194980 and -3671 erg/cm^{3}.

The value of K changes from positive to negative as the thickness of CZF increases (Figure 8). A negative value of K indicates in-plane anisotropy [30]. Thus with increasing CZF thickness, the direction of magnetization is confined to the plane of the sample surface. The thickness value at which the anisotropy changes sign is 56 nm. The perpendicular anisotropy for small values of x may be due to the lattice mismatch at the interfaces [31].

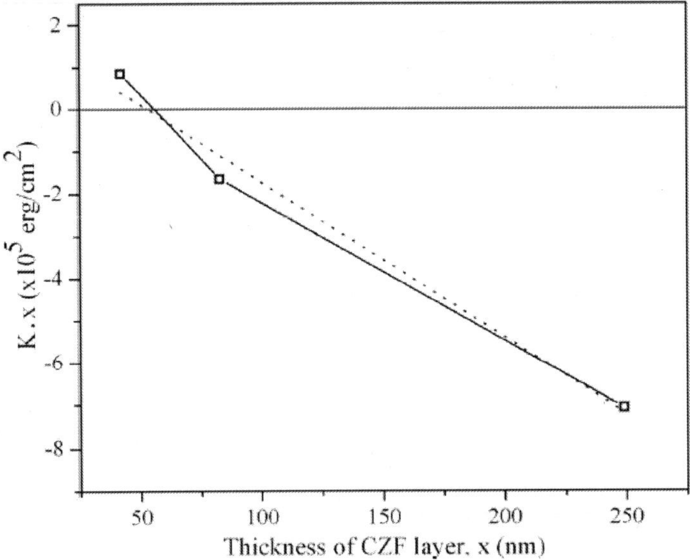

Figure 8 (K..x) vs x plots for [SnO_2(46 nm)CFZ(x nm)]$_5$ ML for x = 43, 83 and 249 nm.

Gilbert damping is a spin relaxation phenomenon in magnetic systems which controls the rate at which the spins reach equilibrium. Spin-orbit coupling [32], non-local spin relaxations like spin wave dissipation [33,34] and disorder present in the materials are the major factors causing Gilbert damping. The damping parameter α values are 0.0096, 0.0083 and 0.0108 for x = 42, 83 and 249 nm respectively. The increase in α value with CZF thickness indicates the increase in the damping of spins with increasing thickness.

Table 2. Damping and anisotropy parameters for or [SnO_2(46 nm)/ CZF (x)]$_5$ ML for various values of x.

x (nm)	α		K_1 (10^3 erg/cm^3)
	$\theta_H = 90°$	$\theta_H = 0°$	
42	0.049	0.057	2.027
86	0.041	0.045	-1.989
249	0.053	0.146	-2.845, -1.254

The temperature dependent FMR studies in the temperature range 133- 473K were carried out on the ML. Figure 9 shows the FMR spectra of the ML at various temperatures. The change in line shape indicates change in the magnetic interaction with increasing temperature. Figure 10 shows the variation of Ipp with temperature in the parallel configuration. Ipp increases with increasing temperature for all the ML. Ipp of the ML with x= 42 nm exhibits a linear trend. Whereas for ML with x=83 and 249 nm, the increase is not linear. It increases slowly from 133K to 298K (RT), followed by a rapid increase with temperature in the temperature range RT- 473K. From 133K to 250K there is no significant difference in amplitude for ML with x= 83 and 249 nm. At temperatures above RT the amplitude of ML with x= 83 nm is higher than that of ML with x = 249 nm.

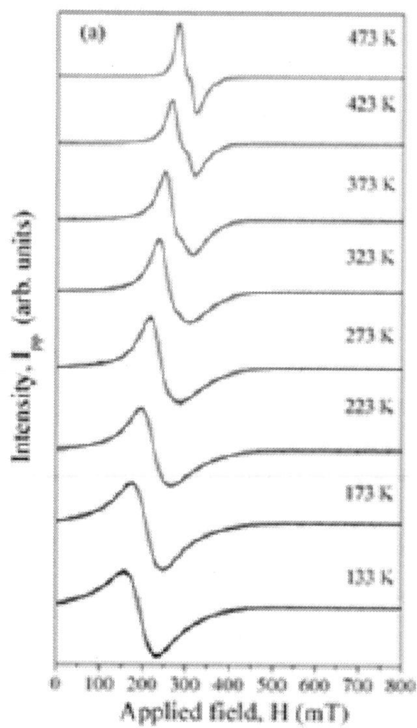

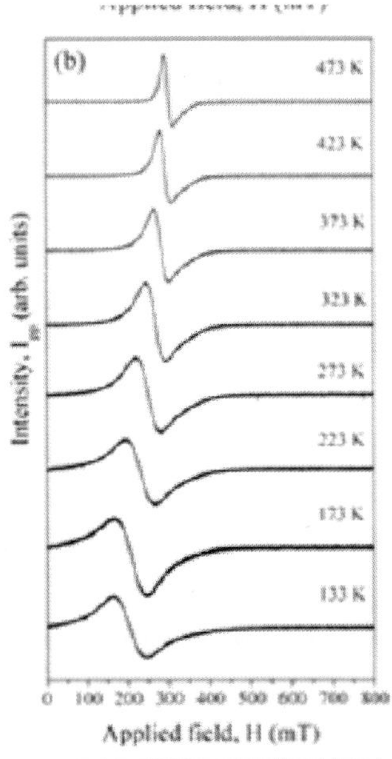

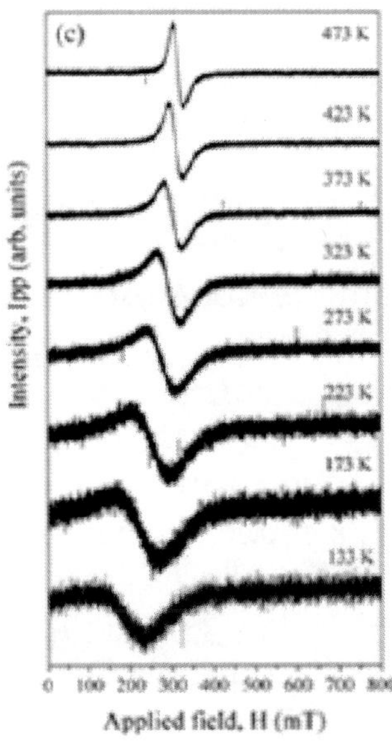

Figure 9. FMR spectra of [SnO_2(46 nm)CFZ(x nm)]$_5$ ML at various temperatures for x = 249 (a), 83 (b) and 42 nm (c)in the parallel configuration.

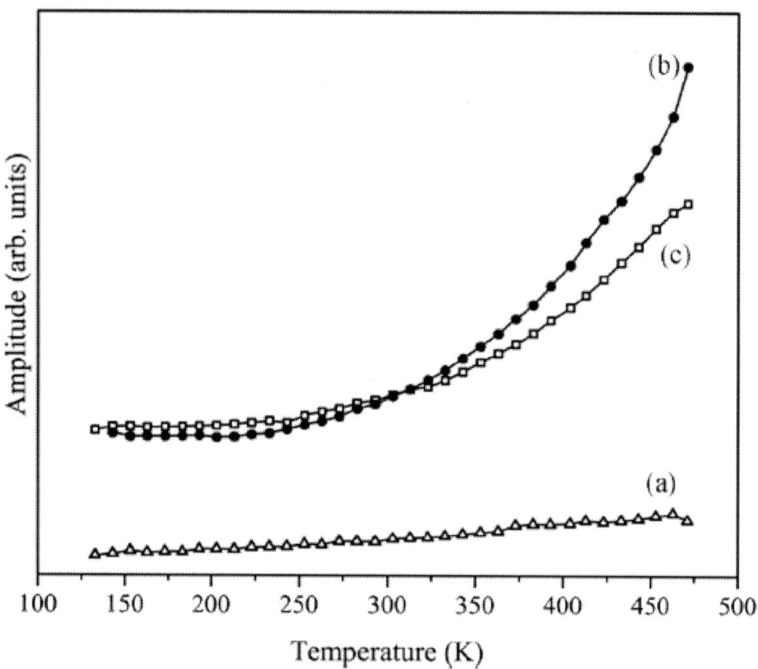

Figure 10. Temperature dependent amplitude of $[SnO_2(46\ nm)CFZ(x\ nm)]5$ for x 249(a), 83(b) and 42 nm (c).

Variation of ΔH with temperature in the parallel configuration is displayed in figure 11. The width of the FMR spectra increases continuously with decreasing temperature for x= 42 nm. For x = 83 nm it decreases with temperature and saturates at ~ 215 K . Whereas for x = 249 nm it begins saturating at ~ RT presumably due to freezing of spins. The H_{res} increases with increasing temperature (Figure 12) in all the ML presumably due to increasing internal field at low temperatures. The ML with x = 249 nm exhibits a linear dependence in the entire temperature range whereas the ML with x = 83 and 42 nm exhibit saturating tendency at higher temperatures in the parallel configuration. This may be an interfacial effect. For x = 42 nm the ratio of spins at the interfaces to the spins in the bulk is high. Whereas for x = 83 and 249 nm the ratio is lower. Hence the contribution from the bulk of the film becomes significant than that from the interfaces. The saturating tendency at high temperatures for the ML with x = 83 and 42 nm shows that even at higher temperatures, the thermal energy is not sufficient to unlock the spins at the interfaces.

Similar behavior is observed for CZF thin films [12]. The increase in saturation temperature with increase in CZF thickness could be due to the interlayer coupling. The change of slope of ΔH indicates a change in the kind of magnetic interactions in the ML. The line shape of the FMR spectra also changes around the same temperature. The curvature of the temperature dependent of ΔH plot changes as x increases from 42 to 249 presumably due to increase in active to dead layer ratio. Decreasing trend in Ipp and H_{res} and increasing trend in ΔH with decreasing temperature is a signature of super para magnetism [35]. The resonance field of a given particle includes contribution from magneto crystalline anisotropy field and demagnetizing field.

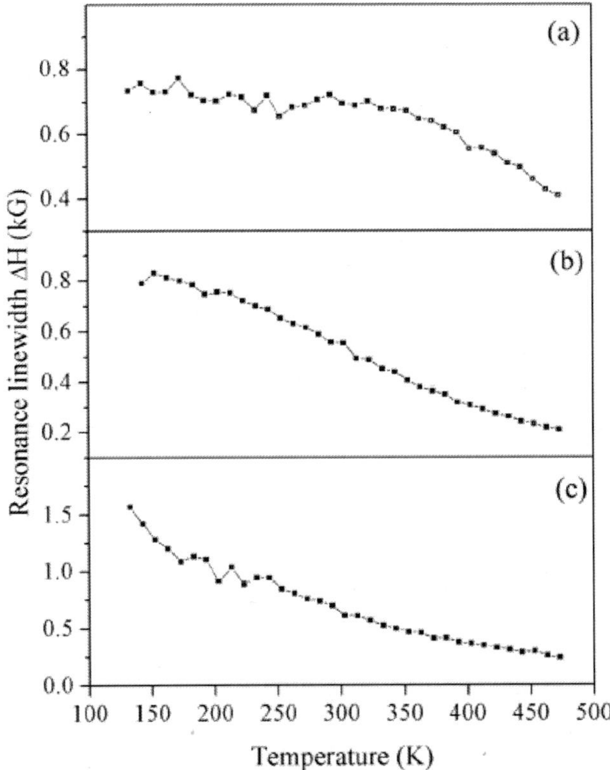

Figure 11. Temperature dependent resonance linewidth of [SnO2(46 nm) CFZ(x nm)]5 for x = 249(a), 83(b) and 42 nm (c).

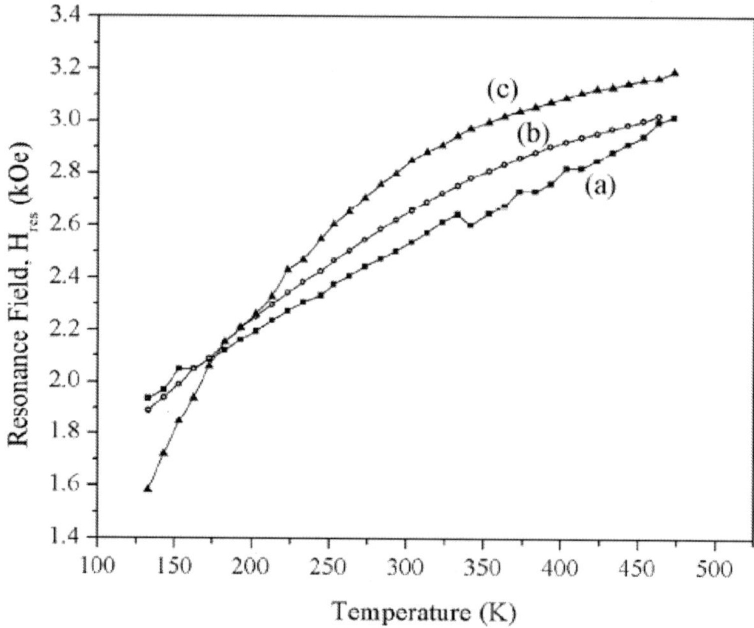

Figure 12. Temperature dependent resonance field, H_{res}, in the parallel and perpendicular configuration of [SnO_2(46 nm)CFZ(x nm)]5 for x = 249(a), 83(b) and 42 nm (c).

For a system containing randomly oriented particles, the magnetic resonance signal is broad. At high temperatures where kbT is much greater than the anisotropy barriers the thermal fluctuations of the magnetic moments reduces the angular anisotropy of the resonance fields and the linewidth of individual nanoparticle [36] resulting in a superparamagnetic resonance (SPR) spectra in the case of nanoparticles. At low temperatures the particle moments are unable to overcome the local anisotropy barriers and, thus, become trapped in metastable states (blocking phenomenon). This results in broad resonance spectra with linewidth comparable to that of the resonance field.

A characteristic feature of SPR is that at low temperatures, ΔH is almost equal to the value of Hres. For the ML with x = 42 nm around 40% of CZF layer is dead. The diffused region consists of fine particles of CZF dispersed in SnO2 matrix. Hence the SPR phenomenon is pronounced for this ML. Whereas for ML with x = 83 and 249 nm, the thickness of the diffused region is small compared to that of the bulk CZF layer. Hence the value of ΔH at low temperatures is not same as that of Hres.

FMR susceptibility is estimated by calculating the double integrated intensity i.e. area under the absorption curve in FMR is proportional to the concentration of spins. Figure 13 shows temperature dependence of ESR susceptibility in the parallel configuration for *[SnO_2(46 nm)CFZ(x nm)]$_5$* ML. It initially increases with decreasing temperature, reaches a maximum and then decreases with further decrease in temperature evidencing the super paramagnetic behavior. The width of the transition is large due to the wide spread of blocking temperatures.

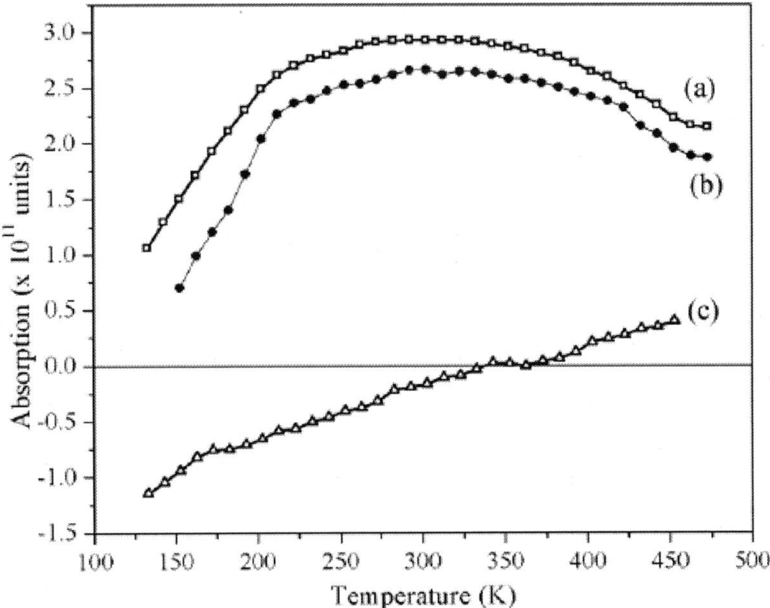

Figure 13. Variation of FMR susceptibility, χFMR with temperature of [SnO_2(46 nm)CFZ(x nm)]5 ML for x= 249 (a), 83 (b) and 42 nm (c).

For ML with x = 42 nm, χFMR decreases continuously with decrease in temperature down till 183 K. The negative absorption is due to the fact that the area under the lower part of the FMR curve is greater than that of the upper part. The Curie- Weiss law does not fit 1/ χ FMR vs T plot due to the continuous curvature in the entire temperatures range.

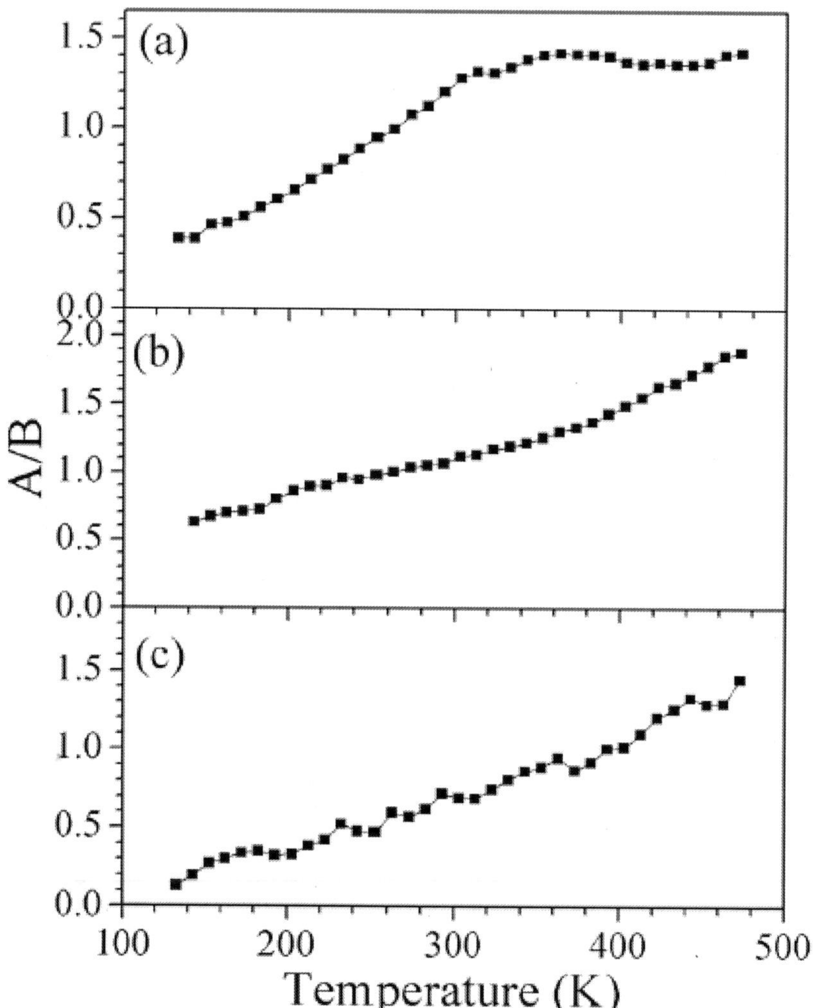

Figure 14. Variation A/B ratio with temperature of [SnO2(46 nm)CFZ(x nm)]5 ML for x = 249(a), 83(b) and 42 nm (c).

Figure 14 shows the A/B ratio with temperature for these ML. It decreases almost linearly with decreasing temperature from 400 to 200 K for the ML with x = 42 nm. Below 200 K there is a change in slope. The temperature dependence of A/B ratio exhibits different slopes in different temperature region for the ML with x = 83 nm. Whereas for the ML with x = 249 nm, there is no significant change in the A/B ratio in the temperature range between 500 and 350 K. Below 350 K the ratio decreases continuously. The temperature at which the slope for A/B ratio changes matches with that of temperature dependence of linewidth.

Temperature dependence of FMR spectra of these ML in the perpendicular configurations is displayed in Figure 15. The resonance field shifts towards higher values with decreasing temperature. This is accompanied with decrease in intensity and increase in line width. This indicates that AFM interactions dominate as temperature decreases in the perpendicular configuration.

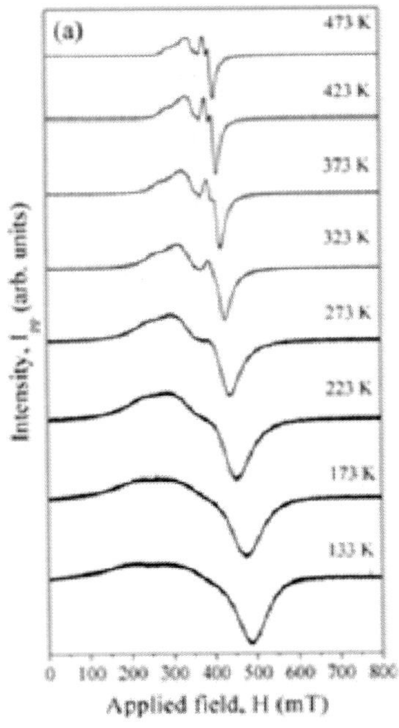

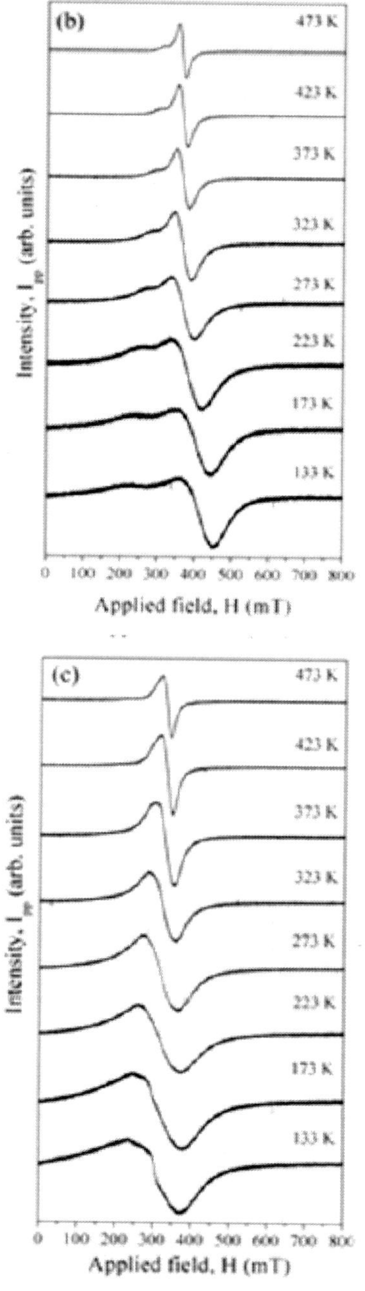

Figure 15. FMR spectra of [SnO2(46 nm)CFZ(x nm)]5 at various temperatures in the perpendicular configuration for x = 249 (a), 83 (b) and 42 nm (c).

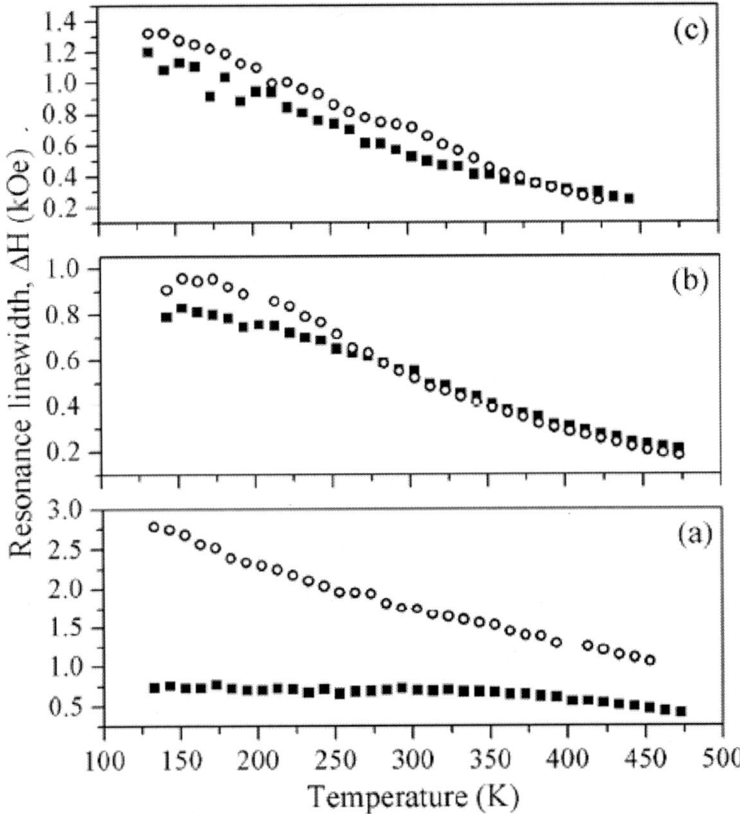

Figure 16 Temperature dependent resonance linewidth in the parallel (full circle) and perpendicular (open circle) configuration of [SnO_2 (46 nm)CFZ(x nm)]5 for x = 249(a), 83(b) and 42 nm (c).

Figure 17 shows the variation of Hres in the parallel and perpendicular configuration. Unlike the parallel configuration, the resonance field increases with decreasing temperature in the perpendicular configuration for ML with x = 249 and 83 nm. Hres remains insensitive to temperature in the range 300- 473K. Below 300 K it increases with further decrease in temperature.

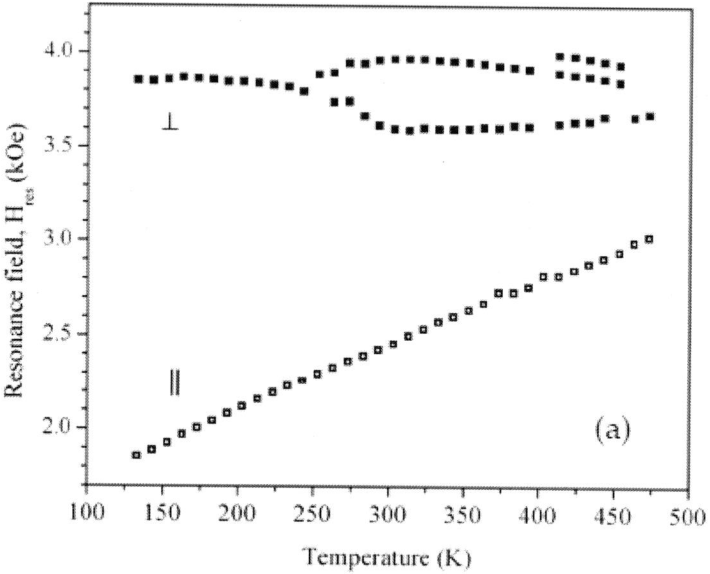
Resonance field, H_{res} (kOe)
Temperature (K)
⊥
||
(a)

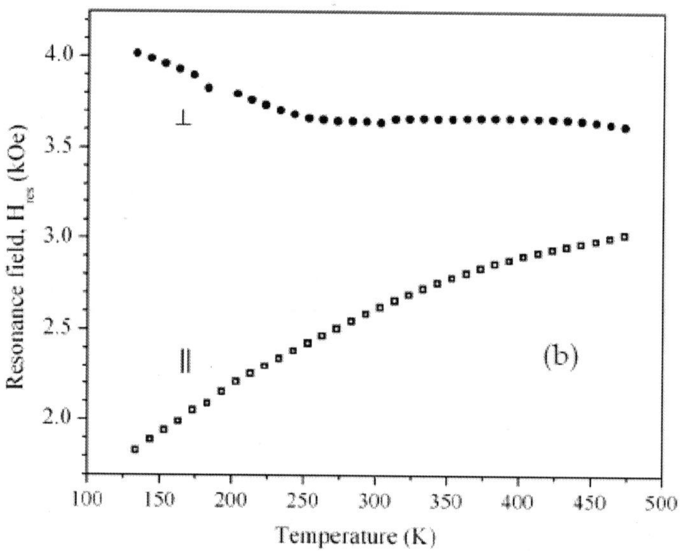
Resonance field, H_{res} (kOe)
Temperature (K)
⊥
||
(b)

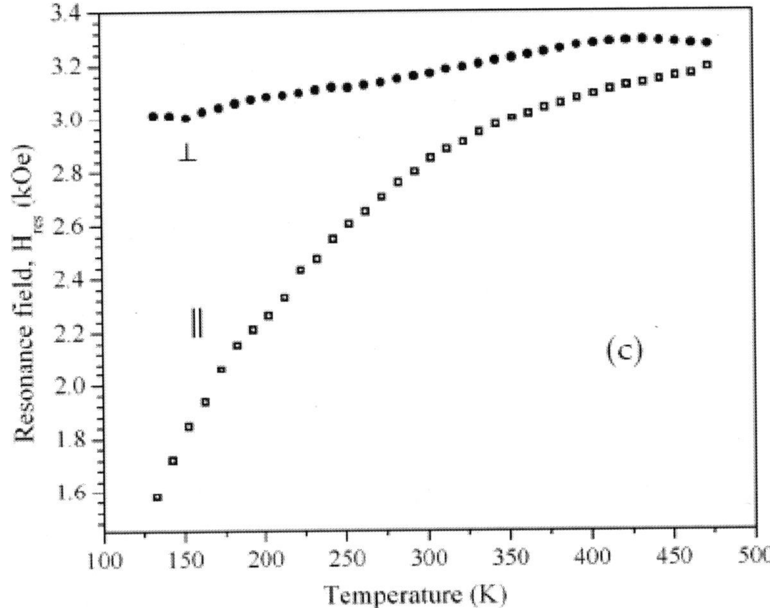

Figure 17 Temperature dependent resonance field, H_{res}, in the parallel and perpendicular configuration of $[SnO_2(46 \text{ nm})CFZ(x \text{ nm})]_5$ for x = 249 (a), 83 (b) and 42 nm (c).

Whereas for the ML with x = 42 nm, initially there is a small increase in H_{res} as temperature decreases from 473 to 433 K. Below 433 K it decreases continuously. At all temperatures, $_{Hres}$ is less sensitive to temperature in the perpendicular configuration than in the parallel configuration.

For all the ML, the FMR susceptibility increases with decreasing temperature in the perpendicular configuration. The effective magnetization is a measure of net magnetization within the ML. Figure 18 shows the variation of effective magnetization with temperature. It increases with decreasing temperature for all ML.

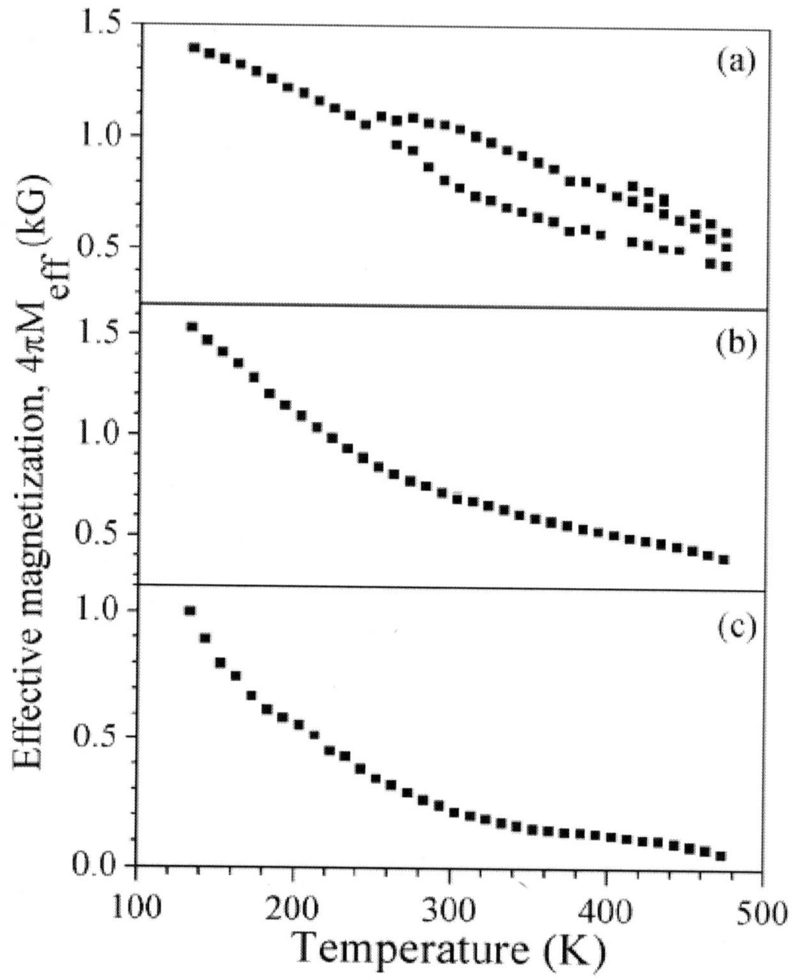

Figure 18. Variation of effective magnetization with temperature for [SnO2 (46 nm)CFZ(x nm)]5 ML for x = 249(a), 83(b) and 42 nm (c).

Figure 19 is a representative graph of [SnO2 (46 nm)/ CZF(249 nm)]$_5$ ML showing the variation of FMR susceptibility with temperature in both parallel and perpendicular configurations. Unlike the parallel configuration, χFMR in the perpendicular configuration increases continuously with decreasing temperature. The FMR spectra were recorded at different orientations of the normal to the sample with the applied field (θH). The rotational dependence of FMR spectra was measured at 133, 298 (room temperature) and 473 K. At all temperatures the resonance field increases continuously as θH decreases exhibiting uniaxial anisotropy. For the ML, the difference between ΔH values in the parallel and perpendicular configuration increases with decreasing temperature. This shows increasing anisotropy with decreasing temperature. The multiple resonance for the ML with x = 249 is prominent at 298 and 473 K (Figure 20(a)). For

ML with x < 249 nm, the absence of multiple resonances even at high temperatures indicates absence of interlayer coupling. At all temperatures the ΔH_{res} is highest for the ML with x = 249 nm. These multiple resonances disappear with decreasing temperature.

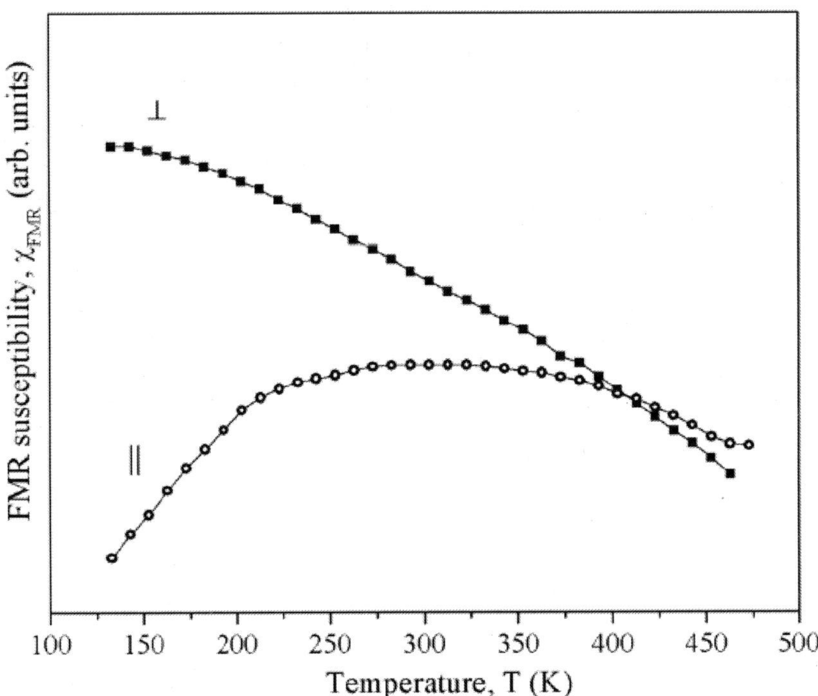

Figure 19. Representative plot of the FMR susceptibility, FMR of [SnO2 (46 nm) CFZ(249 nm)]5 ML in parallel and perpendicular configuration.

At room temperature, ΔH shows a minimum ~ 30-40° (figure 21). Ipp shows maximum around the same value. Hence the sample has an easy axis of magnetization at that inclination. At 473 and 298 K its intensity initially increases, reaches a maximum and then decreases with decreasing θH. Whereas at 133 K, Ipp values remains constant initially then decreases continuously with decreasing value of θH.

Another evidence for decreasing anisotropy with decreasing x is shown in figure 21. At 473 K, ΔH for the ML with x = 249 nm initially increases, exhibits a minimum ~ 40° and then increases continuously with further increase in θH . Whereas for the ML with x = 83 nm, with increasing θH, ΔH value remains constant till certain angle then decreases continuously. On the other hand the ΔH value for the ML with x = 42 nm, the ΔH value continuously decrease reaches a minimum and then increases with increasing θH. The occurrence of minimum in ΔH may be due to existence of regions with different effective magnetization values due to inhomogeneous nature of the sample [37]. The minimum in ΔH shifts towards higher angle with decreasing x. Similar behavior is observed at 298 K for all the ML. With decreasing temperature, the curvature of the angle dependent linewidth changes. At 133 K, for x = 249 nm, the minima vanishes and ΔH increases continuously with increasing θH. For the ML with x = 83 nm, the curvature is reversed at 133 K. It increases with increasing θH, attains a maximum value at ~ 40° and then decreases. On the contrary there is no significant change in ΔH for the ML with x = 42 nm. Thus at 133 K, x = 249 nm ML is more anisotropic than the ML with x = 42 and 83 nm.

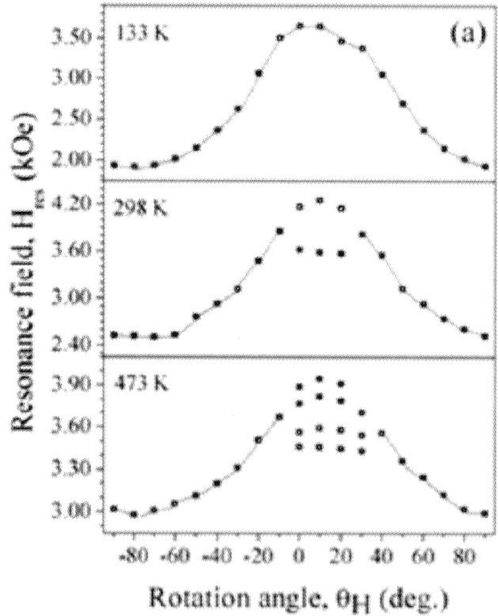

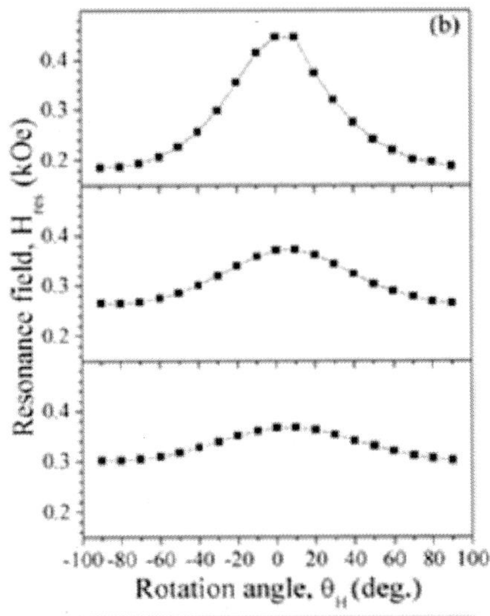

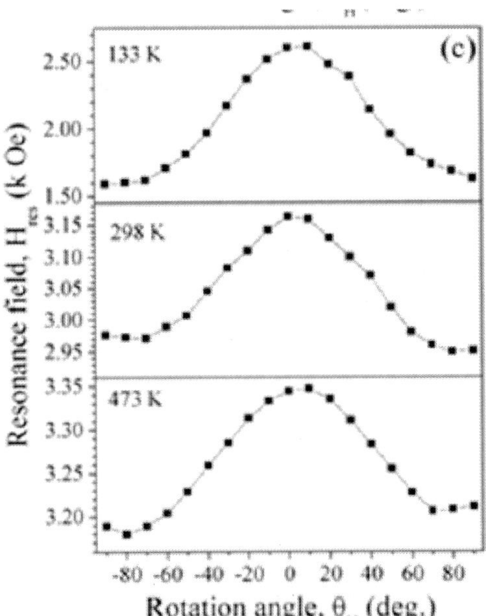

Figure 20. Rotational dependence of resonance field, Hres, at various temperatures for [SnO_2 (46 nm)CFZ(x nm)]5 ML where x = 249(a), 83(b) and 42 nm (c).

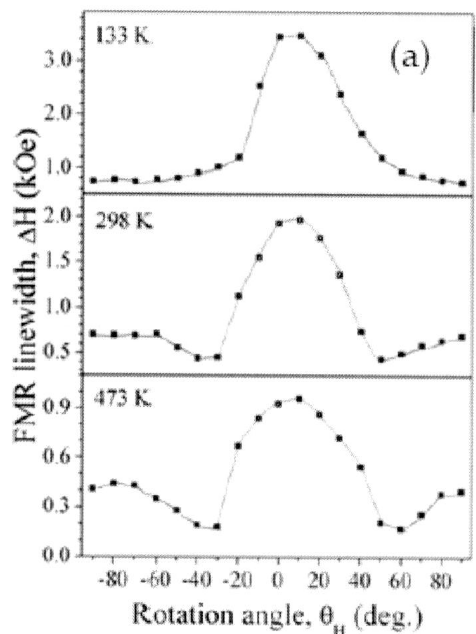
133 K
(a)
298 K
473 K
FMR linewidth, ΔH (kOe)
Rotation angle, θ_H (deg.)

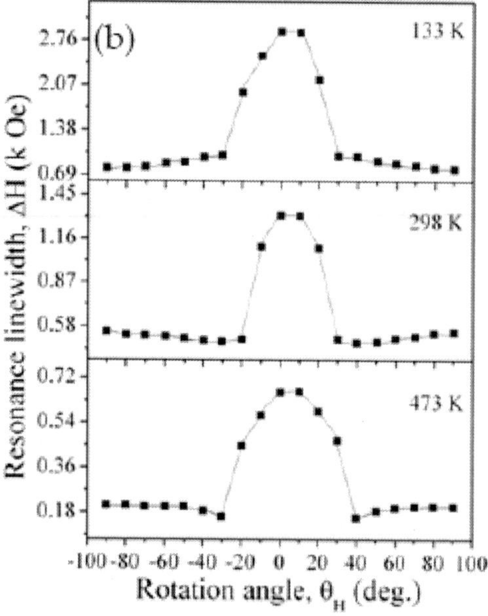
(b)
133 K
298 K
473 K
Resonance linewidth, ΔH (k Oe)
Rotation angle, θ_H (deg.)

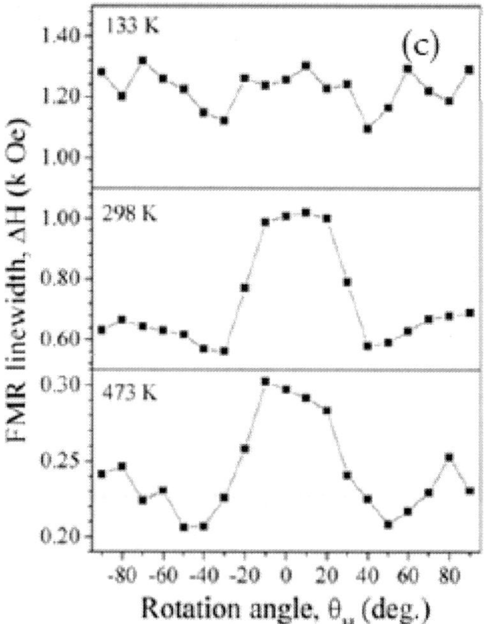

Figure 21. Rotational dependence of resonance linewidth, ΔH, at various temperatures for [SnO_2 (46 nm)CFZ(x nm)]$_5$ ML for x = 42(a), 83(b) and 249 nm (c).

Temperature, the curvature of the angle dependent linewidth changes. At 133 K, for x = 249 nm, the minima vanishes and ΔH increases continuously with increasing θH. For the ML with x = 83 nm, the curvature is reversed at 133 K. It increases with increasing θH, attains a maximum value at ~ 40° and then decreases. On the contrary there is no significant changein ΔH for the ML with x = 42 nm. Thus at 133 K, x = 249 nm ML is more anisotropic than the ML with x = 42 and 83 nm.

Effect of Spacer Layer Thickness

The effect of thickness of spacer (SnO2) layer was studied on 3 sets of samples with varying thickness of the SnO2 layer. The total number of bilayers is 5 and individual CZF layer thickness is 83 nm for all the samples. The ML samples were [SnO2(x nm) CZF(83 nm)]5 ML for x = 46 , 115 and 184 nm .

Stack of alternate dark and light layers are evident from the FESEM images (Figure 22). The dark columnar layers correspond to CZF and the light layers correspond to SnO2. The variation in the thickness ratio between SnO2 and CZF is also clearly visible from these images. The absence of fringes in XRR spectra may be due to large thickness of the ML stack.

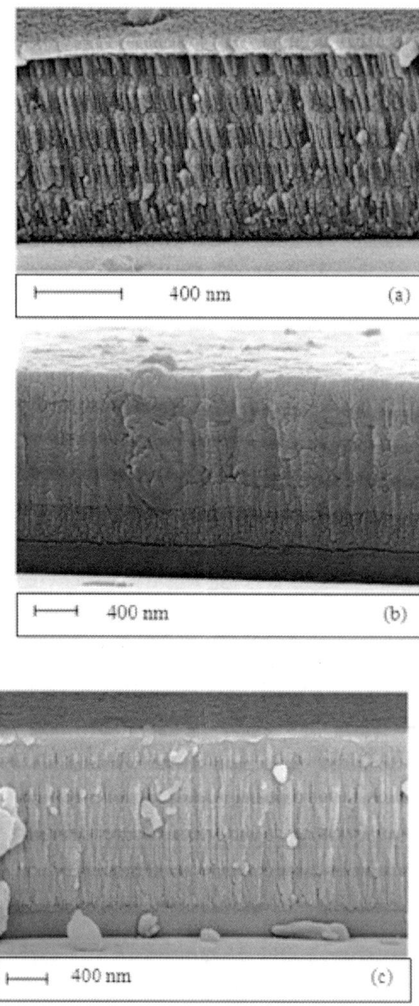

Figure 22. FESEM images of [SnO2(x nm) CZF(83 nm)]5 ML for x = 46 (a), 115 (b) and 184 nm (c).

FMR studies were carried out at different temperatures and by varying the angle θH between the film normal and the direction of the applied field. Figure 23 shows the first derivative of the FMR absorption spectra of the ML at room temperature in parallel (θH = 90°) and perpendicular (θH = 0°) geometry, normalized to the volume of the CZF layers. Ipp increases as the spacer (SnO2) thickness increases from x = 46 to 115 nm followed by a decrease for x = 184 nm. The peak to peak linewidth (ΔH) is sensitive to inhomogenieties, surface roughness, internal field etc. The χ FMR is estimated by calculating double integrated intensity (DI). Both ΔH and DI follow a trend similar to that of Ipp with increase in spacer layer thickness. This behavior may be attributed to the oscillation in the exchange coupling across the NM layer due to quantum interference of the electron waves reflected from the interfaces. The A/B ratio is 0.82, 0.74 and 0.79 for x = 46, 115 and 184 nm respectively. The asymmetry ratio is not very different from each other. For the ML with x = 115 nm, the upper part of the FMR signal rises steeply whereas the rise is relatively slow for the lower part indicating a negative anisotropy.

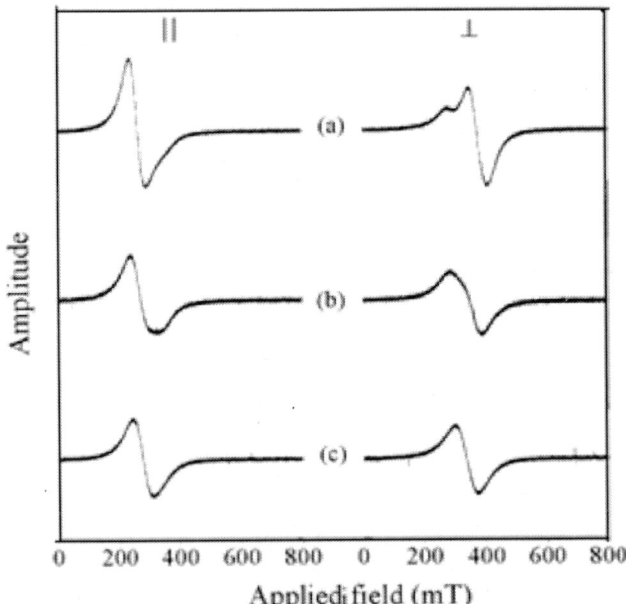

Figure 23. Room temperature FMR spectra of [SnO_2(x nm) CZF(83 nm)]5 ML in the parallel and perpendicular configuration for x = 46 (a), 115 (b) and 184 nm (c).

The resonance field (H_{res}) increases from 2.65 to 2.78 kG and the g value decreases from 2.47 to 2.36 with increase in SnO_2 layer thickness in parallel geometry. The shift in Hres to higher values might be due to the decrease in interlayer coupling strength. The Hres decreases and g values increases from 1.77 to 1.96 with increasing spacer layer thickness in the perpendicular geometry. This might be attributed to lower perpendicular anisotropy for well separated films [38]. The FMR parameters in the parallel and perpendicular configurations are listed in Table 3. The effective magnetization decreases with increasing SnO_2 thickness.

Table 3. FMR parameters of [SnO2(x nm) CZF(83 nm)]5 multilayers for various spacer layer thickness

x (nm)	H_{res} (kG)		ΔH_{res} (kG)	g (±0.01)		$4\pi M_{eff}$ (Oe)	ΔH (G)		Ms (emu/cc CZF)
	$\theta_H = 90°$	$\theta_H = 0°$		$\theta_H = 90°$	$\theta_H = 0°$		$\theta_H = 90°$	$\theta_H = 0°$	
46	2.652	3.721	1.069	2.48	1.77	710	541	595	50
115	2.733	3.452	0.719	2.41	1.90	485	693	733	100
184	2.784	3.356	0.576	2.36	1.96	406	675	736	30

The ML with x = 46 nm exhibits multiple peaks in the FMR spectrum recorded in the perpendicular configuration. This may be due to the interlayer coupling of the CZF layers mediated by the nonmagnetic spacer layer. When two spins are coupled across the spacer layer, they tend to process at a frequency different from that of the rest. Hence the resonance for the coupled spins occurs at a different field. The absence of multiple splitting in the other two samples indicates lack of interlayer coupling in the magnetic layers. Apart from the multiple resonances, the asymmetry in the FMR signal in the perpendicular configuration also decreases with the increase in SnO2 layer thickness. As the thickness of the spacer increases, the distance between the CZF layers increases and hence the coupling between them decreases leading to symmetrical FMR spectra. The effective magnetization value at room temperature was calculated using the Kittel's relations and are 710, 485 and 406 Oe for x = 46, 115 and 184 nm respectively.

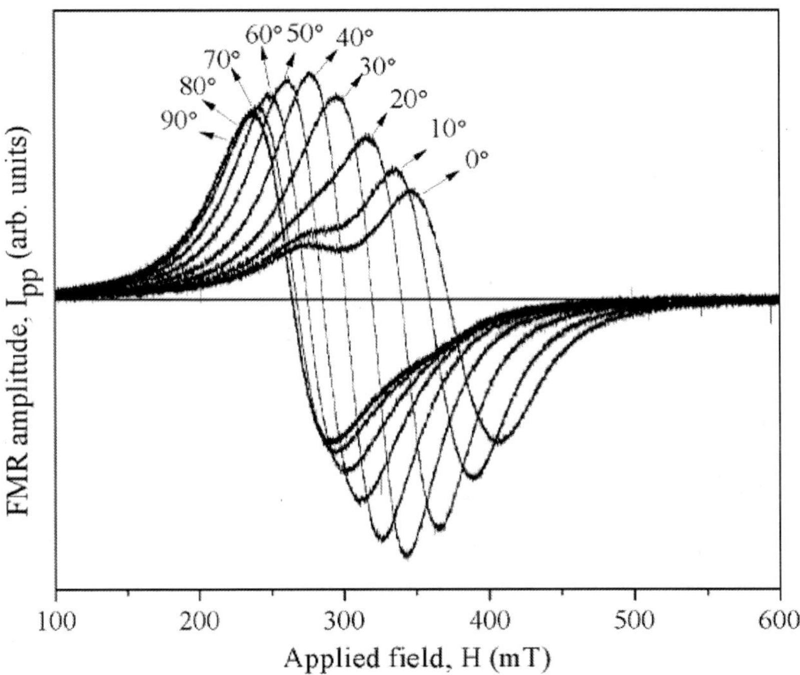

Figure 24. Room temperature FMR spectra of [SnO_2(46 nm)CFZ(83 nm)]5 at different orientations.

The angle dependent FMR spectra are shown in figure 24. The symmetry in the FMR spectra increases as θH decreases from 90° to 40°. For θH > 40° the spectra becomes asymmetric and the single FMR peak splits into two resonance peaks. With decreasing θH the separation of the peaks increases.

Figure 25 is a representative graph of the angular dependence of ΔH and Ipp. For all ML, ΔH value initially decreases till θH = 40° and then increases till θH = 0°. Whereas, Ipp increases till

θH = 40° and then decreases till θH = 0°. This indicates that at θH = 40°, the susceptibility of the film increases resulting in higher magnetization. This may be due to existence of regions with two different effective magnetizations. For all the samples Hres shifts to higher fields as the angle between the sample normal and the direction of the field (θH) changes from 90° to 0°. This indicates that the ML exhibits uniaxial anisotropy.

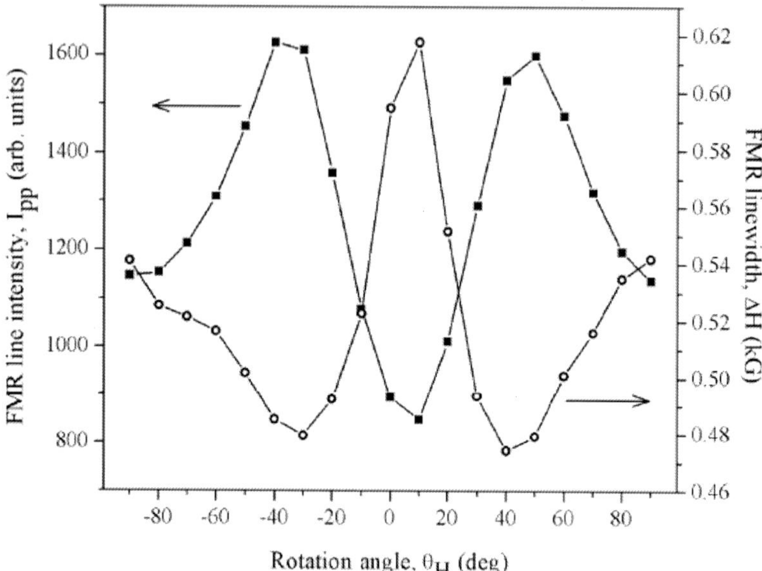

Figure 25. Variation of room temperature FMR linewidth (ΔH) and intensity (Ipp) of [SnO_2 (46 nm) CZF(83 nm)]$_5$ ML at different orientations of the film.

FMR spectra recorded at various temperatures between 133 and 473 K are shown in figure 26. The change in the line shape with decreasing temperature evidences an evolution of a different type of magnetic interaction. The FMR signal is asymmetric at high temperatures and the symmetry increases with decreasing temperature. In the high temperature region A/B is greater than 1 indicating that the upper part of the FMR spectrum is of higher intensity than the lower half and indicates the presence of dispersive component in the ML. Figure 27 shows the variation of A/B with temperature.

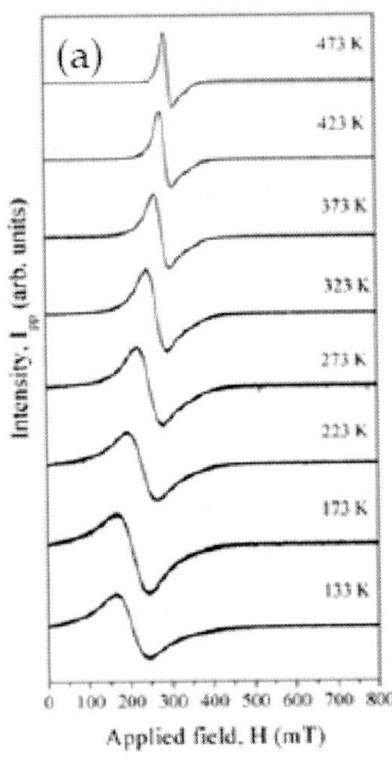
(a)
473 K
423 K
373 K
323 K
273 K
223 K
173 K
133 K
Intensity, I_{pp} (arb. units)
0 100 200 300 400 500 600 700 800
Applied field, H (mT)

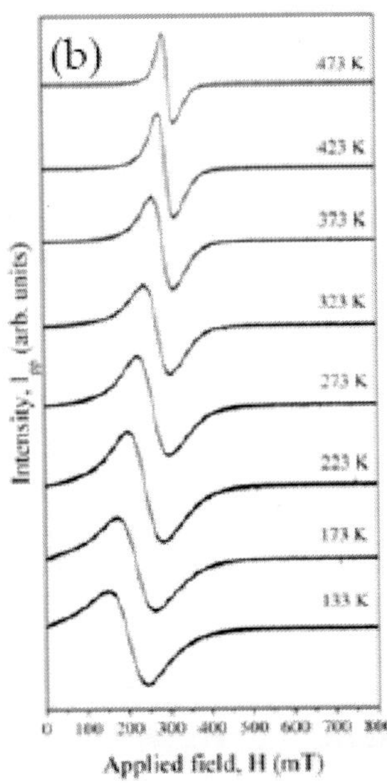
(b)
473 K
423 K
373 K
323 K
273 K
223 K
173 K
133 K
Intensity, I_{pp} (arb. units)
0 100 200 300 400 500 600 700 800
Applied field, H (mT)

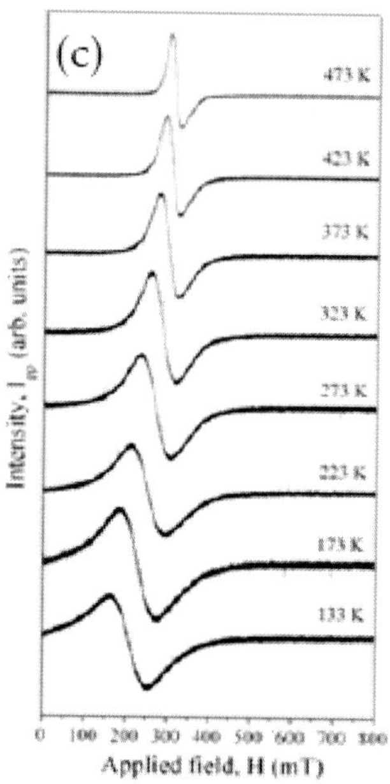

Figure 26. FMR spectra of [SnO_2 (x nm) CZF(83 nm)]$_5$ ML recorded at various temperatures in parallel configuration for x = 46 (a), 115 (b) and 184 (c).

The Ipp value decreases with decreasing temperature. The increase in ΔH at lower temperatures indicates the large spin relaxation times and hence large coercivity at low temperatures (Figure 28(a)).

The decreasing trend in Ipp and Hres and increasing trend in ΔH with decreasing temperature is a signature of superparamagnetism. Figure 29 shows variation of FMR susceptibility, χ FMR as a function of temperature. It initially increases, reaches a maximum and then decreases with decreasing temperature confirming the superparamagnetic nature of the ML.

Figure 30 shows the temperature dependent FMR spectra in the perpendicular configuration. All the ML samples exhibit multiple resonances. The resonances broaden and the separation between them increases with decreasing temperature. The intensity of the peak at lower field side increases and that at higher field side decreases with decreasing temperature.

The behavior of ΔH and Ipp is same as that of θH = 90°. The difference between ΔH in both the configurations is negligible in the

HT region for all the samples. The substantial difference occurs at low temperatures (Figure 31(a)). This may be due to the presence of in homogeneities in the ML [39]. In the parallel configuration H_{res} decreases with decrease in temperature indicating strengthening of FM interactions. Whereas in the perpendicular configuration it is less sensitive to temperature in the HT- RT range, then increases slowly with decrease in temperature (Figure 31 (b)).

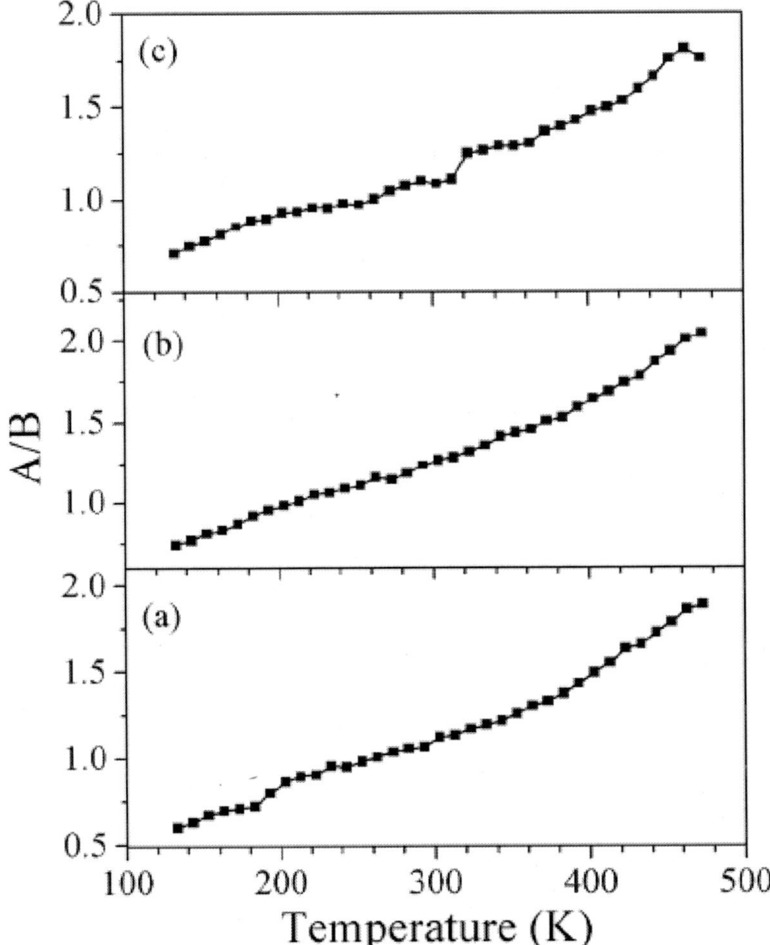

Figure 27. Variation of A/B ratio of [SnO_2 (x nm) CZF (83 nm)]$_5$ with temperature for x = 46 (a), 115 (b) and 184 (c).

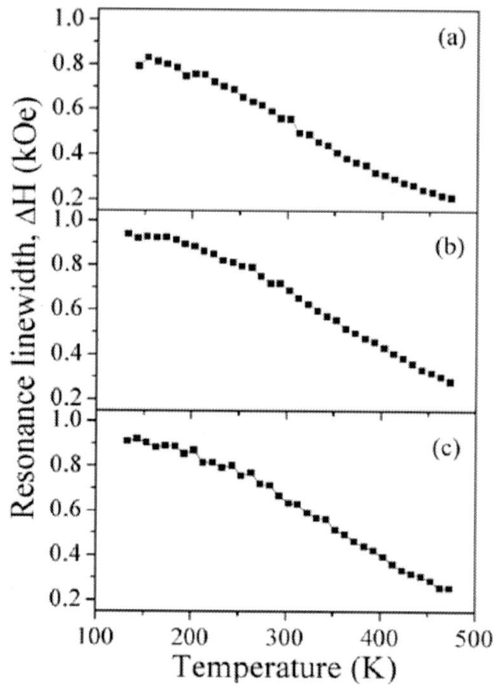

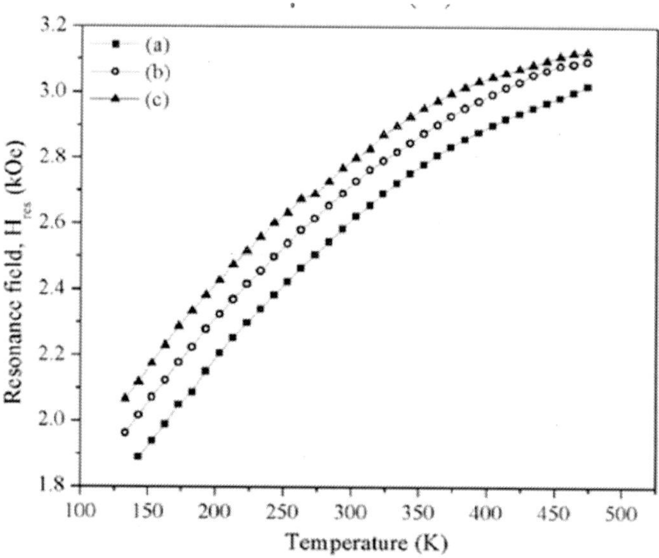

Figure 28. Variation of FMR resonance linewidth (top) and resonance field (bottom) with temperature of [SnO_2 (x nm) CZF(83 nm)]$_5$ for x = 46 (a), 115 (b) and 184 nm (c).

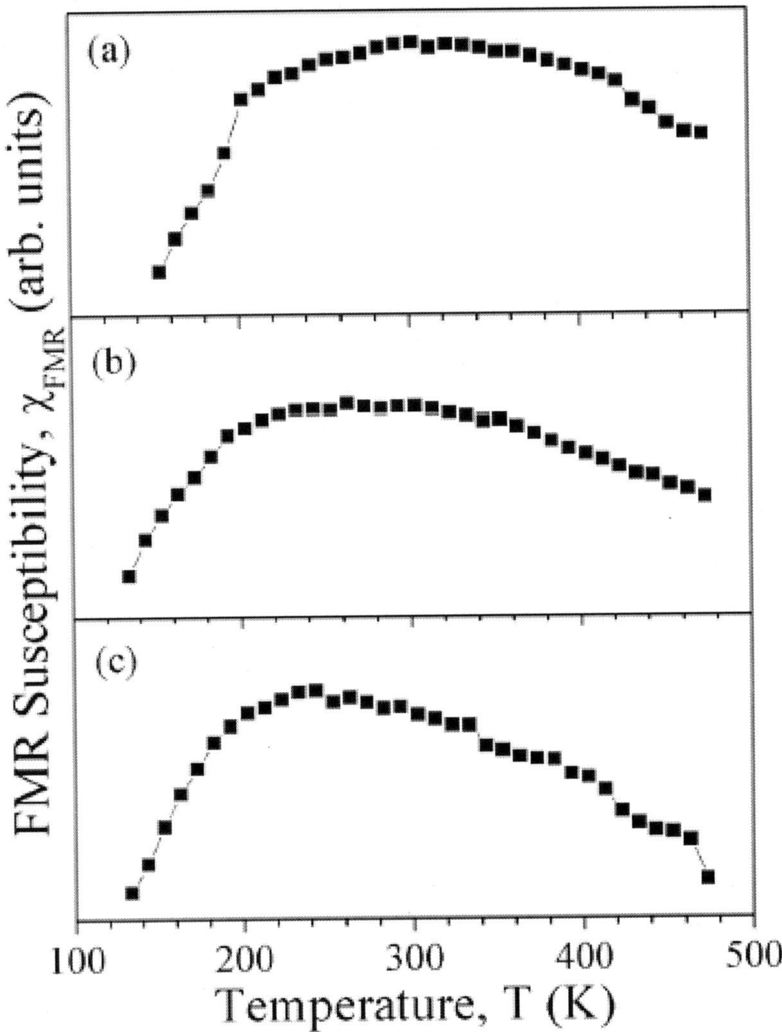

Figure 29. Variation of FMR susceptibility, χFMR with temperature of [SnO2(x nm)CFZ(83 nm)]5 ML for x = 46 (a), 115 (b) and 184 nm (c).

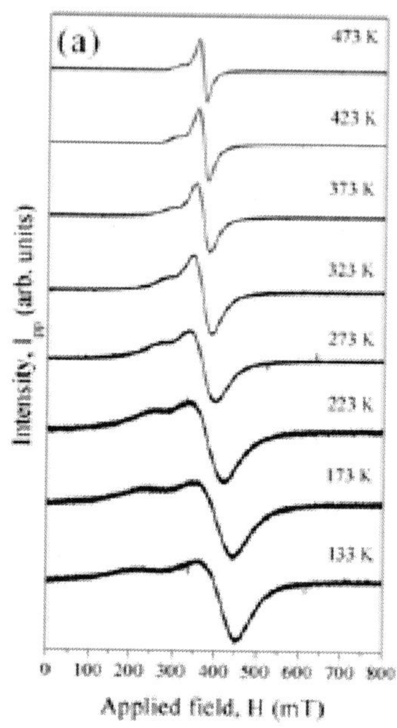
(a)
473 K
423 K
373 K
323 K
273 K
223 K
173 K
133 K
Intensity, I_{pp} (arb. units)
0 100 200 300 400 500 600 700 800
Applied field, H (mT)

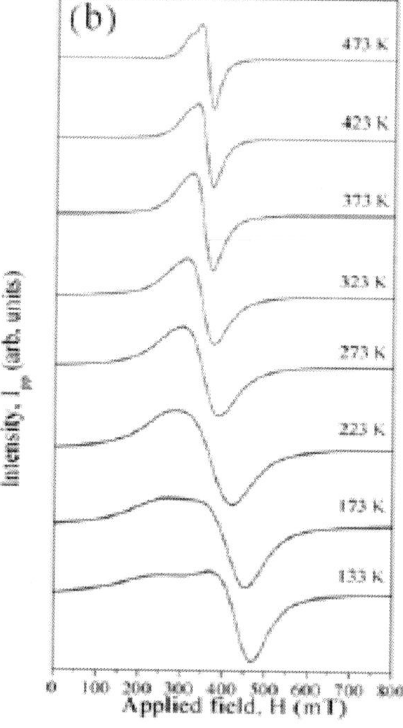
(b)
473 K
423 K
373 K
323 K
273 K
223 K
173 K
133 K
Intensity, I_{pp} (arb. units)
0 100 200 300 400 500 600 700 800
Applied field, H (mT)

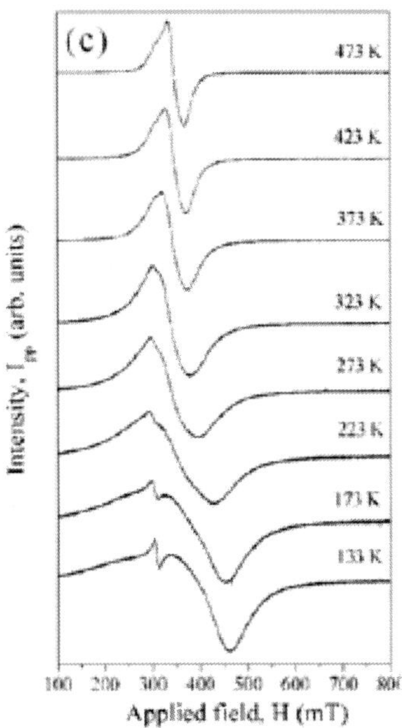

Figure 30. FMR spectra of [SnO_2 (x nm) CZF(83 nm)]$_5$ ML recorded at various temperatures in perpendicular configuration for x = 46 (a), 115 (b) and 184 (c).

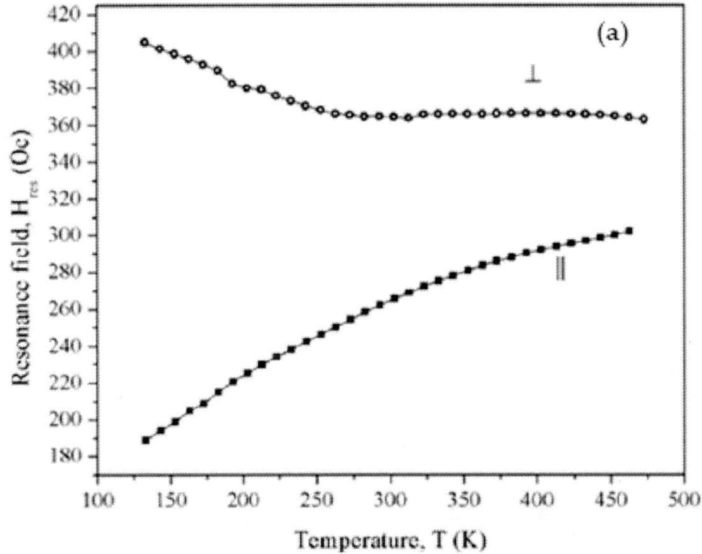

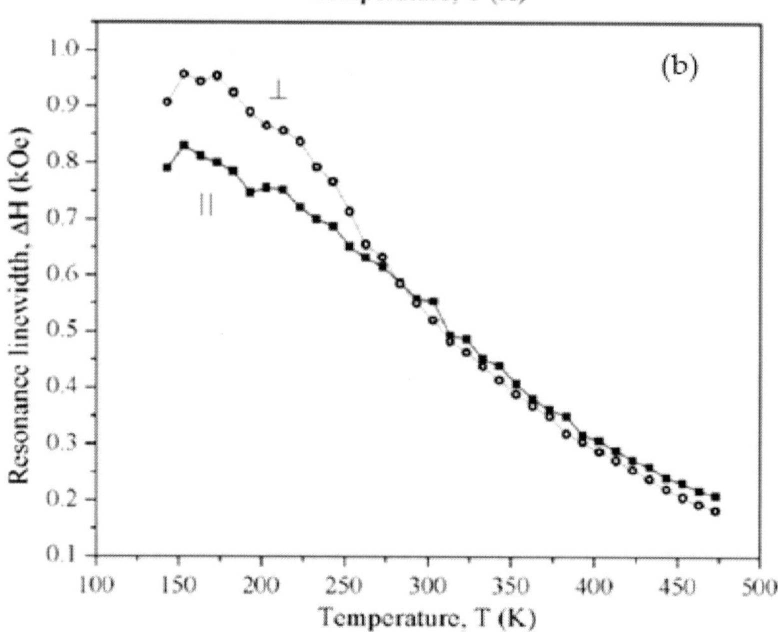

Figure 31. Variation of resonance field, Hres (a) and resonance linewidth, ΔH (b) with temperature in parallel and perpendicular configurations for [SnO2 (46 nm) CZF(83 nm)]5 ML.

The effective magnetization is found to increase with decreasing temperature. Major contribution to the effective magnetization comes from the perpendicular anisotropy. This implies that as the temperature decreases the perpendicular anisotropy relaxes. χFMR shows a different behavior as a function of temperature in the perpendicular configuration. Figure 32 is a representative plot of variation of χFMR with temperature in parallel and perpendicular configuration. It increases continuously with decreasing temperature for all the ML in the perpendicular configuration.

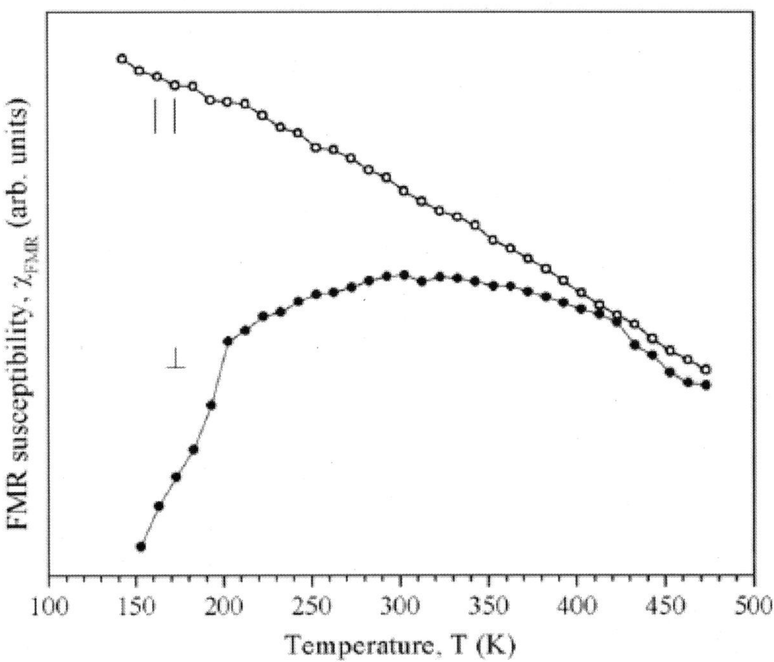

Figure 32. Representative plot showing variation of FMR susceptibility with temperature in parallel and perpendicular configurations for [SnO_2 (46 nm) CZF(83 nm)]$_5$ M.

CONCLUSIONS

The temperature dependent ferromagnetic resonance (FMR) studies on [SnO_2/ Cu-Zn ferrite]n as function of magnetic layer and spacer layer thickness were carried out in the temperature range 100- 475 K. The study of temperature dependence of FMR lineshape, peak-to-peak linewidth, peak-to -peak line intensity and resonance field provide an insight into the interfacial effects. The in plane and out of plane FMR studies provide information about the magnetic anisotropies of the ML.

Author Details

R.Singh and S.Saipriya

School of Physics, University of Hyderabad, Central University P.O., Hyderabad, India

REFERENCES

1. P. Bruno and C. Chappert, *Phys. Rev. Lett. B* 67 (1991) 1602
2. D.M.Edwards, J.Mathon, R.B.Muniz, and M.S. Phan, *Phys. Rev. Lett.* 67 (1991) 493
3. C.H.Marrows, Nathan Wiser, B. J. Hickey, T.P.A Hase and B.K. Tanner, *J. Phys.: Cond.Matter* 11 (1999) 81
4. K. R. Nikolaev, A. Yu. Dobin, I. N. Krivorotov, W. K. Cooley, A. Bhattacharya, A. L.
5. Kobrinskii, L. I. Glazman, R. M. Wentzovitch, E. Dan Dahlberg, and A. M. Goldman Phys. Rev. Lett. 85 (2000) 3728
6. D. M. Lind, S. D. Berry, G. Chern, H. Mathias, and L. R. Testardi, *Phys. Rev. B* 45 (1992) 1838.
7. M. A. James, F.C. Voogt, L. Niesen, O.C. Rogojanu and T. Hibma, *Surf. Sci.* 402 (1998) 332
8. P. J. Van der Zaag, R.M. Wolf, A.R. Ball, C. Bordel, L.F. Feiner and R. Jungblut, *J. Magn. Magn. Mater.*148 (1995) 346
9. S. Saipriya, Joji Kurian and R. Singh, *IEEE Trans Mag.*147 (2011) 10
10. S. Saipriya, Joji Kurian and R. Singh, AIP Conf. Proc. 1451 (2012) 67
11. S. Saipriya and R. Singh, *Mater. Lett.* 71 (2012) 157
12. S. Saipriya, Joji Kurian and R. Singh, J. Appl. Phys 111 (2012) 07C110
13. M. Sultan and R. Singh, J. Appl. Phys. 107 (2010) 09A510
14. S. Saipriya, M. Sultan and R. Singh, Physica B 406 (2011) 812
15. M. Sultan and R. Singh, *J. Phys. D: Appl. Phys.* 42 (2009) 115306
16. G. S. Bales and A. Zangwill, J. Vac. Sci. Technol A 9 (1991) 145
17. O. S. Josyulu and J. Shobanadri, *J. Mat. Sci.* 20 (1985) 2750
18. Janhavi P. Joshi, Rajeev Gupta, A. K. Sood and S. V. Bhat, *Phys. Rev. B* 65 (2001) 024410
19. Y. Gong, Z. Cevher, M. Ebrahim, J. Lou, C. Pettiford, N. X. Sun and Y. H. Ren, *J. Appl. Phys.* 106 (2009) 063916
20. S. S. Yan, *Acta Metullurgica Sinica* 9 (1996) 283
21. A. Ohtomo, D. A. Muller, J. L. Grazul and H. Y. Hwang, *Nature* 419 (2002) 378
22. S.S. Kang, J.W. Feng, G.J. Jin, M. Lu, X.N. Xu, A. Hu, S.S. Jiang and H. Xia, *J. Magn.Magn. Mater.* 166 (1997) 277
23. C. Kittel, *Phys. Rev.* 110 (1958) 1295
24. A. M. Portis, *Appl. Phys. Lett.* 2 (1963) 69
25. S. J. Yuan, L. Wang, R. Shan and S.M. Zhou, *Appl. Phys. A* 79 (2004) 701

26. R. D. McMichael, M. D. Stiles, P. J. Chen, and W. F. Egelhoff, Jr., *Phys. Rev. B* 58 (1998) 8605
27. M. Sultan and R. Singh, J. Phys: Conf. Ser. 200, 072090 (2010).
28. U. Gradmann and A Mueller, *Phys. Status. Solidi.* 27 (1968) 313
29. M. T. Johnson, P. J. H. Bloemen, F. J. A. den Broeder and J. J. de Vriesy, *Rep. Prog. Phys.*59 (1996) 1409
30. J. H. Jung, S. H. Lim and S. R. Lee, *J. Appl. Phys.* 108 (2010) 113902
31. J. Pelzl et al, *J. Phys.: Condens. Matter* 15 (2003) S451
32. M. C. Hickey and S. J. Moodera, *Phys. Rev. Lett.* 102 (2009) 137601
33. Y. Tserkovnyak, Arne Brataas, Gerrit E. W. Bauer and Bertrand I. Halperin, *Rev. Mod.Phys.* 77 (2005) 1375
34. G. Eliers, M. Lüttich and M. Münzenberg, *Phys. Rev. B.* 74 (2006) 054411
35. R. Rai, K. Verma, S. Sharma, Swapna S. Nair, M. A. Valente, A. L. Kholkin and N. A.
36. Sobolev, *J. Phys. Chem. Solids* 72 (2011) 862
37. M. Marysko and J. Simsova, *Phys. Stat. Sol. A,* 33 (1976) K133-K136
38. M. R. Diehl, J-Y Yu, J. R. Heath
39. R. Topkaya, M. Erkovan, A. Öztürk, O. Öztürk, B. Aktaş and M. Özdemir, *J. Appl.Phys.*108 023910
40. B. X. Gu and X. Wang, *Acta Metalurgica Sinica(English letters)*12 (1999) 181

Chapter 3

NEW METHODS AND NEW TYPES OF FUNCTIONALIZE NANOCOMPOSITES INTENT-ED FOR THE ECOLOGICAL DEPOLLUTION OF WATERS

Nicoleta Petrea, Petrişor Zamora Iordache, Rodica Mihaela Lungu, Ioan Safta, Razvan Petre and Andrada Pretorian

Scientific Research Center for CBRN Defense and Ecology Romania

INTRODUCTION

Nanotechnology as techniques for the removal of organic pollutants implies the use of certain types of nanostructures or nanostructured materials as a carrier host for the reticulation, encapsulation and degradation of pollutants. The main mechanisms through which Nano composite structures separate or destroy organic pollutants are those of the catalytic oxidation, reduction, sorption or reticulation types (Lu, Zhao & Wan, 2010, Choi et al., 2009). Most materials and methods which are based on nanotechnologies have high selectivity regarding specific organic pollutants, and their applicability rely on the economic factor. The limited spectrum of separated or degraded pollutants, the decontamination time and the lack of control over degradation products are key factors which limit the widespread use of most nanostructured materials in organic decontamination processes. Nanotechnologies offer promising solutions to the depollution field, thanks to their remarkable properties: depollution process control ability, structural and functional modeling flexibility, with a view to improving depollution parameters, high specific surface, etc. In addition, nanostructured materials offer solutions for the obtaining of new ecological depollution biodegradable materials. Microstructured and nanomembranary materials, as well as functionalized nanostructured materials have the most promising applicative potential. Functionalized nanomaterials are carriers of chemical functions capable of reticulating, incorporating and removing organic pollutants. Functionalized composite materials have depollution properties which are similar to those of functionalized materials. As compared to functionalized nanomaterial, composites materials have several possibilities of structural and functional modelling, bearing direct effect on the yield, efficiency and spectrum of reticulated pollutants. Ion exchangers or natural adsorbents (zeolites, synthetic resins, functionalized polymers, etc.) (Cheremisinoff, 2002; Evanghelou 1998) have organic depollution properties, which are similar to those of functionalized composites. Degradation residues of organic pollutants, resulting from depollution processes, as well as from secondary pollution processes caused by the latter are one of the decisive factors that limit current depollution technologies (Rosenfeld & Feng, 2011; Bayliss & Langley, 2003). Most depollution waste fractions have lower molecular weights, as compared to the initial pollutants, which can induce carcinogenic effects on humans or changes on the structural and functional properties of the biodiversity and the environment. Membrane nanostructured materials and functionalized composites have pronounced limiting effects over the dispersion effects of secondary degradation pollutants by their retention, encapsulation and separation. As a consequence of the major social and economic implications (environmental protection, natural resource conservation, medicine, Nano medicine), organic depollution has imposed itself as a priority research field, as a source which generates solutions and implementable technologies. This chapter presents the experimental results acquired as a result of the obtaining and testing of new biodegradable functionalized composite materials, intended for the removal of a wide range of organic pollutants. The composites we obtained were tested on four categories of polluted water, coming from milk processing industry, medicine industry, the obtaining and processing of polymers and from the sewage waters in the city of Bucharest.

FUNCTIONALITY RELATIONS BETWEEN ENVIRONMENTAL ELEMENTS AND POLLUTANTS

The environment can be defined as the manifestation of dependence and functionality relations established between physicochemical and biological structures characteristics of soil, water and air, as its fundamental structural macroentities (Tolgyessy, 1993). Overall, the environment represents a complex and dynamic structure, on the evolutionary process, due to the way and type of interaction established between its structural elements, and material and energy mediation vectors (Fig. 1). Ecosystems represent manifestations of the local distribution manner of basic structural entities of the environment. Ecosystems represent organized systems, well defined in relation to their physical, chemical and biological structure of their structural elements, among which are established structural and functional relations of their own (Socolow et al., 1994). Each set of functions is characteristic for a particular ecosystem, reflecting its biochemical structure, mediation and transport vectors, the dynamics of internal evolution, as well as the rate, role and functions of each structural component. Any changes to global environmental factors or local macro structural distribution of an ecosystem may irreversibly affect the structural and functional evolution. From this perspective, ecosystems can be defined as stable phases of local equilibrium, established between the biochemical structure, compositional structure and functional structure of their macro- and micro components.

Transport and mediation vectors (TMV) influence local and global dynamics of transformation and transport, established between the local structural components of the environment and its ecosystems. The main TMV, which form the evolution and transformation of the environment and of its ecosystems are air, water and climatic factors. The distribution and local physical and chemical characteristics of TMV model the distribution and evolution of biodiversity models, as well as the distribution and evolution of the elements which form the composition of the soil, water, rocks, sediments and minerals. The modeling process leads to the establishment of relationships and dynamic biochemical balances, always subject to the evolutionary processes of transformation and structural and functional remodeling. The development and the establishment of equilibrium states within and between the local structures of the environment, reflect ongoing relationship of mass and energy transfer established at the level of micro structured elements, with TMV as mediating factor. The morph chemical and morph functional structuring manner of the soil influence pollutant absorption, modeling and conditioning applicative technologies and depollution resources (Reddy, 2010; Lal & Shukla, 2005).

Organic fractions resulting from degradation processes of biodiversity (humic and fulvinic acids, organic derivatives and their macro complex, carbon, nitrogen, oxygen, water, etc.) constitute the foundation of chemical and biochemical transformation processes that model host ecosystems, and are factors which catalyse and support chemical and biochemical functionalities (Evanghelou, 1998; Haider & Schäffer, 2009).

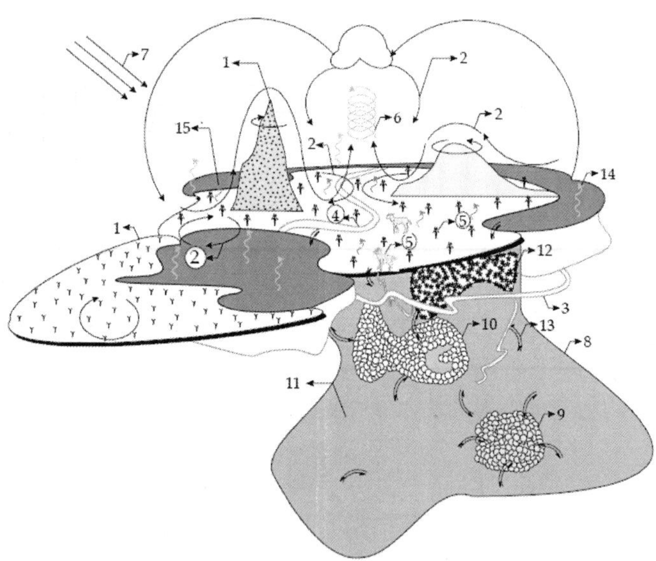

Figure 1: Ecosystems and their structural and functional organization.

(1 - surface form relief; 2 – flow of air and water vapor currents; 3 – underground water flows; 4 – surface water flows; 5 – biodiversity; 6 – energy fluxes; 7 – light; 8 – underground; 9, 10, 12 – underground mineral and rock distributions; 11 - other underground fine particulated matter; 13 – flow of matter and energy carried by fluids; 14 - sea water vapor; 15 – land water vapor)

Material and chemical structures resulting from natural processes of evolution are not a threat to biodiversity and the environment, as they support and model local exchange mechanisms and energy and material balance mechanisms. In terms of chemical functionality, the material structures resulting from natural processes of degradation can be considered inert structures, as they do not have the ability of modeling the morphochemical structure and the functions of flora, fauna and microorganisms of ecosystems. Pollutants can be defined as factors that induce sudden steps in the structural and functional evolution of an ecosystem, the biodiversity being the fastest element of the environment that responds and adapts to changes. The emergence, suppression or evolution of certain species or biological structures is conditioned by the relations established between the functionality structure of the host ecosystem and the genetic profile specific to the targeted bio structures, as a structural conservation factor. Most dependence relations established between the functionality of ecosystems and the genetic profile of their bio structures are of the evolving type, in which the ecosystem, as a material and energy source, ensures the transition of available resources in higher biological forms of organization. The suppression of material and energy resources is equivalent to the suppression of the development and evolution of biodiversity. The factors that may suppress the dependency relations established between the functionality and the structure of the environment and the biodiversity hosted by it can be factors of organic and inorganic pollution, as well as natural equilibrium factors (air, light, temperature, humidity, etc.). These factors model and provide energetic and catalytic support, as well as the dynamics and complexity of structural and material organization forms of the soil, water and air. The dynamics and complexity of the structural and material organization forms of the soil, water and air constitute the structure that supports the entire evolution and the material and functional structuring of biodiversity. Most artificial chemical compounds are material structures with high functional and chemical reactivity, small molecular weight and high diffusion coefficients, which affect the morphochemical structure and the functionality of the environment biodiversity. For this reason, artificial chemical compounds are considered and act as poisons upon microorganisms or biochemical structures with nonexplicit biological functionality (amino acids, proteins etc.). The morph functional uniformity and the broad spectrum of natural, synthetic organic pollutants, or those resulting from industrial and domestic activities, contained in the surface waters are the main reasons that restrain the methods, materials and current organic depollution techniques. The constraints bearing upon functionalized nanostructured materials or nanostructured composites are due to the non-selective charging with chemical compounds, which are not toxic for the environment. The limitation of the type of functionality of nanostructured materials is the second major cause that limits the depollution processes, by separately constraining the range of pollutants. Considering the above mentioned statements, the most suitable retention methods of organic pollutants are the mixed ones, which use as active principles of pollutant separation: the mechanical filtration, chemical reticulation, reverse osmosis, degradation, encapsulation and controlled extraction of depollution products.

URGENT NEEDS IN THE FIELD OF DEPOLLUTION AND DECONTAMINATION

Inorganic depollution involves relatively low complexity processes, related to organic depollution processes. In the field of inorganic depollution there are thoroughly-studied applicative depollution methods and technologies dedicated to each class of pollutants. Inorganic depollution can be considered as a controlled process, as pollutants may be eliminated in a controlled manner, directly in the stage of generation, limiting the dispersion and secondary contamination processes. In comparison with the organic pollution, the inorganic pollution has a much simpler complexity of the spectrum of pollutants contained in surface water, extending the possibilities of separation, storage, management, treatment and reassessment. Inorganic pollution represents a priority problem of depollution, which has a high complexity level, and whose resolution implies the utilization of some partial, limited applicative solutions. Both organic pollutants and the pollutants having a mixed morphochemical structure (of the organic – inorganic type) are compounds with high and average, slightly degradable reactivity. Most organometallic pollutants and of the compounds having a mixed morphochemical structure (adsorbent inorganic matrices, host matrices, fibres, macro polymers, active suspensions, functionalized inorganic mixtures, etc.) come from food industry, medicine industry, polymer industry, wood processing industry, etc. The slightly degradable character of organic pollutants changes the organic pollution into a degenerative process that generates in time new types of degradation compounds. Most degradation compounds are toxic in relation to the environment, being able to influence, on long and short terms, the functionality and structure of the component elements of the environment. The accumulation and degradation processes are factors that support and amplify the secondary processes of pollution and biological contamination, by favoring the development and adaptation of some new biological structures, extraneous to the host ecosystems. Also, the biological pollution can influence, on long and short terms, the human health and chemical, biological and functional balance, established at the biodiversity level of the ecosystems, by favoring the uncontrollable evolution of the structure and functionality of the local and global environmental factors. By the nature of the sources, due to the complexity of industrial and domestic activities, as well as the needs and exigencies of the consumer society, huge quantities of residue result that require suitable management for collection, storage, reassessment and disposal. The identification of optimum solutions to residue management represents a priority field of depollution as a preventive and control factor. The main function of the waste management is that of controlling and limiting the pollution, by isolating and treating the resulted residue. Due to the exponential increment of the collected residue quantity, as well as to the financial and environmental implications, the depollution methods and technologies have been oriented in the direction of finding some alternative solutions of unitary depollution, disposal, reassessment and ecological reutilization of the residue (Asano et al., 2007). The aim of alternative solutions is that of minimizing the collected quantities of residue and reducing their impact upon the environment. Residue disposal technologies are limited by the implementation costs, the complexity of the type of contained contaminants, as well as the large quantity of collected residue (Cheremisinoff, 2003; Cheremisinoff & Cheremisinoff, 2005; Harrison & Hester, 2002). This type of technologies provides partial solutions, being dedicated especially to the disposal of certain classes of pollutants with high toxicity level or easily separable contaminants (paper, plastics, metals, wood, etc). Reassessment and reutilization of the collected residue have the highest applicative potential, favoring the preservation and regeneration of the natural resources and environmental factors (Baud, Post & Furedy, 2004). Research in the field of waste recovery follows more directions, among them: reassessment -reutilization (plastics, metals, precious metals, etc.), production of new biodegradable depollution materials, the recovery of wastes having organic substrates (building materials, semi-artificial fertilizers, etc.). Waste recovery has some direct active functions within depollution processes, having as direct objects: the simplification of management procedures, the reduction of pollution factors, minimizing the effects of residue and pollutants upon the environment, the preservation and regeneration of natural energy resources. Finding viable and applicable solutions of management and waste recovery, which might have a low impact upon the environment and which might be financially sustainable represents a top priority and a necessity of environment science.

The soil represents the fundamental structural element of the environment, highly exposed to pollutants, acting like an absorbent, due to its own morphological structure and due to the morphochemical and compositional structure of its component elements. The soil presents in its structure an amorphous mineral component (95 - 99 %) (acid, basic and neutral rocks and particles containing: O 46.6%, Si 27.7%, Al 8.1%, Fe 5.0%, Ca 3.7%, Na 2.8%, K 2.6%, Mg 2.1%, etc.; mechanical, chemical, biogenical sediments or their mixtures) a fluid component (characterized by: osmotic pressure, pH, oxido-reduction potential, colloidal structure, buffering capacity, absorbing capacity), a gaseous component (CO2 - up to 1%, O2 - 10-20%, N2, NH3, water vapours, H2S, H2, CH4, SO2, etc), a living matter component (bacteria, fungi, algae, protozoa, insects, arachnoides, molluscs, earthworms, etc.) and non-living matter components (humic acids, fulminic acids, humins, simple saccharides, fatty acids, alcohols, esters, starch, proteins, complex proteins, pectins, hemicellulose, cellulose, lignin, wax, bitumen, etc) (Frank & Tolgyessy, 1993). The chemical and morphochemical structure of the soil shows the complexity of its morph functional structure. The morph functional complexity is expressed by the active chemical functionality variation, capable of absorbing, encapsulating and reticulating the pollutants. Excepting the persistent compounds, in most cases of diffusion of the inorganic pollutants in the soil, the local morphochemical composition is irreversibly modified (Yu & Wang, 1997).

Organic pollutants (especially the surfactants, fatty acids, gasoline, petroleum and its derivatives) diffuse in the surface and depth structure of the structural elements of the soil, deactivating, masking or isolating the structures having functions in establishing and adjusting the chemical and biological balance mechanisms (minerals, organic matter, microorganisms, colloids, etc.) (Perk, 2007; Evanghelou, 1998). Organic pollution has a deep persistent character and in most situations of this kind, the morphochemical, functional and biochemical soil structure is irreversibly altered. Biotechnologies remain the only solutions that have potential for organic depollution and remediation of the soil. Soil depollution and remediation biotechnologies depend on the type of removed pollutants, they effect on short-term the local biochemical balance of the ecosystems, the action mechanisms taking place in time and being noninvasive (Evanghelou, 1998). The substantiation and development of new technologies of soil depollution and remediation is a fundamental priority of environmental science, with major implications to: human health, environmental factors, human habitat distribution, redistribution of natural resources, economy.

Radiological pollution differs from the other types of pollution sources through its generating mechanisms, methods and decontamination-depollution technologies and through the effects that it induces to biodiversity and environmental factors (Bayliss & Langley, 2003). The only known radiological depollution methods are the preventive measures, aiming at the complete isolation of radiological materials and compounds and at preventing their dispersion in the structure of environmental factors. Radiological pollution generators are made up of different radioactive chemicals that have the ability to emit radiation (charged radiation, γ, neutrinos, etc.), independent of the chemical structure that hosts the unstable nucleus. The carrier support of radioactive elements in the environment occurs through external factors (aerosols, encapsulating matrices, suspensions, oxides, halides, contaminated materials, etc.) called carrier vectors. In the case of nuclear accidents, the separation of carrier vectors is a delicate problem, due to the risk of staff contamination and due to their low chemical reactivity, which does not allow the separation and encapsulation processes. Considering the physical and chemical characteristics of the carrier vectors, the only solutions with applicative potential in the separation of radiological contaminants are those based on the use of externally stimulable highly functionalized materials, in the form of sorbents or nanostructured suspensions with high specific surface, that will be able to reticulate, encapsulate, separate and store the contaminants safely, without exposing human staff (Ojovan & Lee, 2005). Radiological pollution and its effects model radically and on long term the structure and the overall functionality of the environment, human health and factors influencing macroeconomy. This type of pollution has an unpredictable nature, its implications on the environment and biodiversity, and the evolution orientation towards local and global equilibrium factors being difficult to estimate. Radiological depollution represents another prioritary research direction, due its deep implications on energy resources and on the needs of the consumer society (medicine, research, material science).

NEW TYPES OF FUNCTIONALISED MATERIALS AND NEW APPLICATIVE SOLUTIONS CONCERNING THE ECOLOGICAL ORGANIC DEPOLLUTION OF WASTE-WATERS

This paper presents the way a new class of functionalized nanocomposite materials, intended for the organic depollution of surface water from industrial and domestic activities, was obtained, investigated and tested. The means of obtaining the composite has already been presented in the previous chapter, "Nanocomposite materials with oriented functionalized structure - The modeling of structure and functionality of nanocomposites intended for ecological depollution". Considering the limitations of current methods and techniques for the separation of pollutants, research and structural and functional modelling investigations have been oriented with a view to obtaining a nanocomposite with an oriented functionality, able to: a. undifferentiatedly reticulate a large number of organic pollutants; b. encapsulate and degrade pollutants and their degradation products; c. separate and extract pollutants and depollution waste under controlled stimulability.

Amorphous natural metal-oxide structures are the most suitable class of materials to be used in processes of chemical modeling and functionalization, intended for the removal of organic pollutants. Their functional, morphological and morphochemical structure, resulting from functionalizing processes, is compact and able to fully saturate its functionalized surface, with different pollutants which have been collected from the depolluted environment. Following the processes of chemical modeling and functionalization, the deep structure of the chemically modified metal-oxide components remains unchanged, giving the material a pronounced ecological and biodegradable character. In the first stage, the pollutants reticulated on the surface and in the depth of the composite, are degraded and encapsulated, resulting in degradation products with modified functionality and toxicity. It is likely that the toxicity of the resulting degradation products might not be substantially modified, but it is very important that, following the reticulation process, the pollutants and the degradation products should remain reticulated on the surface and in the depth of the material, allowing a slow selective biodegradation of the pollutant layer collected on its surface. Equally, the stable reticulation of the pollutants and of the degradation products limits the effects of secondary pollution and contamination by dispersing the resulting degradation compounds. Such depollution materials have an increased application potential in relation to the applications of depollution under the dynamic conditions of polluted surface water flows (rivers, lakes, etc.), as they minimize the dispersion processes and favor local biodegradation processes by local sedimentation of pollutant-charged functionalized suspensions in bottom water. Due to the macromolecular character and the micrometric structure of depollution suspensions, nanostructured materials cannot be assimilated at cell level by the biodiversity they encounter, as it has already been demonstrated in the previous chapter that the surface and the depth of the composite are able to host biological structures, without affecting their morphochemical structure and metabolism.

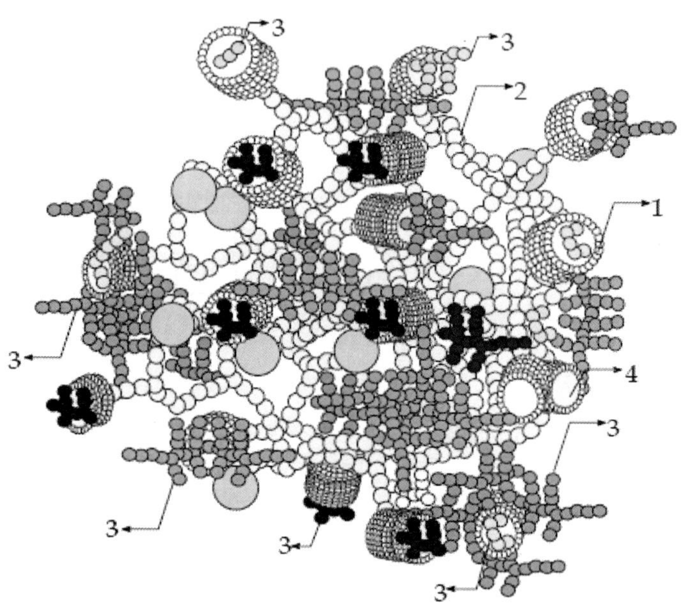

Figure 2. (1 – mineral metal oxide; 2 – internal retaining metal-oxidic cage – mixed oxidic morphostructure; 3 – different types of organic and inorganic pollutants; 4 – external retaining metal-oxidic cage – specific and mixed oxidic component morphostructure).

The metal-oxide mixture (Su) used as a structural component of compositing of the depollution material (MOF-DP) has a natural origin and has some remarkable morphostructural (Figure. 4), morphochemical and functional (Figure.3) features. Using the basic segmentation method proposed by Iordache et al. (Iordache et al., 2010), the main morphostructural parameters of Su were determined and quantified (fig.4b): area (11.41 nm2) compactness (1.29), perimeter (16.48 nm). Investigations carried out by X-rays dispersion show that Su contains, apart from oxidic fractions of the following elements Na, Mg, Al, Si, P, S, K, Ca, Ti, V, Cr, Mn, Fe, Ni, Cu, Zn , As, Rb, Sr, Y, Zr, Ba, Pb, B, O (45.76%), oxidic fractions of other elements (Sn, Ge etc. - Fig.3), C (~ 4%), halogens (Cl (~ 0.019%), Br) and nitrogen. All metal-oxide structures identified in the Su structure form nanostructured elements having a compact amorphous structure (Fig.5), whose topological distribution is most likely of the type modelled in Figure 2. This was confirmed by the numerical data obtained by elementary morphological segmentation (fig. 4b) for the compactness parameters. The compactness range of values (Round) has values in the interval (1.29 ÷ 9), representing a measure of deviation from the sphericity of agglomeration domains and of the nanostructured elements which enter their structure (1 corresponds to the perfect spherical structures) (Jain, 1989). The perimeter (Round = P2/4πS) has values on the domain (~ 10.3-107.12 nm), a measure of the size of nanoparticles and their agglomeration domains.

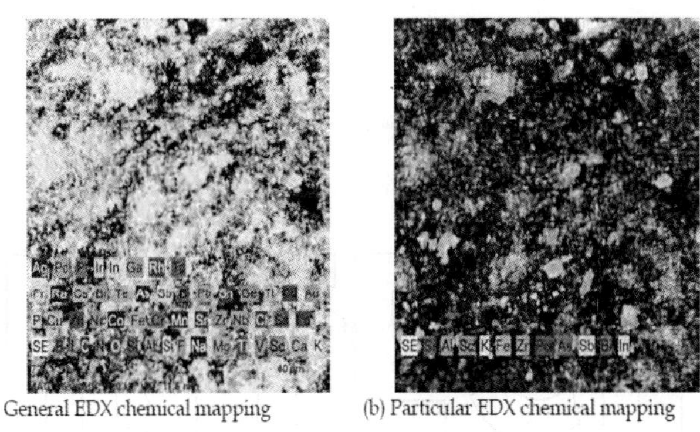

(a) General EDX chemical mapping (b) Particular EDX chemical mapping

Figure 3: Su EDX mapping (Su laid-down on ordinary paper).

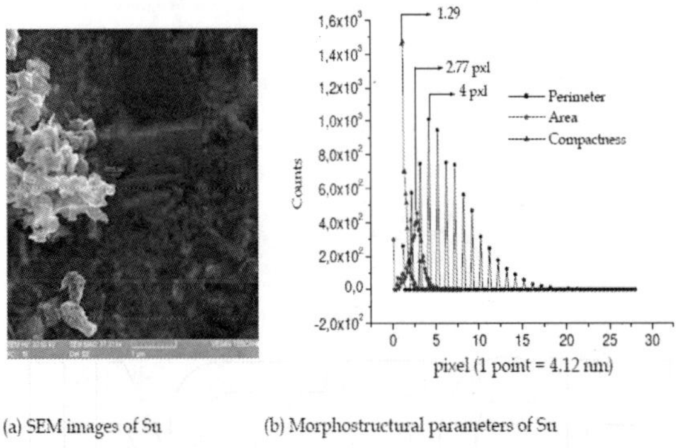

(a) SEM images of Su (b) Morphostructural parameters of Su

Figure. 4: SEM morphotopological investigation of Su (VEGA II LMU microscope).

The identification of chlorine and bromine in the Su structure implies that the metal-oxide mixture has a complex chemical functionality, which includes metal salts (MX, X - Cl, Br, etc.), halide oxides (MOX), metal-oxide-halogenated mixtures ((Mi-O-X)n(Mj-O-X)m; i, j - different types of metals; m, n - compositing ratio), H and -OH chemical reactive groups, electrical charged microsites, etc. MOF-DP was obtained by the regeneration of the compositing mixture (functionalised magnetite, functionalised Su and solubilised celulose) in distilled water. As an applicative and testing form, MOF-DP was obtained as a powder by mild thermic dehydration at 800C (fig.4A.a), and as a membrane (fig.4A.b) concentrated suspension.

(a) Fine powder of dehydrated Su

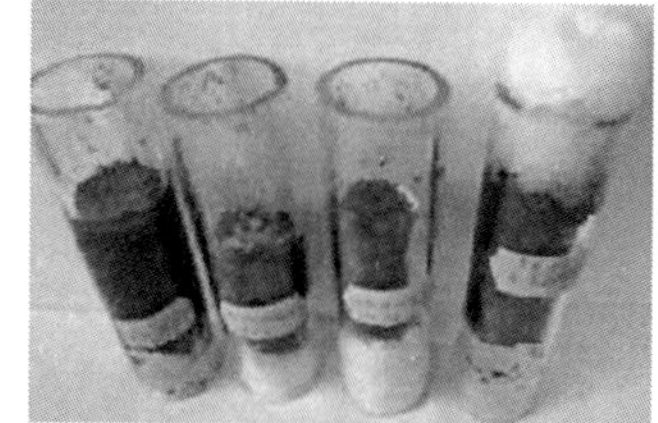

(b) Membranary layer of hydrated Su

Figure. 4A: Applicative forms of Su.

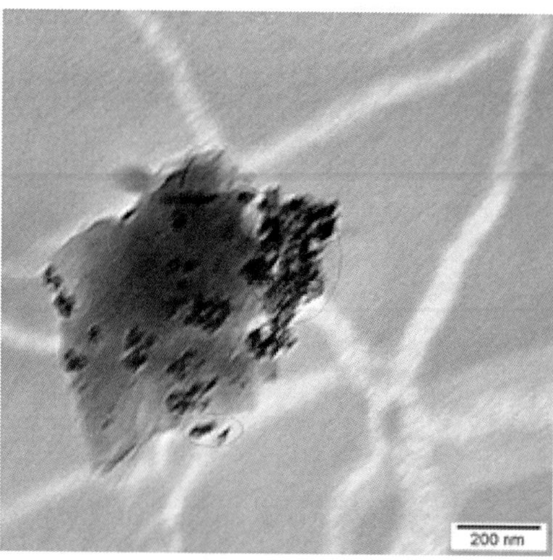

Figure. 5. TEM images of Su embedded in the structure of MOF-DP -uncharged pollutants.

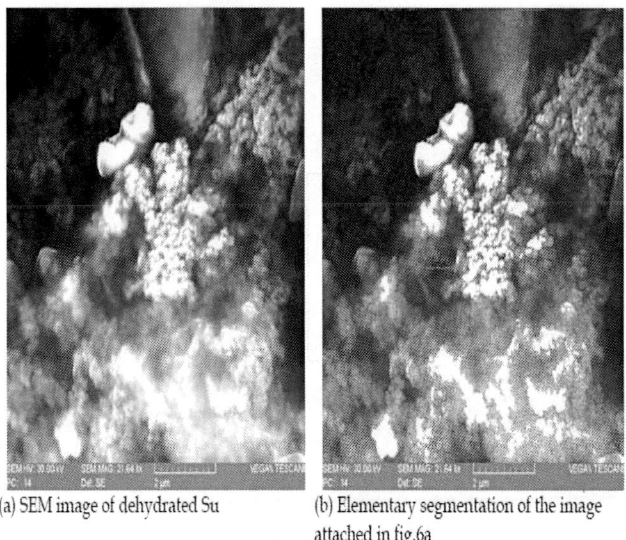

(a) SEM image of dehydrated Su

(b) Elementary segmentation of the image attached in fig.6a

Figure. 6: SEM investigation of dehydrated MOF-DP (VEGA II LMU microscope).

Morphostructural investigations carried out by scanning electron microscopy (SEM) show that dehydrated MOF-DP has a smaller and more irregular specific surface, as compared to the hydrated MOF-DP. The distributions of the agglomeration domain surfaces of the dehydrated MOF-DP have values between 194 nm2 and 12 µm2 (fig.8). From the investigation of the distribution of the domain compactness of the agglomerated structures and of the nanoparticles within their structure, taking into account the analytical relationship Round = P2/4πS, it follows that the dehydrated material suffered internal degradation processes that led to its severe discretization and to the increasing of agglomeration domains. It is likely that the increasing of the agglomeration domains may be due to secondary internal morphochemical modeling processes under the influence of temperature, as a result of which, uncontrolled reticulation processes might be established between the structural elements of the composite. It is very likely that, following the dehydration processes, MOF-DP may lose and partially modify its functionality.

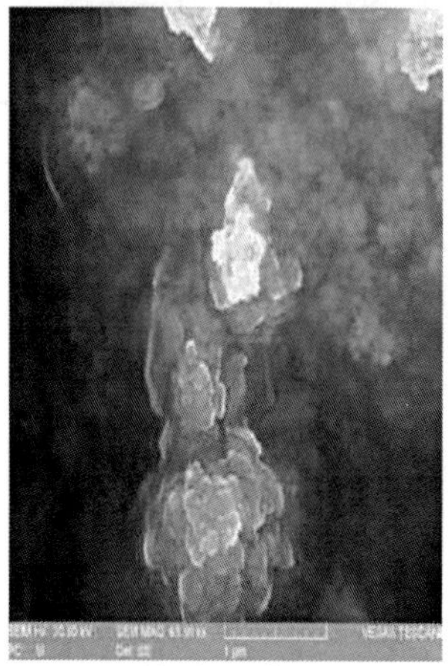

(a) SEM image of hydrated M_{OF-DP}

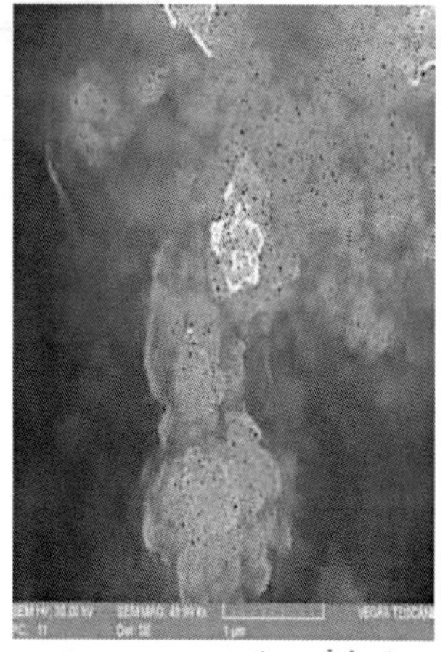

(b) Elementary segmentation of the image attached in fig.7a

Figure 7: SEM investigation of hydrated MOF-DP (acquired with VEGA II LMU microscope).

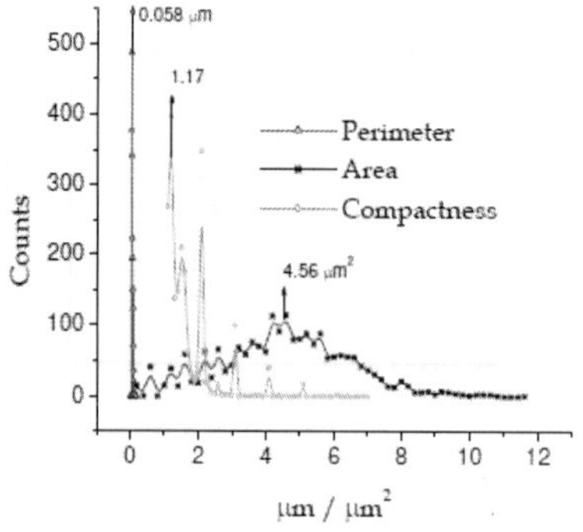

(a) Dehydrated M_{OF-DP} (attached to fig.6b)

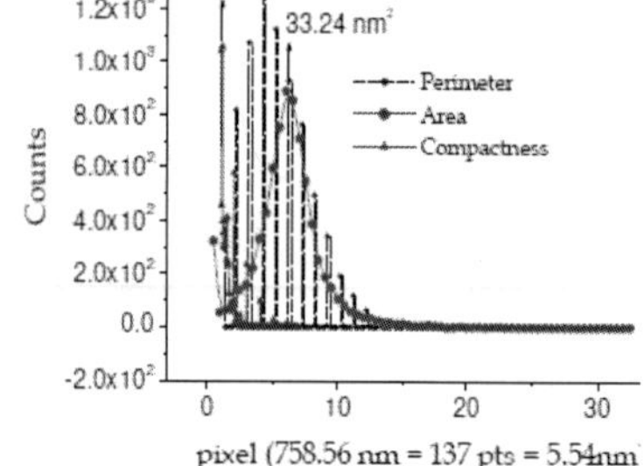

(b) Hydrated M_{OF-DP} (attached to fig.7b)

Figure. 8: Elementary morph structural parameters of MOF-DP.

Depollution of Sewage Waters using functionalized MOF-DP

This type of investigations had as main objective the determination of depollution parameters (yield and depollution efficiency) for the type of investigated water, as well as for the morph structural and morphochemical stability of MOF-DP. Investigations have been configured in accordance to the diagram in figure 9.

Samples were investigated by gas chromatography coupled with mass spectrometry (GC-MS), using a cromatograph of the GC Focus type (AI300 autosampler, split / splitless injector, chromatograph column of the TR5MS - 30 m x type with Φext = 0,25 mm and 0.25 mm stationary film thickness) and a mass spectrometer of the DSQII type. Previously, samples were concentrated by solvent extraction and evaporated in nitrogen flow.

To check the saturation point of MOF-DP with pollutants, 50 ml of water charged with pollutants were passed through the filtering layer (fig. 9), in four steps, the (P1, P2, P3, P4) samples were properly prepared and investigated. To determine the analytical correspondence holding between the separated pollutant quantity and the real pollutant quantity in the contaminated water (Rp), the quantity of pollutants present in the non-filtered water was determined before starting the investigating processes.

The separated pollutant quantity was estimated by determining the pollutant fractions found in P1, P2, P3 and P4, after filtering (c% (pollutants) + c% (separated) = 1). The investigated samples were taken from the sewerage system of Bucharest municipality. With a view to determining the morphochemical and morphostructural stability of MOF-DP, the composite charged with pollutants was investigated by scanning electron microscopy (SEM), energy-dispersive X-ray spectroscopy (EDX) and infrared spectroscopy (IR).

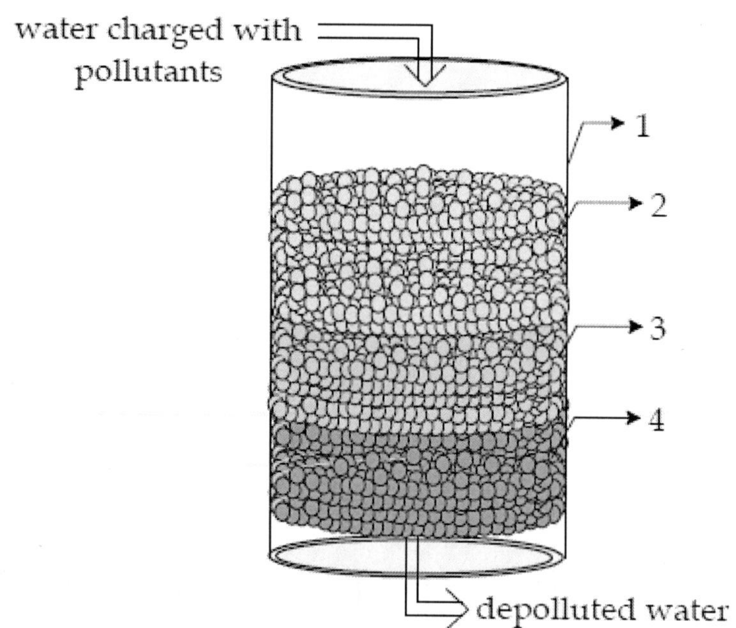

Figure 9: Experiment configuring for the investigation of water depollution yield and efficiency of MOF-DP (1 – fixing column; 2 – polluted water; 3 - MOF-DP layer; 4 – fibre glass).

The results of the investigations carried out by GC-MS are presented in table 1 and they demonstrate the fact that the tested water had a high level of organic pollution, containing a large range of organic pollutants (n-tridecane, propanoic acid, 2-methyl-,1-(2-hydroxy-1- methylethyl)-2,2-dimethylpropyl ester propanoic acid, 2-methyl-,3-hydroxy-2,4,4- trimethylpentyl ester, diethyl benzene-1,2-dicarboxylate, phthalic acid, isobuthyl-2- penthylester, butyl octyl phthalate, dibutyl phthalate, ((Z)-9-octadecenenitrile).

Table 1: The relative abundance (Ar) of decelable and partially decelable pollutants in filtered water – case of sewage waters (where: PL1 - n-tridecane; PL2 - propanoic acid,2- methyl-,1-(2-hydroxy-1-methylethyl)-2,2-dimethylpropyl ester; PL3 - propanoic acid,2- methyl-, 3-hydroxy-2,4,4-trimethylpentyl ester; PL4 - diethyl benzene-1,2-dicarboxylate; PL5 - phthalic acid, isobuthyl-2-penthylester; PL6 - butyl octyl phthalate; PL7 - dibutyl phthalate; PL8 - (9Z)-9-octadecenenitrile).

Pollutant/ undecelable peaks	A..-Rr	A..-P1	A..-P3	A..-P4
TSp	2848678635	15047488950	10533922491	8526288052
8.33	5094541	0	2272703	7299050
PLl	977320	0	0	0
10,75	5268711	0	0	5547832
Pl2	2465265	0	0	0
PL3	1405799	0	0	0
12,49	1740771	0	0	0
12,93	2828747	0	0	3710438
14,85	3347278	0	1841992	2755159
PIA	1680227	0	0	0
16,13	523552	0	0	0
16,51	1704027	0	0	0
17,23	1278456	0	0	0
17,98	901907	0	0	1129834
PL5	132278	1730996	0	0
18,27	267140	0	0	0
PL6	507889	0	0	0
PL7	10830818	8652426	0	0
PL8	1859244	0	0	0
24,00	5434244	2712903	4682044	2903562
26,56	11791893	8461410	6699962	4400335

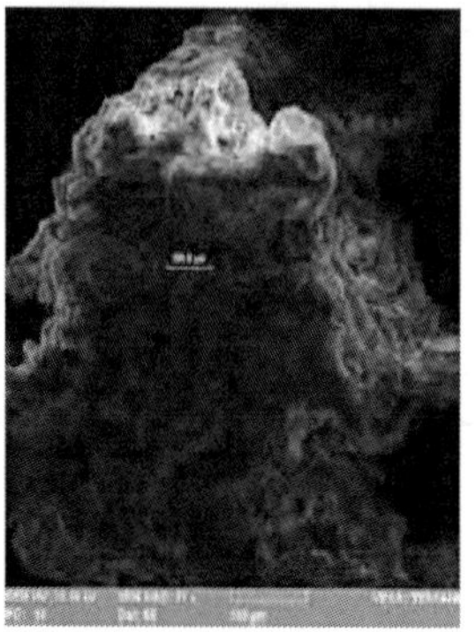

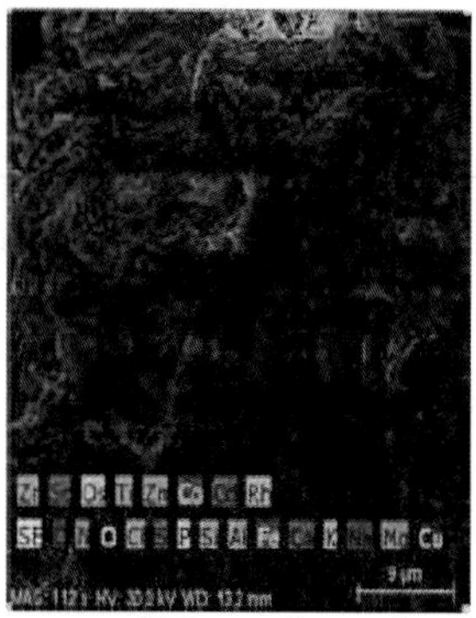

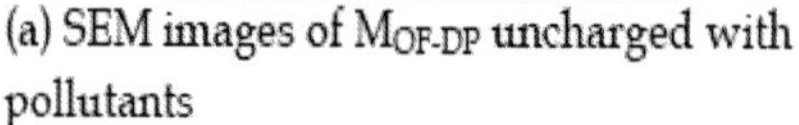

(a) SEM images of $M_{OF\text{-}DP}$ uncharged with pollutants

(b) EDX images of $M_{OF\text{-}DP}$ charged with pollutants (attached to fig.10a)

Figure 10: SEM investigation of MOF-DP charged and uncharged with pollutants – case of sewage waters (images acquired with VEGA II LMU SEM microscope).

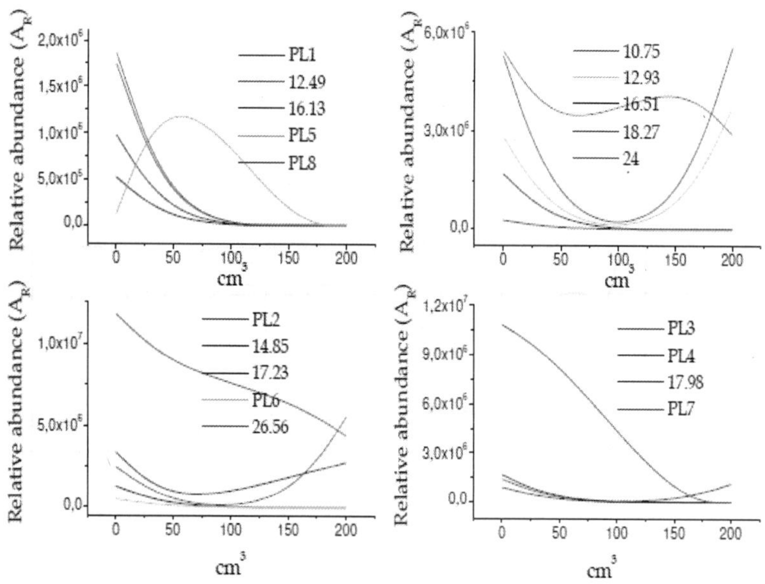

Figure 11: Relative yield of decelable and partially decelable pollutants retained by 71.43 grams of MOF-DP (according to Table.1)

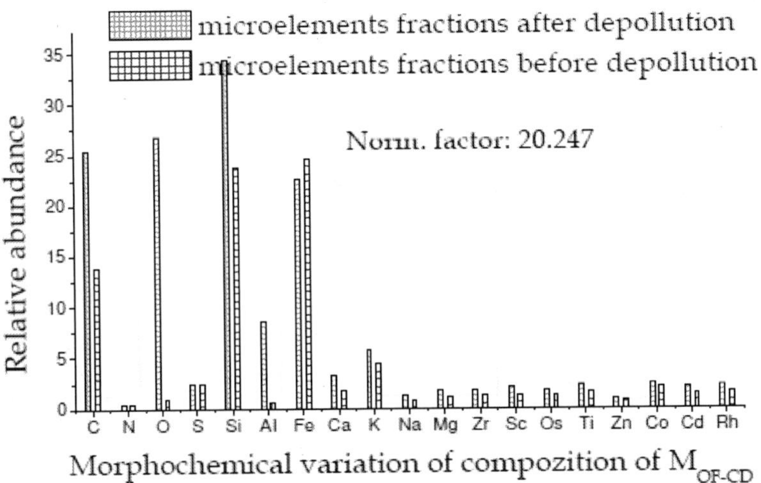

Figure. 12: Relative variation of microelement fractions in the depth and on the surface of MOF-DP charged with pollutants.

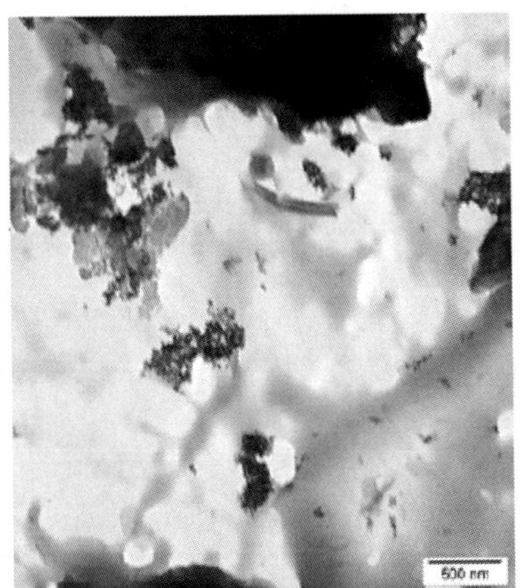

(a) TEM images of M_{OF-DP} - uncharged pollutants

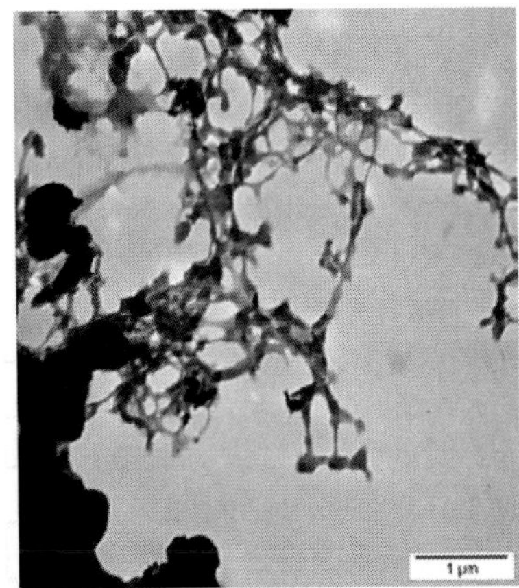

(b) TEM images of M_{OF-DP} - charged pollutants

Figure.13: TEM morph structure investigation of MOF-DP charged with pollutants - case of sewage waters (images acquired with Philips S208 TEM microscope).

The GC-MS investigations revealed the fact that the filtered water (the water charged with pollutants) contains a wide range of organic pollutants, a significant fraction of them being partially decelable. Partially decelable pollutants have clear, distinct peaks, whose variation can be quantified depending on the quantity of filtered water (table 1). A great part of the chemical compounds present in the investigated water could not be sensitized in relation to the retention time and the specific mass fragments. The quantification of these types of pollutants was entirely determined, by quantifying the variation degree of the whole sensitization area. It has been noticed that the concentrations of decelable pollutants and of the partially decelable ones decrease rapidly with the increase of the filtered water quantity (figure.11). Each pollutant attains a certain value for which the membrane is saturated through specific reticulation. After attaining the saturation peak, the filtering layer loses its ability to separate the pollutants which saturated it, thus becoming permeable in relation to them.

The saturation processes are most likely due to the neutralization through reticulation of the functional groups with which the pollutants come into contact. This fact was pointed out indirectly by SEM (fig.10 a), TEM (fig.13) and EDX (fig.10b, fig.12) investigations. Thus, we observed the severe degeneration of the specific reticulation surface of MOF-DP (fig.13b). The degradation and reticulation processes have also been confirmed by EDX investigations, during which we noticed the fact that the microelement fractions (especially C and O) found in the depth and on the surface of MOF-DP increase significantly (fig.12). This fact can be explained by admitting the fact that organic pollutants reticulated on the surface or in the depth of MOF-DP form covering layers which lead to the masking of the detection of the structural elements of the composite. The determinations in fig.12 have been carried out by supposing that the statistical quantity of silicon in the composite structure is approximately constant. Thus, to obtain the same quantity of silicon on the investigated micro surfaces, both before and after the depollution process, a normalization factor of 20.247 was added (20.247 x c%Si(after) = c%Si(before)), to minimize the errors of the statistical distribution of microelements. In order to obtain results as accurately as possible, the normalization microelement must have variations of the concentration distributions only on one of the investigated microsurfaces. These errors may be due to the different distribution probabilities of morphochemical patterns on distinct microsurfaces, as well as to the different periods of time of data acquisition.

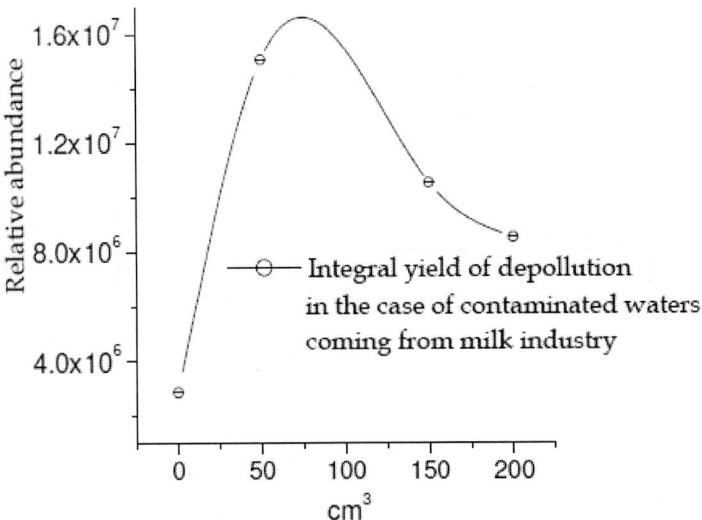

Figure 14: Integral yield of sewage water depollution in the presence of 71.43 grams of MOF-DP.

The integral retention spectrum of the pollutants has been defined as the whole recorded sensitization area, minus the area of the background signal. This represents an indirect measure of the pollutant retention spectrum size, offering relative information regarding the variation of the quantity of organic compounds, which are generated or disappear from the collected depolluted water, before and after the depollution process. The integral retention curve shown in figure 14 presents the maximum variation at approximately the same value as that of the saturation point of MOF-D with decelable and partially decelable pollutants. This suggests that during the depollution process, the reticulated pollutants are partially or completely degraded, at the level of functionalized micro surfaces of the structural components of MOF-D. Degradation processes can influence the dynamics of the depollution processes, as well as the achievement of specific saturation points. This explains the analytical structure of the depollution curves of pollutants 12.93 (specific retention time), 24 and PL2 (Figure.11). The same tendency is also shown by the decontamination curve of the entire spectrum of pollutants (Figure.14).

CONCLUSIONS

In this chapter we have presented and investigated the specific mechanisms underlying the reticulation, separation and encapsulation processes of organic pollutants. Specific depollution mechanisms, as well as the main depollution parameters (yield, efficiency), have been investigated by testing a polyfunctionalized nanocomposite material (MOF-DP) on three types of water charged with organic pollutants. The polluted waters on which the depollution features of MOF-DP have been investigated came from various pollution sources: milk processing industry, medicine industry, sewage waters. The investigations' secondary aim was to determine the effects of organic pollutants on the morphological and morphochemical structure of the obtained MOF-DP, as well as to determine how they influence the depollution parameters. It was intended, in particular, to check the range of pollutants that MOF-DP is able to separate, in order to determine the type of chemical functionality of pollutants capable to establish reticulation relations with the functional structure of the composite. The reticulated functionalities are a direct measure of the depollution efficiency of the obtained material, by defining the range of separated pollutants and the extent to which the depollution was carried out. The investigation processes included various analytical techniques, in order to confirm, through complementarity (SEM, TEM, EDX, FTIR, GC-MS), the depollution parameters and the morphostructural characteristics of MOF-DP, both before the testing and after its completion. The investigation results confirm the reticulation, encapsulation and degradation mechanisms of the pollutants that the authors proposed in chapters „New methods and new types of functionalised nanocomposites intended for the ecological depollution of waters" and „Nanocomposite materials with oriented functionalized structure". The acquired experimental data have shown that the obtained MOF-DP presents high pollutant separation yield and efficiency. The concentrations of most decelated pollutants and of those partially decelated in depolluted water were below the detection limit, showing the applicative potential of this type of materials. The aim of the investigations of the morphological and morphochemical structure which we have carried out was to identify and quantify the critical parameters that influence the processes of encapsulation and degradation of pollutants. The acquired experimental data indicate that MOF-DP is able to reduce the level of organic pollution in surface waters by up to two orders of magnitude (the observed typical values have a separation factor of 20 ÷ 50 x). The oxide materials with a functionalised mixed structure represent a viable solution for the removal of organic pollutants in surface waters, having the most favourable applicative potential in terms of economy, depollution yield and efficiency.

ACKNOWLEDGEMENTS

The authors are grateful for the financial support granted by the CNMP (National Research, Development and Innovation Plan II - Program 4) inside project 32-165/2008, as well as for the logistic support granted by the Scientifical Research Center for CBRN Defence and Ecology. Also, for their contribution to the obtaining of results, we are grateful to Dionesie Bojin, Filipescu Laurențiu, Toader Constantin and Mihaela Murașan. For her contributions to text editing and translating, we are grateful to Loredana Iordache.

REFERENCES

1. Lu, A. H.; Zhao, D. & Wan, Y. (2010). Nanocasting - A Versatile Strategy for Creating Nanostructured Porous Materials, RS□C, ISBN 978-0-85404-188-6
2. Cheremisinoff, N. P. (2002). Handbook of water and wastewater treatment technologies, Butterworth-Heinemann, ISBN 0-7506-7498-9, Boston
3. Tolgyessy, J. (2003). Water, air and soil - Fundamental sources of the biosphere, In: Chemistry and biology of water, air and soil - Environmental Aspects, J. Tolgyessy, (Ed.), 3-13, Elsevier, ISBN 0-444-98798-3, Amsterdam-London-New York-Tokyo
4. Evanghelou, V. P. (1998). Environmental soil and water chemistry - Principles and applications, John Wiley & Sons, ISBN 0-471-16515-8 (cloth: alk. paper), New York
5. Rosenfeld, P. E. & Feng, L. G. H. (2011). Risks of Hazardous Wastes, Elsevier, ISBN 978-1-4377- 7842-7, Amsterdam

6. Bayliss, C. R. & Langley, K. F. (2003). Nuclear Decommissioning, Waste Management, and Environmental Site Remediation, Elsevier, ISBN 978-0-7506-7744-8, Amsterdam

7. Socolow, R. H.; Andrews, C.; Berkhout, F. & Thoma, V. (2003). Industrial ecology and global change, Cambridge University Press, ISBN 0-521-47197-4, New York

8. Lal, R. & Shukla, M. (2005). Principles of soil physics, Marcel Dekker, ISBN 0-8247-5324-0, New York, Basel

9. Haider, K. & Shäffer, M. (2009). Soil Biochemistry, Science Publishers, ISBN 978-1-57808-579- 8, Enfield (NH), Jersey, Plymouth

10. Asano, T.; Burton, F. L.; Leverenz, H. L.; Tsuchihashi, R. & Tchobanoglous, G. (2007). Water Reuse - Issues, Technologies, and Applications, McGrow-Hill, ISBN 978-0-07-145927-3, Enfield (NH), New York

11. Cheremisinoff, N. P. (2003). Handbook of solid waste management and waste minimization technologies, Butterworth-Heinemann, ISBN 0-7506-7507-1 (alk. paper), Amsterdam, Boston, London

12. Cheremisinoff, N. P. & Cheremisinoff, P. N. (1995). Hazardous materials and waste management, Noyes Publications, ISBN 0-8155-1372-0, Park Ridge, New Jersey, U.S.A.

13. Hester, R. E. & Harrison, R. M. (2002). Environmental and health impact of solid waste management activities, RS.C, ISBN 0-85404-285-7, Manchester

14. Baud, I.; Post, J. & Furedy, C. (2004). Solid waste management and recycling, Kluwer Academic, ISBN 1-4020-1975-0, New York, Boston, Dordrecht, London, Moscow

15. Yu, T. R. & Wang, E. J. (1997). Chemistry of variable charge soils, In: Chemistry of variable charge soils, T. R. Yu, (Ed.), 309-338, Oxford University Press, ISBN 0-19-509745-9, New York, Oxford

16. Ojovan, M. I. & Lee, W. E. (2005). An introduction to nuclear waste immobilisation, Elsevier, ISBN 0-080-44462-8, Amsterdam, Boston, Heidelberg

17. Iordache, P. Z.; Lungu, R. M.; Epure, G.; Mureşan, M.; Petre, R.; Petrea, N.; Pretorian, A.; Dionezie, B.; Mutihac, L. & Ordeanu, V. (2011). The determination of the nanostructurated materials' morphology, by applying the statistics of the structural element maps, Journal of Optoelectronics and Advanced Materials, Vol. 13, No. 5, pp. 550-559, ISSN 1454-4164

Chapter 4

IRON OXIDE NANOPARTICLES

Mohammed M. Rahman[1], Sher Bahadar Khan[1,2], Aslam Jamal[3], Mohd Faisal[3] and Abdullah M. Aisiri[1,2]

[1]The Center of Excellence for Advanced Materials Research, King Abdulaziz University, Jeddah

[2]Chemistry Department, Faculty of Science, King Abdulaziz University, Jeddah

[3]Centre for Advanced Materials and Nano-Engineering (CAMNE),
Department of Chemistry, Faculty of Sciences and Arts, Najran University, Najran Kingdom of Saudi Arabia

INTRODUCTION

A semiconductor is a material with electrical conductivity owing to the electron flow (as opposed to ionic conductivity) intermediate in magnitude between that of a conductor and an insulator. This means conductivity roughly in the range of 103 to 10−8 siemens per centimeter. It is well known that semiconductor has governed a significant role in progressing research in nanoscience and nanotechnology, leading to novel classes of semiconductor nanomaterials which are capable of nano-particles and wide range of synthesis and applications. Semiconductor materials are also the foundation of modern electronics, including radio, computers, telephones, and many other devices. Such devices include transistors, solar cells, many kinds of diodes including the light-emitting diode, the silicon controlled rectifier, and digital and analog integrated circuits. Similarly, semiconductor solar photovoltaic panels directly convert light energy into electrical energy.

In a metallic conductor, current is carried by the flow of electrons. In semiconductors, current is often schematized as being carried either by the flow of electrons or by the flow of positively charged "holes" in the electron structure of the material. Actually, however, in both cases only electron movements are involved. Common semiconducting materials are crystalline solids, but amorphous and liquid semiconductors are known. These include hydrogenated amorphous silicon and mixtures of arsenic, selenium and tellurium in a variety of proportions. Such compounds share with better known semiconductors intermediate conductivity and a rapid variation of conductivity with temperature, as well as occasional negative resistance. Such disordered materials lack the rigid crystalline structure of conventional semiconductors such as silicon and are generally used in thin film structures, which are less demanding for as concerns the electronic quality of the material and thus are relatively insensitive to impurities and radiation damage. Organic semiconductors, that is, organic materials with properties resembling conventional semiconductors, are also known. Silicon is used to create most semiconductors commercially. Dozens of other materials are used, including germanium, gallium arsenide, and silicon carbide. A pure semiconductor is often called an "intrinsic" semiconductor. The electronic properties and the conductivity of a semiconductor can be changed in a controlled manner by adding very small quantities of other elements, called "dopants", to the intrinsic material. In crystalline silicon typically this is achieved by adding impurities of boron or phosphorus to the melt and then allowing the melt to solidify into the crystal.

An iron oxide nanoparticle, in nanotechnology, a particle is defined as a small object that behaves as a whole unit in terms of its transport and properties. Particles are further classified according to size: in terms of diameter, fine particles cover a range between 100 and 2500 nanometers. On the other hand, ultrafine particles are sized between 1 and 100 nanometers. Similar to ultrafine particles, nanoparticles are sized between 1 and 100 nanometers. Nanoparticles may or may not exhibit size-related properties that differ significantly from those observed in fine particles or bulk materials (Buzea et al., 2007).

Although the size of most molecules would fit into the above outline, individual molecules are usually not referred to as nanoparticles. Nanoclusters have at least one dimension between 1 and 10 nanometers and a narrow size distribution. Nanopowders (Fahlman, 2007) are agglomerates of ultrafine particles, nanoparticles, or nanoclusters. Nanometer-sized single crystals, or single-domain ultrafine particles, are often referred to as nanocrystals.

Nanoparticle research is currently an area of intense scientific interest due to a wide variety of potential applications in biomedical, optical and electronic fields. Iron oxides are chemical compounds composed of iron and oxygen. Altogether, there are sixteen known iron oxides and oxyhydroxides (Cornell & Schwertmann, 2003). The uses of these various oxides and hydroxides are tremendously diverse ranging from pigments in ceramic glaze, to use in thermite.

Oxides:

- Iron(II) oxide, wüstite (FeO)
- Iron(II,III) oxide, magnetite (Fe3O4)
- Iron(III) oxide (Fe_2O_3)
    - alpha phase, hematite (α-Fe2O3)
    - beta phase, (β-Fe2O3)
    - gamma phase, maghemite (γ-Fe2O3)
    - epsilon phase, (ε-Fe2O3)

Hydroxides:

- iron(II) hydroxide ($Fe(OH)_2$)
- iron(III) hydroxide ($Fe(OH)_3$), (bernalite)

Oxide/hydroxide:

- goethite (α-FeOOH),
- akaganéite (β-FeOOH),
- lepidocrocite (γ-FeOOH),
- feroxyhyte (δ-FeOOH),
- ferrihydrite ($Fe_5HO_8 \cdot 4H_2O$ approx.), or $5Fe_2O_3.9H_2O$, better recast as $FeOOH.0.4H_2O$ high-pressure FeOOH

- schwertmannite (ideally $Fe_8O_8(OH)_6(SO)\cdot nH_2O$ or $Fe^{3+}_{16}O_{16}(OH,SO_4)12\text{-}13\cdot 10\text{-}12H_2O$)

Beta phase iron oxide: Cubic face centered, metastable, at temperatures above 500 °C converts to alpha phase. It can be prepared by reduction of hematite by carbon, pyrolysis of iron(III) chloride solution, or thermal decomposition of iron(III) sulfate.

Iron(III) oxide or ferric oxide is the inorganic compound with the formula Fe2O3. It is of one of the three main oxides of iron, the other two being iron(II) oxide (FeO), which is rare, and iron(II,III) oxide (Fe_3O_4), which also occurs naturally as the mineral magnetite. As the mineral known as hematite, Fe_2O_3 is the main source of the iron for the steel industry. $Fe_2O_3$is paramagnetic, reddish brown, and readily attacked by acids. Rust is often called iron(III) oxide, and to some extent, this label is useful, because rust shares several properties and has a similar composition. To a chemist, rust is considered an ill-defined material, described as hydrated ferric oxide.

LITERATURE SURVEY

Semiconducting nano-materials have attracted much attention because of their unique properties and potential applications in all areas of science (Kumar & Singhal, 2007). The simplest synthetic route to nanomaterials is probably self aggregation, in which ordered aggregates are formed in a spontaneous process (Whitesides & Boncheva, 2002). However, it is still a big challenge to develop simple and reliable synthetic way for low-dimensional metal oxide nanostructures with designed chemical components and controlled morphologies which strongly affect the properties of nano-materials (Dale & Huber, 2009).

In recent years, nanosized and nanostructures of super-paramagnetic iron oxides have been investigated extensively because of their wide applications in nano-fields such as ferrofluids (Raj et al., 1995), magnetocaloric refrigeration (McMichael et al., 1992), biotechnology, and in vivo bio-medical field (Rahman et al., 2011). These materials offer several potential biomaterial (Rahman et al., 2011) as well as biomedical (Rahman et al., 2009) applications in magnetically controlled drug delivery, magnetic resonance imaging as contrast agent, tissue repair, and detoxification of biological fluids (Garcia et al., 2004; Music et al., 1997). It has been extensively studied in diverse fields including catalysis (Brown et al., 1998) environment protection (Chen et al., 2005), magnetic storage media (Zeng et al., 2002), clinical diagnosis, and treatment (Jordan et al., 2003). Nanomaterials may also be utilized in different technological requests, viz. refrigeration systems, medical imaging, drug targeting, other biological applications, and catalysis (Kesavan et al., 1999). Reducing the crystal dimension and enhancing the surface area of the sensing materials are an optional approach to improve the responses, since the reduction/oxidation reactions are mainly activated by the active surfaces area.

Removal of organic pollutants (especially AO dye) in water has been a significant issue in wastewater treatment because of the non-biodegradable nature of these pollutants. The degradation methods, such as, adsorption using activated carbon and coagulation using coagulants, actively convert the aqueous dyes to the solid state leaving the contaminant intact. During the past decades, many researchers have put focus on searching for a direct and effective method to solve this problem (Rahman et al., 2006; Saquib et al., 2008).

Recently the ecological contamination and inadvertent leakage of detrimental gases and liquids, semiconducting metal oxide chemical sensors have attracted the interest of ecologists, technologists, environmentalist, and others (Pare et al., 2008). One of the industrial gases in liquid form of interest is ammonia as an ammonium hydroxide because of its toxic and polluting nature. Because of its wide variety of applications including the production of nitrogenous fertilizers and other nitrogenous chemicals, as well as an industrial refrigerant, its global production is in excess of 100 million tons per annum. If highly concentrated it might lead to severe burns on our skin, eyes, throat, or lungs causing permanent blindness and lung disease. In addition, ammonia, hydroxide aerosols have a sun blocking function, corrosive natures, and the fume produced results in temperature reduction. Hence early detection and monitoring of presence of ammonium hydroxide in a wide range of industrial applications is desired. Solution phase ammonia sensors based on optical, electrical, and chemical detection have focused significant attention due to the possibility to operate at room temperature, to measure low level concentrations with fast response time and applicable to detect trace level ammonia solution in environments. In order to detect NH4OH, there have been lots of efforts in developing a variety of chemical sensors such as electrochemical sensor and chemi-resistive sensors (Ballun et al., 2003), and optical sensors (Christie et al., 2003).

Many sensing research works have been performed with the metal oxide nanostructures for the detection of various chemicals such as hydrazine, acetone, ethanol etc. The chemical sensing by metal oxide thin films utilize mainly the properties of thin film formed by the physi-sorption and chemisorptions techniques. The chemical detection is based on the current changes of the fabricated thin films caused by the chemical components of the reacting system in aqueous medium (Ansari et al., 2008). Here, the main efforts are focused on detecting the minimum ammonium hydroxide concentration necessary for electrochemical detection. Low-dimension of NPs in metal oxide nanostructures allows very sensitive transduction of the liquid/surface interactions into a change in the electrochemical properties. The possibility to form a variety of structural morphologies metal oxide NPs offers many opportunities of tuning the chemical sensing properties. Taking into account their applications, low-dimensional NPs with iron oxide is prepared by hydrothermal technique to fabricate a simple and efficient chemical sensors consisting on a side-polished gold surface; and to test the chemical sensing performance on ammonium hydroxide at room temperature. To best of our knowledge, this is the first report for detection of ammonium hydroxide (in liquid phase) using simple and reliable I-V technique in short response time. Additionally we investigated the photo-catalytic degradation of organic dye named AO (water soluble) with as-grown iron oxide NPs in aqueous system under 250W mercury lamp.

EXPERIMENTAL SECTIONS

Materials and Methods

Ferric chloride, urea, ammonia solution (25%), acridine orange, butyl carbitol acetate, ethyl acetate, monosodium phosphate, disodium phosphate, and all other chemicals used were of analytical grade and purchased from Sigma-Aldrich Company. The λ_{max} (404 nm) of asgrown Iron particles was measured using UV/visible spectroscopy Lamda-950, Perkin Elmer, Germany. FT-IR spectra were recorded on a Spectrum-100 FT-IR spectrophotometer in the mid-IR range in KBr media purchased from Perkin Elmer, Germany. Raman station 400 was used to measure the Raman shift of as-grown iron NPs using radiation source (Ar+ laser line, λ; 513.4 nm), which was purchased from Perkin Elmer, Germany. Morphology of iron oxide NPs were recorded on FE-SEM instrument (FESEM; JSM-7600F, Japan). Elemental analysis was investigated using EDS from JEOL, Japan. The morphologies, sizes, and structures of NPs were executed by HR-TEM (TEM; JEM-2100F, Japan). HR-TEM sample were prepared as follows; the synthesized as-grown iron NPs were dispersed into ethanol under ultrasonic vibration for 5 minutes. Then the HR-TEM film is dipped in the solution and dry for investigation. The powder X-ray diffraction (XRD) patterns were taken on a Xray diffractometer (XRD; X'Pert Explorer, PANalytical diffractometer) equipped with Cu- $K\alpha_1$ radiation (λ =1.5406 nm) using a generator voltage of 45 kV and a generator current of 40 mA were applied for the determination. I-V technique is employed by using Electrometer (Kethley, 6517A, Electrometer, USA).

Hydrothermal Methods

Hydrothermal synthesis includes the various techniques of crystallizing substances from high-temperature aqueous solutions at high vapor pressures; also termed "hydrothermal method". The term "hydrothermal" is of geologic origin. Geochemists and mineralogists have studied hydrothermal phase equilibria since the beginning of the twentieth century. George W. Morey at the Carnegie Institution and later, Percy W. Bridgman at Harvard University did much of the work to lay the foundations necessary to containment of reactive media in the temperature and pressure range where most of the hydrothermal work is conducted. Hydrothermal synthesis can be defined as a method of synthesis of single crystals that depends on the solubility of minerals in hot water under high pressure. The crystal growth is performed in an apparatus consisting of a steel pressure vessel called autoclave, in which a nutrient is supplied along with water. A gradient of temperature is maintained at the opposite ends of the growth chamber so that the hotter end dissolves the nutrient and the cooler end causes seeds to take additional growth. Possible advantages of the hydrothermal method over other types of crystal growth include the ability to create crystalline phases which are not stable at the melting point. Also, materials which have a high vapor pressure near their melting points can also be grown by the hydrothermal method. The method is also particularly suitable for the growth of large good-quality crystals while maintaining good control over their composition. Disadvantages of the method include the need of expensive autoclaves, and the impossibility of observing the crystal as it grows (O'Donoghue, 1983). A large number of compounds belonging to practically all classes have been synthesized under hydrothermal conditions: elements, simple and complex oxides, tungstates, molybdates, carbonates, silicates, germanates etc.

Hydrothermal synthesis is commonly used to grow synthetic quartz, gems and other single crystals with commercial value. Some of the crystals that have been efficiently grown are emeralds, rubies, quartz, alexandrite and others. The method has proved to be extremely efficient both in the search for new compounds with specific physical properties and in the systematic physicochemical investigation of intricate multicomponent systems at elevated temperatures and pressures.

Synthesis of Iron Oxide NPs

Low dimensional iron oxide NPs have been synthesized by adding uni-molar concentration of ferric chloride and urea into hydrothermal cell (Teflon line autoclave). $FeCl_3$ and urea were slowly dissolved into the de-ionized water separately to make 0.5 M concentration at room temperature. Then the solution was mixed gently and stirred until mix properly. The solution pH was slowly adjusted using ammonia solution drop wise to approximately 9.66. Then the mixture was put in hydrothermal cell (Teflon line autoclave) to put in the oven for 6 hours heated up and maintained at 150oC. Then the solution was washed with acetone and kept for drying at room temperature. The as-grown iron oxide products were characterized in detail in terms of their structural and optical properties.

The development of Fe_2O_3 nanoparticles can be well explained stranded on the chemical reactions concerned and crystal growth behaviors of iron oxide. For the synthesis of Fe_2O_3 NPs, ferric chloride ($FeCl_3$) and NH_4OH (in presence of urea) were mixed under continuous stirring at 150oC. During the reaction method, the NH_4OH performs in major rules, like control the pH value of the solution as well as resource to supply hydroxyl ions to the solution. The $FeCl_3$ reacts with NH4OH and forms FeOOH, which, upon heating, further produce into Fe^{2+} and OH- ions, which consequently assists in the development of Fe_2O_3 ions according to the chemical reactions (i)-(iii).

$$NH_4OH_{(aq)} \rightarrow NH_4^+{}_{(aq)} + OH^-{}_{(aq)} \quad \text{(i)}$$

$$FeCl_{3(s)} \rightarrow Fe^{2+}{}_{(aq)} + 3Cl^-{}_{(aq)} \quad \text{(ii)}$$

$$Fe^{2+}{}_{(aq)} + 2OH^-{}_{(aq)} \rightarrow FeOOH_{(aq)} + H^+ \quad \text{(iii)}$$

The FeOOH finally, however dissociates to the formation of Fe_2O_3 nuclei according to the reactions (iv)-(v).

$$FeCl_{3(aq)} + 2NH_4OH_{(aq)} \rightarrow FeOOH_{(aq)} + 2NH_4^+{}_{(aq)} + H^+{}_{(aq)} + 3Cl^-{}_{(aq)} \quad \text{(iv)}$$

$$FeOOH_{(aq)} \rightarrow Fe_2O_{3(s)} + H_2O_{(l)} \quad \text{(iv)}$$

The initially formed Fe_2O_3 nuclei perform as building blocks for the development of final products. With reaction time under the appropriate heating conditions in hydrothermal method, the Fe_2O_3 nuclei concentration enlarges which escorts the construction of desired nanoparticle products. As the Fe_2O_3 NPs are prepared by the well accretion of lowdimensional particles, therefore it is supposed that the fundamental entity for the configuration of Fe_2O_3 structure is nanoparticles.

Photocatalyst

In chemistry, photocatalysis is the acceleration of a photoreaction in the presence of a catalyst. In catalyzed photolysis, light is absorbed by an adsorbed substrate. In photogenerated catalysis, the photocatalytic activity (PCA) depends on the ability of the catalyst to create electron-hole pairs, which generate free radicals (hydroxyl radicals: OH) able to undergo secondary reactions. Its comprehension has been made possible ever since the discovery of water electrolysis by means of the titanium dioxide. Commercial application of the process is called advanced oxidation process (AOP). There are several methods of achieving AOP's that can but do not necessarily involve TiO2 or even the use of UV light. Generally the defining factor is the production and use of the hydroxyl radical.

Photocatalytic degradation of AO with NPs

Photo-catalytic degradation of AO using Iron oxide NPs was executed by optical absorption spectroscopy. The catalytic reaction was carried out in a 250.0 ml beaker, which contain 150.0 ml of AO dye solution (0.03 mM) and 150.0 mg of catalyst. Prior to irradiation, the solution was stirred and bubbled with oxygen for at least 15 min in the dark to allow equilibrium of the system so that loss of compound owing to the adsorption can be taken into account. The suspension was continuously purged with oxygen bubbling throughout the experiment. Irradiation was carried out using 250 W Mercury lamps. Samples (5.0 ml) were composed before and at usual intervals during the irradiation and AO solution were isolated from the photo-catalyst by centrifugation before analysis. The degradation was investigated by measuring the absorbance using UV-visible spectrophotometer (Lambda 950). The absorbance of AO (0.03 mM) was followed at 491 nm wavelength. All solutions were prepared and organized with de-ionized water in room conditions.

Fabrication of gold electrode using NPs

Gold electrode (surface area 0.0216 cm^2) is fabricated with as-grown iron NPs where butyl carbitol acetate (BCA) and ethyl acetate (EA) as a fabricating agent. Then it is kept in the oven at 50oC for 2 hours until the film is completely uniform, stable, and dried. Phosphate buffer solution (PBS, 0.1M, pH 7.0) is prepared by mixing 0.2M Na_2HPO_4 and 0.2M NaH_2PO_4 solution in 100.0ml de-ionize water. The as-grown iron oxide NPs were employed for the detection of ammonia in liquid phase. The thin film of NPs were made with conducting binders and embedded on the gold electrode, which is presented in the Scheme

The Pd and gold electrodes are used as counter and working electrodes, which are presented in Scheme 1A and Scheme 1B respectively. The fabricated electrode was kept in the oven at low temperature (60.0oC) for 2 hours to dry and uniform the film completely (Scheme 1B). The ammonium hydroxide was used as a target chemical in the liquid phase for the total measurement. The electrical response of target compound has been measured using I-V method, which is shown in Scheme 1C.

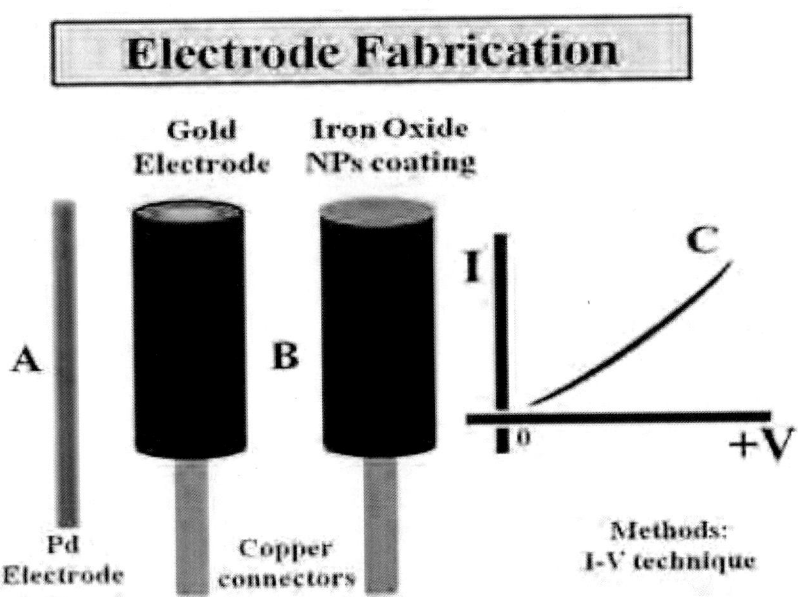

Scheme 1. Fabrication of chemical sensors using as-grown β-Fe_2O_3 NPs and its detection methodology.

Lower detection limit (LOD)

In analytical chemistry, the detection limit, lower limit of detection, or LOD (limit of detection), is the lowest quantity of a substance that can be distinguished from the absence of that substance (a blank value) within a stated confidence limit (generally 1%)

(MacDougall & Crummett, 1980). The detection limit is estimated from the mean of the blank, the standard deviation of the blank and some confidence factor. Another consideration that affects the detection limit is the accuracy of the model used to predict concentration from the raw analytical signal. There are a number of different "detection limits" that are commonly used. These include the instrument detection limit (IDL), the method detection limit (MDL), the practical quantification limit (PQL), and the limit of quantification (LOQ). Even when the same terminology is used, there can be differences in the LOD according to nuances of what definition is used and what type of noise contributes to the measurement and calibration (Long & Winefordner, 1983).

Linear dynamic range (LDR)

The detector response is said to be linear if the difference in response for two concentrations of a given compound is proportional to the difference in concentration of the two samples. Such response appears as a straight line in the calibration curve. The linear dynamic range of a detector is the maximum linear response divided by the detector noise. Most detectors eventually become non-linear as sample size is increased and this upper point is usually well defined. The chromatographer should know where this occurs to avoid errors in quantification.

Detection of ammonium hydroxide by I-V technique

A cell is constructed consisting of NPs coated gold electrode as a working electrode and Pd wire is used a counter electrode. Ammonium hydroxide solution is diluted at different concentrations in DI water and used as a target chemical. Amount of 0.1M phosphate buffer solution was kept constant as 20.0 ml throughout the investigation. Solution is prepared with various concentrations of ammonium hydroxide as 77.0 μM to 7.7 M (25% ammonia solution). The sensitivity is calculated from the ratio of voltage versus current of the calibration plot. Electrometer is used as a voltage sources for I-V measurement in simple two electrode system.

PROPERTIES OF NANOPARTICLES

Optical properties of β-Fe_2O_3 NPs

The optical property of the β-Fe_2O_3 nanoparticles is one of the important characteristics for the evaluation of its optical and photocatalytic activity. UV/visible absorption are a method in which the outer electrons of atoms or molecules absorb radiant energy and undergo transitions to high energy levels. In this procedure, the spectrum obtained owing to optical absorption can be analyzed to acquire the energy band gap of the semiconductor nanomaterials. The optical absorption measurement was carried out at room conditions. Therefore, the nanomaterial may be helpful for the development of non-linear optical sensors in this wavelength region, as the lack of absorption peaks is the major prerequisite for the nanomaterials to confirm non-linear properties. The absorption spectrum of as-grown iron oxide NPs solution is presented in Fig. 1(A).

It displays an onset of absorption maxima at 404.0 nm in visible range between 200 to 800 nm wavelengths. The lambda maxima of as grown NPs are quite different to those observed earlier due to β-Fe_2O_3 and β-Fe_2O_3(Cherepy et al., 1998) but are very similar to those reported for β-Fe_2O_3 (Cornell & Schwertmann, 2003). It shows a broad absorption band around 404.0 nm indicating the formation of low dimensional β-Fe_2O_3NPs having reddish colors. Band gap energy is calculated on the basis of the maximum absorption band (404nm) of β-Fe_2O_3 NPs and obtained to be 3.06931 eV, according to following equation (vi).

$$E_{bg} = \frac{1240}{\lambda}(eV) \qquad \text{(vi)}$$

Where Ebg is the band-gap energy and λ_{max} is the wavelength (404.0 nm) of the nanoparticles.

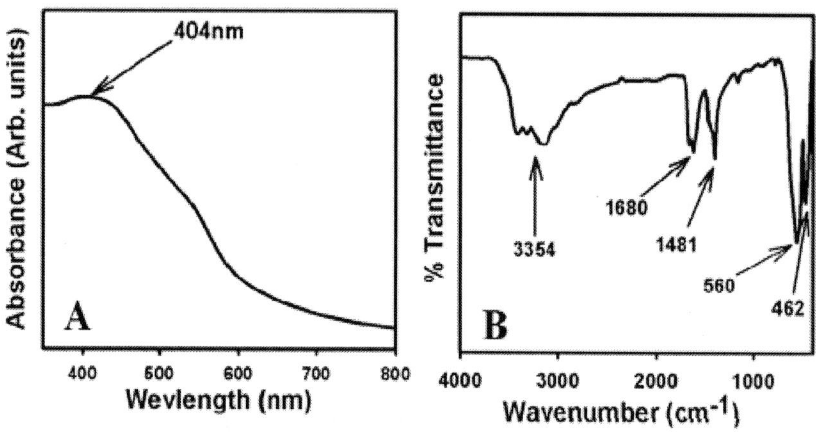

Figure 1 (A) UV-vis absorption and (B) FT-IR spectra of as-grown Fe2O3 NPs.

Infrared spectroscopy (IR spectroscopy) is the spectroscopy that deals with the infrared region of the electromagnetic spectrum that is light with a longer wavelength and lower frequency than visible light. It covers a range of techniques, mostly based on absorption spectroscopy. As with all spectroscopic techniques, it can be used to identify and study chemicals. A common laboratory instrument that uses this technique is a Fourier transform infrared (FT-IR) spectrometer. The infrared portion of the electromagnetic spectrum is usually divided into three regions; the near-, mid- and far- infrared, named for their relation to the visible spectrum. The higher energy near-IR, approximately 14000–4000 cm^{-1} (0.8- 2.5 μm wavelength) can excite overtone or harmonic vibrations. The mid-infrared, approximately 4000–400 cm−1 (2.5–25 μm) may be used to study the fundamental vibrations and associated rotational-vibrational structure [Fig. 2]. The far-infrared, approximately 400–10 cm−1 (25–1000 μm), lying adjacent to the microwave region, has low energy and may be used for rotational spectroscopy. The names and classifications of these subregions are conventions, and are only loosely based on the relative molecular or electromagnetic properties. Fourier transform infrared (FT-IR) spectroscopy is a measurement technique that allows one to record infrared spectra. Infrared light is guided through an interferometer and then through the sample (or vice versa). A moving mirror inside the apparatus alters the distribution of infrared light that passes through the interferometer. The signal directly recorded, called an "interferogram", represents light output as a function of mirror position.

A data-processing technique called Fourier transform turns this raw data into the desired result (the sample's spectrum): Light output as a function of infrared wavelength (or equivalently, wavenumber). There is an alternate method for taking spectra (the "dispersive" or "scanning monochromator" method), where one wavelength at a time passes through the sample. The dispersive method is more common in UV-Vis spectroscopy, but is less practical in the infrared than the FTIR method. One reason that FTIR is favored is called "Fellgett's advantage" or the "multiplex advantage": The information at all frequencies is collected simultaneously, improving both speed and signal-to-noise ratio. Another is called "Jacquinot's Throughput Advantage": A dispersive measurement requires detecting much lower light levels than an FTIR measurement. There are other advantages, as well as some disadvantages, but virtually all modern infrared spectrometers are FTIR instruments.

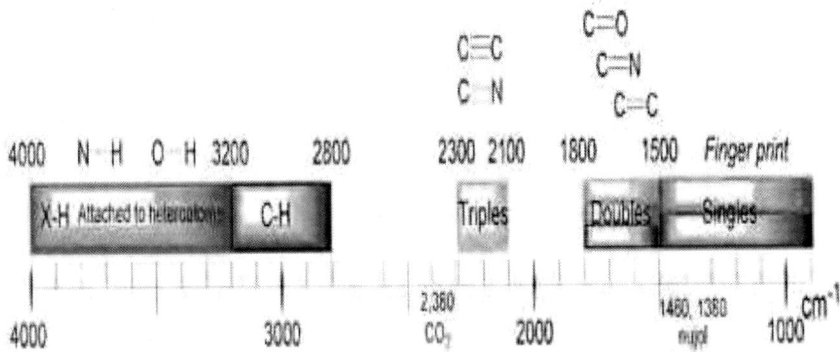

Figure 2 FT-IR wave range for fundamental vibrations and associated rotational-vibrational frequencies.

Fig. 1(B) represents the FT-IR spectrum of the as- grown NPs. It displays several bands at 462, 560, 1481, 1680, and 3354 cm-1. The observed vibration bands may be assigned to Fe–O– Fe stretching vibration (462 and 560 cm^{-1}), O=C=O stretching vibration (1481 cm−1), O-H stretching (3354 cm−1), and O-H bending vibration (1680 cm−1). The absorption bands at 1481, 1680, and 3354 cm-1 normally comes from carbon dioxide and water which generally nanomaterials absorbed from the environment due to their mesophorous structure. The observed vibration bands at low frequencies regions suggest the formation of β-Fe_2O_3.

Structural properties of of β-Fe_2O_3. NPs

Raman spectroscopy is a spectroscopic method utilized to reveal vibrational, rotational, and other low-frequency phases in a Raman active compounds.

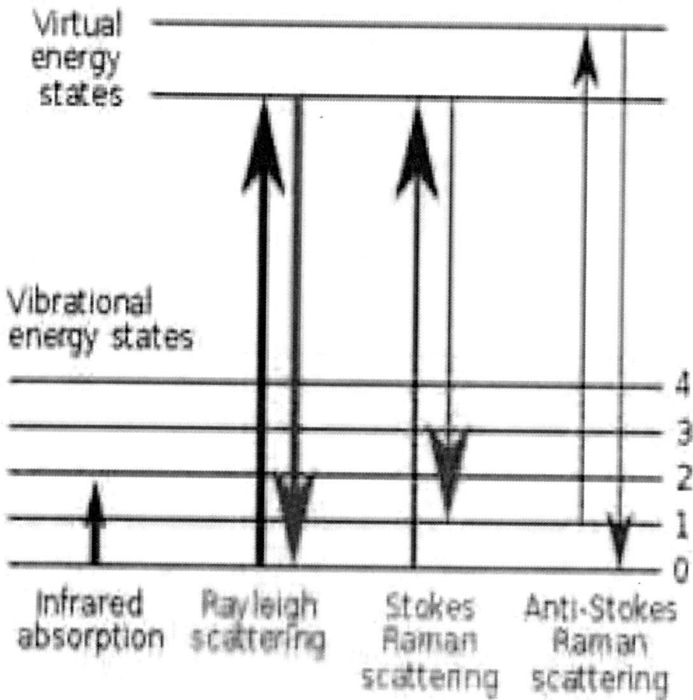

Figure 3 Energy level diagram showing the states involved in Raman signal. The line thickness is roughly proportional to the signal strength from the different transitions.

It depends on inelastic scattering, or Raman scattering, of monochromatic light, generally from a laser in the visible, near infrared, or near ultraviolet range [Fig. 3]. The laser light communicates with molecular vibrations, phonons or other excitations in the modes, showing in the energy of the laser photons being shifted up or down. The shift in energy represents information regarding the phonon modes in the system, where Infrared spectroscopy yields similar, but complementary information. Raman spectroscopy is generally established and utilized in material chemistry, since the information is specific to the chemical bonds and symmetry of metal-oxygen stretching or vibrational modes.

Fig. 4A shows the Raman spectrum where main features of the wave-number are observed at about 221, 290, 410, 500, and 608 cm-1 for Fe-O stretching vibration. These large bands may be assigned to a magnetite phase of iron oxide NPs. At 608 cm-1, higher wave-number shift is observed due to the various dimensional effects of the NPs.

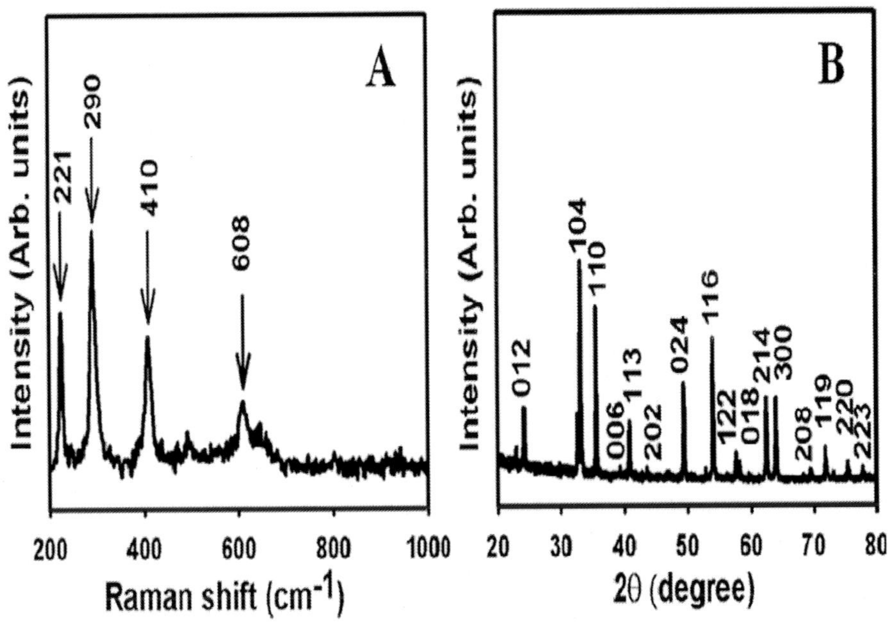

Figure 4 (A) Raman spectra and (B) XRD patterns of as-grown iron oxide NPs.

X-ray scattering techniques are a family of non-destructive analytical techniques which reveal information about the crystallographic structure, chemical composition, and physical properties of materials and thin films. These techniques are based on observing the scattered intensity of an X-ray beam hitting a sample as a function of incident and scattered angle, polarization, and wavelength or energy. X-ray diffraction yields the atomic structure of materials and is based on the elastic scattering of X-rays from the electron clouds of the individual atoms in the system. The most comprehensive description of scattering from crystals is given by the dynamical theory of diffraction (Azároff et al., 1974). X-ray diffraction can be used to determine which iron oxide compounds are present in NPs by calculating or comparing with the standard value of lattice parameters, crystal structures and crystallinity. Fig. 4B shows typical crystallinity of the synthesized as-grown iron oxide NPs and their aggregates. All the reflection peaks in this pattern were found to match with the β-Fe_2O_3 phase having rhombohedral geometry [JCPDF # 089-2810]. The lattice parameters are a = 5.04, c = 13.75, point group: R-3c, and Radiation: $CuK\alpha_1$ (λ= 1.5406). The sample showed the major characteristic peaks for as grown crystalline metallic iron at 2θ values of 24.2(012), 32.2(104), 35.7(110), 39.4(006), 40.9(113), 43.5(202), 49.5(024), 54.1(116), 57.3(122), 57.6(018), 62.5(214), 64.0(300), 69.6(208), 72.0(119), 75.4(220), and 78.0(223) degrees. These indicate that the as-grown iron oxide NPs is well crystalline β-Fe_2O_3. As scanning electron microscopes (SEM) have evolved the electron beam cross section has become smaller and smaller increasing magnification several fold. A field-emission cathode in the electron gun of a scanning electron microscope provides narrower probing beams at low as well as high electron energy, resulting in both improved spatial resolution and minimized sample charging and damage. Advantages of SEM are (a) semiconductor device cross section analyses for gate widths, gate oxides, film thicknesses, and construction details, (b) advanced coating thickness and structure uniformity determination, and (c) small contamination feature geometry and elemental composition measurement. High resolution FE-SEM images of as-grown low dimensional iron oxide NPs are presented in Figure 5.

Figure 5A to 5E shows the low to high magnified images of NPs grown by hydrothermal process. It is clear from the FE-SEM images that the synthesized products are NPs, which grown in a very high-density and possessed almost uniform shape presented in Figure 5A to 5D. Figure 5E exhibits the high-resolution FE-SEM image of the synthesized NPs which reflected that most of the NPs possessing spherical shapes. The diameter of iron NPs is calculated in the range of 50-90 nm where the average diameter of iron oxide NPs is close to 60 ± 10 nm.

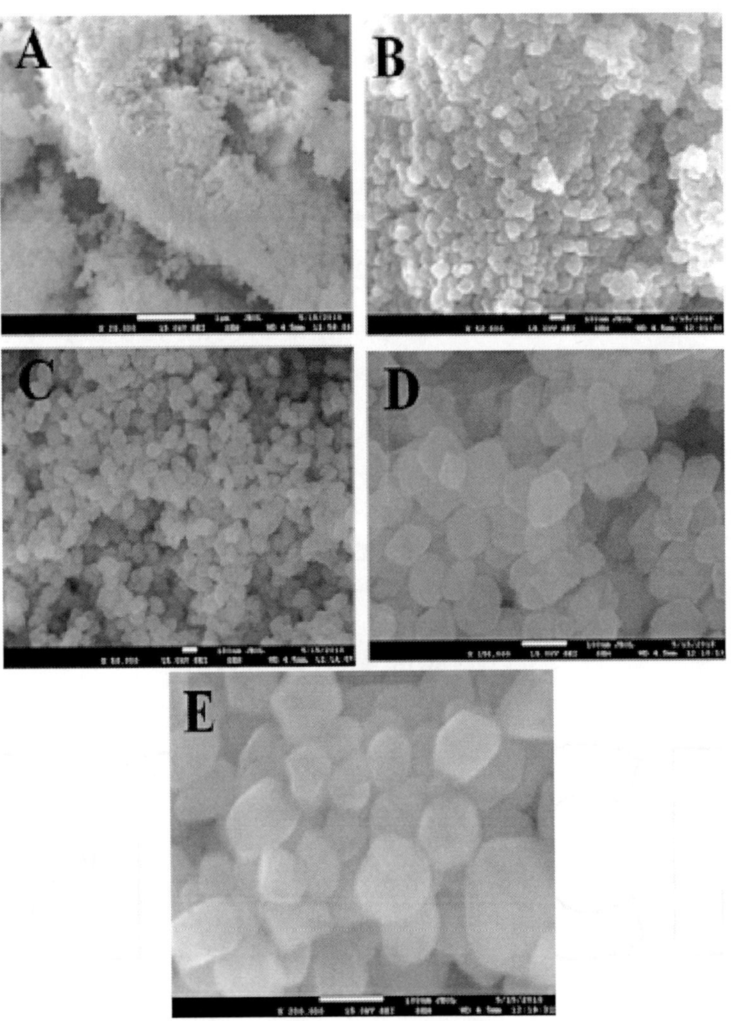

Figure 5 (A)-(E) Low to High resolution FE-SEM images of as-grown iron oxide NPs.

Energy-dispersive X-ray spectroscopy (EDS or EDX) is an analytical technique used for the elemental analysis or chemical characterization of a sample. It is one of the variants of X-ray fluorescence spectroscopy which relies on the investigation of a sample through interactions between electromagnetic radiation and matter, analyzing X-rays emitted by the matter in response to being hit with charged particles. Its characterization capabilities are due in large part to the fundamental principle that each element has a unique atomic structure allowing X-rays that are characteristic of an element's atomic structure to be identified uniquely from one another. To stimulate the emission of characteristic X-rays from a specimen, a highenergy beam of charged particles such as electrons or protons or a beam of X-rays, is focused into the sample being studied. At rest, an atom within the sample contains ground state (or unexcited) electrons in discrete energy levels or electron shells bound to the nucleus. The incident beam may excite an electron in an inner shell, ejecting it from the shell while creating an electron hole where the electron was. An electron from an outer, higherenergy shell then fills the hole, and the difference in energy between the higher-energy shell and the lower energy shell may be released in the form of an X-ray. The number and energy of the X-rays emitted from a specimen can be measured by an energy-dispersive spectrometer. As the energy of the X-rays is characteristic of the difference in energy between the two shells, and of the atomic structure of the element from which they were emitted, this allows the elemental composition of the specimen to be measured. The electron dispersive spectroscopy (EDS) analysis of these particles indicates the presence of Fe and O composition in the pure as-grown iron oxide NPs (Fe_2O_3). It is clearly displayed that asgrown synthesize materials contained only iron and oxygen elements, which presented in Fig. 6. The composition of iron and oxygen is 54.11% and 45.88% respectively. No other peak related with any impurity has been detected in the EDS, which confirms that the asgrown NPs are composed only with iron and oxygen.

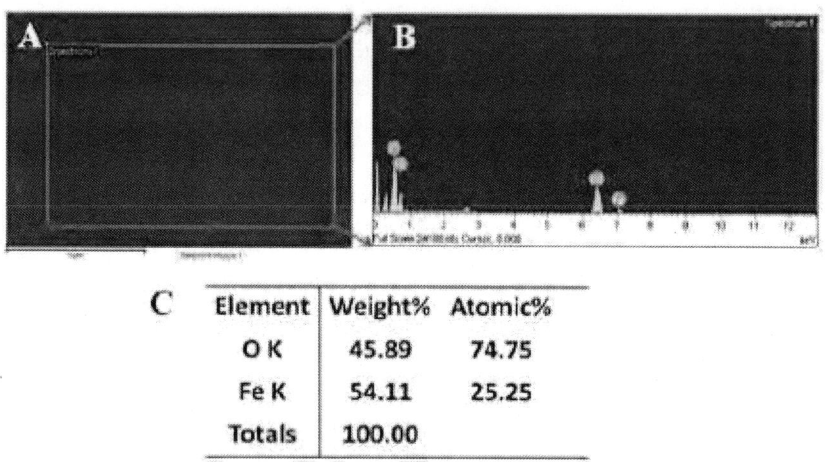

C

Element	Weight%	Atomic%
O K	45.89	74.75
Fe K	54.11	25.25
Totals	100.00	

Figure 6 EDS of as-grown iron oxide NPs. (A) selected area, (B) EDS outcomes, and (C) elemental compositions (weight% and atomic%).

Transmission electron microscopy (TEM) is a microscopy technique whereby a beam of electrons is transmitted through an ultra thin specimen, interacting with the specimen as it passes through. An image is formed from the interaction of the electrons transmitted through the specimen; the image is magnified and focused onto an imaging device, such as a fluorescent screen, on a layer of photographic film, or to be detected by a sensor such as a CCD camera. TEMs are capable of imaging at a significantly higher resolution than light microscopes, owing to the small de Broglie wavelength of electrons. This enables the instrument's user to examine fine detail-even as small as a single column of atoms, which is tens of thousands times smaller than the smallest resolvable object in a light microscope.

TEM forms a major analysis method in a range of scientific fields, in both physical and biological sciences. TEMs find application in cancer research, virology, materials science as well as pollution and semiconductor research. At smaller magnifications TEM image contrast is due to absorption of electrons in the material, due to the thickness and composition of the material. At higher magnifications complex wave interactions modulate the intensity of the image, requiring expert analysis of observed images. Alternate modes of use allow for the TEM to observe modulations in chemical identity, crystal orientation, electronic structure and sample induced electron phase shift as well as the regular absorption based imaging. The first TEM was built by Max Knoll and Ernst Ruska in 1931, with this group developing the first TEM with resolving power greater than that of light in 1933 and the first commercial TEM in 1939.

High-resolution transmission electron microscopy (HRTEM) is an imaging mode of the transmission electron microscope (TEM) that allows the imaging of the crystallographic structure of a sample at an atomic scale (Spence, 1988). Because of its high resolution, it is an invaluable tool to study nanoscale properties of crystalline material such as semiconductors and metals.

At present, the highest resolution realized is 0.8 angstroms (0.08 nm) with microscopes such as the OAM at NCEM. Ongoing research and development such as efforts in the framework of TEAM will soon push the resolution of HRTEM to 0.5 Å. At these small scales, individual atoms and crystalline defects can be imaged. Since all crystal structures are 3-dimensional, it may be necessary to combine several views of the crystal, taken from different angles, into a 3D map. This technique is called electron crystallography. One of the difficulties with HRTEM is that image formation relies on phase-contrast.

In phase-contrast imaging, contrast is not necessarily intuitively interpretable as the image is influenced by strong aberrations of the imaging lenses in the microscope. One major aberration is caused by focus and astigmatism, which often can be estimated from the Fourier transform of the HR-TEM image.

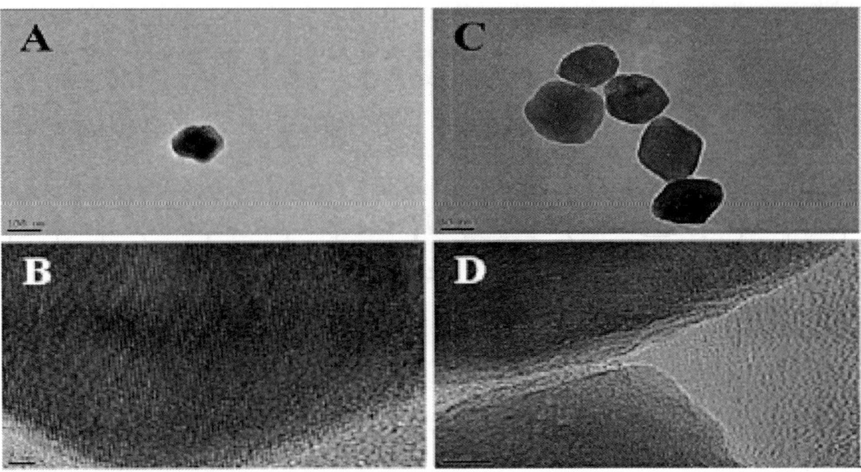

Figure 7 (A and C)TEM and (B and D) HR-TEM images of iron oxide NPs

The morphologies of particles were observed by TEM, which are presented in the Fig. 7A and Fig. 7C. TEM photographs showed spherical NPs of iron oxide which appeared as deep dark spots. The diameter of as-grown iron oxide is in the range of 50 ~ 90 nm. A typical HRTEM image of the iron NPs is shown in Figure 7B and 7D. It appears to be spherical in shape and fairly uniform in size with a mean diameter of 60 ± 10 nm. Most of the NPs are spherical in shape and the dimensions are well consistent with the FE-SEM observations. It is clearly revealed and calculated from the HR-TEM image that the distance between two lattice fringes is about 0.31 nm.

POTENTIAL APPLICATIONS OF DEVELOPED β-Fe_2O_3NPS

Photo-degradation of Acridine Orange (AO)

A dye can generally be described as a colored substance that has an affinity to the substrate to which it is being applied. The dye is generally applied in an aqueous solution, and may require a mordant to improve the fastness of the dye on the fiber. Both dyes and pigments appear to be colored because they absorb some wavelengths of light preferentially. In contrast with a dye, a pigment generally is insoluble, and has no affinity for the substrate. Some dyes can be precipitated with an inert salt to produce a lake pigment, and based on the salt used they could be aluminum lake, calcium lake or barium lake pigments. The first human-made (synthetic) organic dye, mauveine, was discovered by William Henry Perkin in 1856. Many thousands of synthetic dyes have since been prepared (Zollinger, 2003; Hunger, 2003). Synthetic dyes quickly replaced the traditional natural dyes. They cost less, they offered a vast range of new colors, and they imparted better properties to the dyed materials (Garfield, 2000). Dyes are now classified according to how they are used in the dyeing process.

Acridine orange is a nucleic acid selective fluorescent cationic dye useful for cell cycle determination. It is cell-permeable, and interacts with DNA and RNA by intercalation or electrostatic attractions respectively. When bound to DNA, it is very similar spectrally to fluorescein, with an excitation maximum at 502 nm and an emission maximum at 525 nm (green). When it associates with RNA, the excitation maximum shifts to 460 nm (blue) and the emission maximum shifts to 650 nm (red). Acridine orange will also enter acidic compartments such as lysosomes and become protonated and sequestered. In these low pH conditions, the dye will emit orange light when excited by blue light. Thus, acridine orange can be used to identify engulfed apoptotic cells, because it will fluoresce upon engulfment.

The dye is often used in epifluorescence microscopy. Acridine orange is prepared from coal tar and creosote oil. Acridine orange can be used in conjunction with ethidium bromide to differentiate between viable, apoptotic and necrotic cells. Additionally, Acridine orange may be used on blood samples to fluoresce bacterial DNA, aiding in clinical diagnosis of bacterial infection once serum and debris have been filtered. Acridine orange can be used in the assessment of sperm chromatin quality.

Photodegradation is degradation of a photodegradable molecule caused by the absorption of photons, particularly those wavelengths found in sunlight, such as infrared radiation, visible light, and ultraviolet light. However, other forms of electromagnetic radiation can cause photodegradation. Photodegradation includes photodissociation, the breakup of molecules into smaller pieces by photons. It also includes the change of a molecule's shape to make it irreversibly altered, such as the denaturing of proteins, and the addition of other atoms or molecules. A common photodegradation reaction is oxidation. This type of photodegradation is used by some drinking water and wastewater facilities to destroy pollutants. Photodegradation in the environment is part of the process by which ambergris evolves from its fatty precursor.

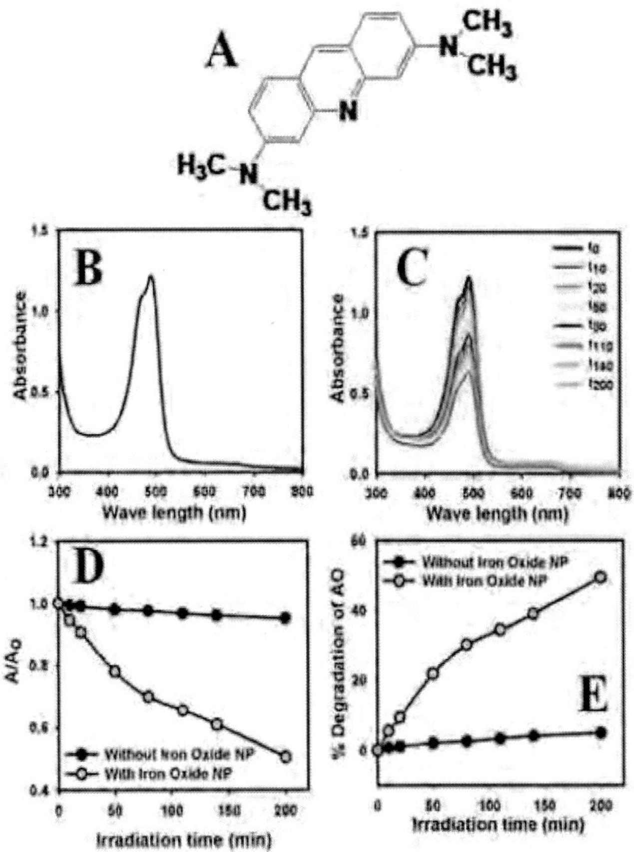

Figure 8 Photocatalytic degradation of AO using iron oxide NPs. (A) Molecular structure of AO; (B) Spectrum of aqueous solution of AO; (C) Spectrum of AO at different time interval; (D) Change in absorbance; (E) % degradation in different time intervals of AO in presence and absence of as-grown iron oxide NPs.

The molecular structure and absorbance spectra of AO are presented in the Figure 8A and 8B respectively. The photo-degradation of the AO dye occurs predominantly on the β-Fe_2O_3 NPs surface. The extent of degradation of the AO dye was measured by monitoring its concentration with and without NP after reaching the degradation. Figure 8C shows the gradual decreasing in absorption spectra for the degradation of AO dye as a function of photo-irradiation time. It is observed that the irradiation of an aqueous suspension of AO dye in the presence of iron oxide NP leads to decrease in absorption spectra at 491 nm.

The absorbance spectra at 491 nm is significantly decreased with increasing in exposure time and gradually decrease until 200 min which concludes that the AO dye has de-colorization property with iron oxide NPs close to 50%. The decrease in absorption intensity vs irradiation time for the AO in the presence and absence of iron oxide NPs is shown in the Figure 8D. Figure 8E shows a plot for the percent degradation versus irradiation time (min) for the oxygen saturated aqueous suspension of AO in the presence and absence of NPs. Figure 8E shows that around 50% degradation of the AO dye takes place after 200 min of irradiation in the presence of catalytic iron oxide NPs. The schematic representation of AO degradation is presented in Scheme 2.

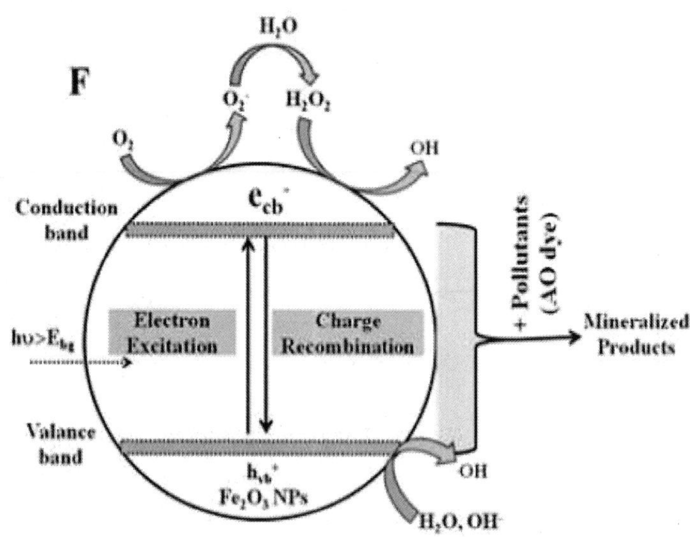

Scheme 2. Schematic representation of AO degradation mechanism

The experimental degradation rate constant is obtained from the initial slope acquired by linear regression from a plot of the natural logarithm (ln) of absorbance of the AO as a function of exposure time, i.e., first-order degradation kinetics. This rate constant is used to evaluate the degradation rates for the decomposition of AO using the formula appended below (vii):

$$-d[A]/dt = kc^n \qquad \text{(vii)}$$

Where, k is the rate constant, c the concentration of the pollutant, and n the order of reaction. The rate of degradation is 9.9x10-5 mole.L-1.Min-1. The degradation of AO clearly demonstrated that prepared iron oxide (MOx) NPs possess significant photocatalytic activity and these NPs could be beneficial photocatalysts for the removal of colored dyes.

Mechanism of heterogeneous photocatalysis has been discussed extensively in literature. When a semiconductor such as MOx absorbs a photon of energy equal to or greater than its band gap, there will be the formation of electron/hole pair (Scheme 2). If charge separation is maintained, the electron and hole may migrate to the metal oxide surface, which can eventually bring about redox reactions of the organic substrates dissolved in water in the presence of oxygen. During the photo-catalytic oxidation processes hydroxyl radicals (.OH) and superoxide radical anions (O.2 -) are supposed to be the main oxidizing species. These oxidative retorts would consequence in the mineralization of the dye. Alternatively, direct absorption of light by the dye named AO, can lead to charge injection from the excited state of the AO to the conduction band of the semiconductor as summarized in the following equations (viii-xii):

$$MO_x + h\upsilon \rightarrow e^-_{cb} + h^+_{vb} \qquad \text{(viii)}$$

$$O_2 + e^-_{cb} \rightarrow \cdot O^-_2 \qquad \text{(ix)}$$

$$H_2O + h^+_{vb} \rightarrow \cdot OH + H^+ \quad (x)$$

$$AO_{ads} + h\upsilon \rightarrow AO_{ads}^* \quad (xi)$$

$$AO_{ads}^* + MO_x \rightarrow AO_{ads}^+ + \cdot MO_2\,(e^-) \quad (xii)$$

The degradation process was initiated by the photolysis of MOx-oxide/hydroxyl species, and accelerated by mercury beam irradiation, due to enhance photolysis of NPs species, which enhances the regeneration of metal precursor with concomitant production of oxide free radicals. Hydroxyl radicals (OH.) and superoxide radical anions (O2.-) contribute to the oxidation process by attacking the dye molecules and would results in the bleaching of the AO dye. The results above clearly indicate that the prepared NPs shows a good photocatalytic activity, thus it can be used as a photo-catalyst for the treatment of wastewater.

Chemical Sensors

A sensor is a device that measures a physical quantity and converts it into a signal which can be read by an observer or by an instrument. For example, a mercury-in-glass thermometer converts the measured temperature into expansion and contraction of a liquid which can be read on a calibrated glass tube. A thermocouple converts temperature to an output voltage which can be read by a voltmeter. For accuracy, most sensors are calibrated against known standards. Nanotechnology can enable sensors to detect very small amounts of chemical vapors or liquids. Various types of detecting elements, such as semiconductor materials, carbon nanotubes, iron oxide nanoparticles, zinc oxide nanowires or palladium nanoparticles can be used in nanotechnology-based chemical sensors. These detecting elements change their electrical characteristics, such as resistance or capacitance, when they absorb a gas/liquid molecule on it. Because of the small size of nanoparticles/nanotubes/ nanowires, a few gas or liquid molecules are sufficient to change the electrical properties of the sensing elements. This allows the detection of a very low concentration of chemical vapors or liquids. This goal is to have small, inexpensive sensors that can sniff out chemicals just as dogs are used in airports to smell the vapors given off by explosives or drugs. The capability of producing small, inexpensive sensors that can quickly identify chemicals provides a king of nano-bloodhound that doesn't need sleep or exercise which can be useful in a number of ways. An obvious application is to mount these sensors throughout an airport, or any facility with security concerns, to check for vapors given off by explosive devices.

The potential application of β-Fe_2O_3 nanoparticles as chemical sensors has been explored for detecting and quantifying hazardous chemicals, which are not environmental friendly as well as safe. Development of this nanostructure material as chemi-sensors is in the primitive stage. The β-Fe_2O_3 NPs chemi-sensors have advantages such as stability in air, non-toxicity, chemical stability, electrochemical activity, ease to fabricate, and bio-safe characteristics. As in the case of chemical sensors, the principle of operation is that the current response in I-V technique of β-Fe_2O_3 NPs drastically changes when aqueous ammonia are adsorbed. The β-Fe_2O_3 nano-materials were used for fabrication of chemical sensor, where aqueous ammonia was considered as target analyte. The emaciated-film of nanomaterials sensor were fabricated with conducting agents and embedded on the gold electrode surface, which is presented in the Scheme 3(A). The fabricated electrode was accumulated in the oven at low temperature (60.0 oC) for 2 hours to dry, stable, and uniforms the film totally. The electrical responses of target analyte were executed using I-V method, which is presented in Scheme 3(B). Hypothetical, experimental, and mechanism of I-V responses of chemical sensor having β-Fe_2O_3 NPs thin film as a function of current versus potential for aqueous ammonia, which is shown in Scheme 3(B), Scheme 3(C), and Scheme 3(D) respectively. The time delaying of electrometer was kept for 1.0 sec. A significant increase in the current value with applied potential is clearly demonstrated.

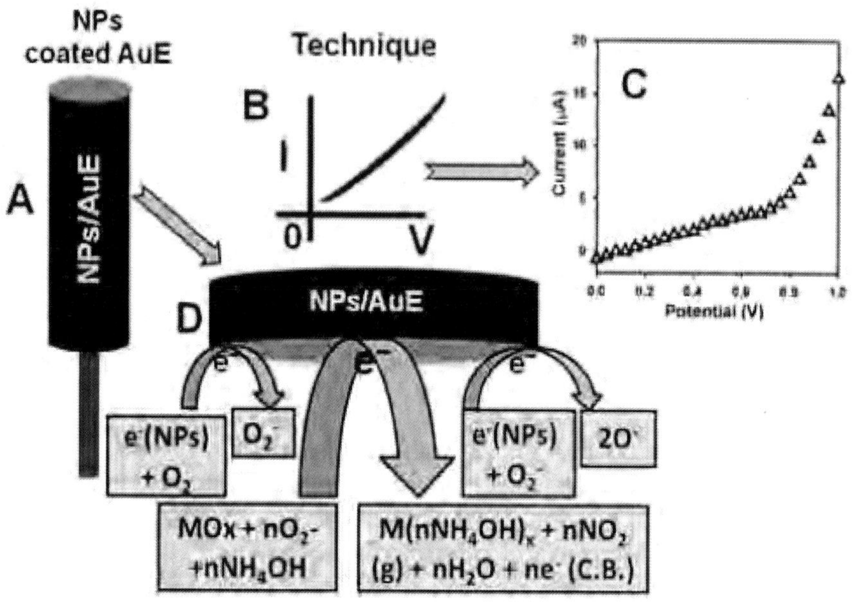

Scheme 3 Schematic view of (A) fabrication of AuE with β-Fe_2O_3 NPs and coating agents, (B) detection I-V method (theoretical), (C) experimental result of I-V plot, (D) proposed mechanisms of aqueous ammonia detection in presence of semiconductor β-Fe_2O_3 nanomaterials.

I-V responses of chemical sensor having β-Fe_2O_3 thin film as a function of current versus potential for the liquid ammonia, which is exposed in Figure 9. A significant increase in the current value with applied potential is clearly demonstrated, where the time delaying for electrometer is kept 1.0 sec. Fig. 9A represents the current changing before (dark-dotted) and after (gray-dotted) coating was made with NPs on gold electrode. The gray-dotted and dark-dotted curves indicated the response of the film before and after injecting 100.0 μl chemicals in bulk solution respectively, which is presented in Fig. 9B. Significant increase of surface film current is measured after every injection of the target component. 77.0 μM concentration of aqueous ammonia is initially taken into the electrochemical cell and added the higher concentration (10 times) in each injection from the stock analyte solution. During I-V measurement, it is injected low to high analyte concentrations (77.0 μM to 7.7 M) gradually into the 20.0 ml phosphate buffer solution in every step. Each I-V response to varying concentration of aqueous ammonia from 7.7 M to 77.0 μM on thin iron oxide NPs coating is presented in the Fig. 9C. It shows the current response changed of the iron oxide NPs as a function of ammonium hydroxide concentration in room conditions. It is observed that at lower to higher concentration of target compound, the current increase gradually. A wide range of analyte concentration was preferred to study the possible analytical parameters, which is calculated in 77.0 μM to 0.77 M. The calibration curve is plotted from the variation of analyte concentrations, which is revealed in the Fig. 9D. The sensitivity is calculated from the calibration curve, which is close to 0.5305 ± 0.02 μAcm-2mM-1. The linear dynamic range of this sensor exhibits from 77.0 μM to 0.77 M and the detection limit was around 21.8 ± 0.1 μM [3×noise(N)/slope(S)].

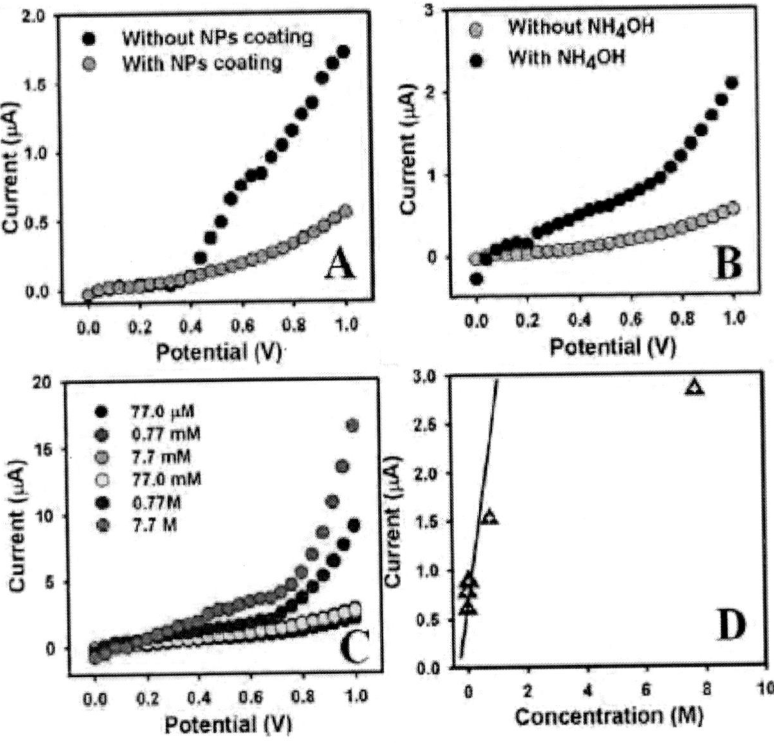

Figure 9 I-V curves of (A) with and without coating of NPs; (B) with and without ammonium hydroxide (77.0 μM) using NPs coated gold electrode; (C) Concentration variation of ammonium hydroxide; and (D) calibration curve of ammonium hydroxide sensors.

The response time was around 10.0 sec for the NPs coated-electrode to reach saturated steady state current. The high sensitivity of film can be attributed to the good absorption (porous surfaces fabricated with coating) and adsorption ability, high catalytic activity, and good biocompatibility of the iron oxide NPs. The estimated sensitivity of the fabricated sensor is relatively higher than previously reported ammonium hydroxide sensors based on other composite or materials modified electrodes (Raj et al., 2010). The aqueous ammonia sensing method of β-Fe_2O_3 NPs sensor is based on the semiconductor oxides, due to oxidation or reduction of the metal-oxide itself, in accordance with the dissolved oxygen (O_2) in bulk-solution or surface-air of the surrounding atmosphere (Scheme 4A), according to the equations (xiii-xiv).

$$\text{electron } (e^-, \beta\text{-}Fe_2O_3 \text{ NPs}) + O_2 \rightarrow O_2^- \quad \text{(xiii)}$$

$$\text{electron } (e^-, \beta\text{-}Fe_2O_3 \text{ NPs}) + O_2^- \rightarrow 2O^- \quad \text{(xiv)}$$

These reactions take place in bulk-solution or air/liquid interface or surrounding air due to the low carrier concentration which increased the resistances. The aqueous ammonia sensitivity headed to β-Fe_2O_3 NPs (i.e. MOx) could be attributed to the high oxygen deficiency and defect density leads to increase oxygen adsorption. Larger the amount of oxygen adsorbed on the MOx NPs surfaces, larger would be the oxidizing capability and faster would be the oxidation of analytes.

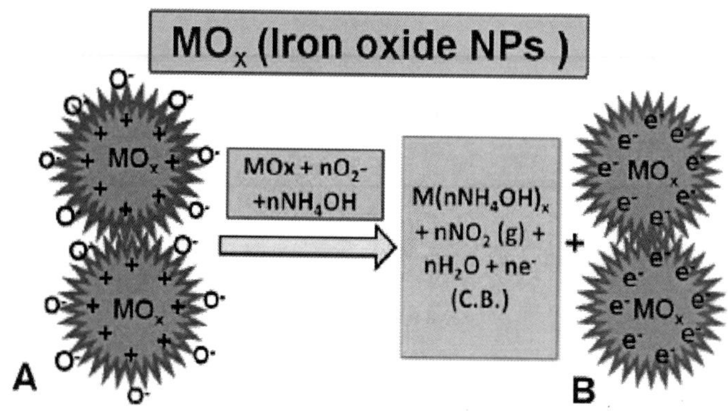

Scheme 4. Mechanism of β-Fe_2O_3 NPs chemical sensors at room conditions.

The reactivity of aqueous ammonia would be very large as compared and contrasted to other chemical with the semiconductor surfaces under the similar condition (Patil et al., 2007). When NH_4OH reacts with the adsorbed oxygen on the surface of the film, it acquires oxidized to nitrogen oxide gas and metal ammonium hydroxide, liberating free electrons in the conduction band (Scheme 4B), which could be articulated through the following reactions (xv-xvi):

$$MO_x + nNH_4OH + nO_2- \rightarrow M(nNH_4OH)x + nNO_2\,(g) + nH_2O + ne^-\,(C.B.) \qquad (xv)$$

$$4NH_4OH + 7O_2- \rightarrow 4NO_2 + 10H_2O + 15e- \qquad (xvi)$$

These reactions corresponded to oxidation of the reducing carriers. These processes increased the carrier concentration and hence decreased the resistance on exposure to reducing liquids/ analytes. At the room condition, the disclosure of metal oxide surface to reducing liquid/analytes results in a surface mediated combustion process. The exclusion of iono-sorbed oxygen amplifies the electron concentration and thus the surface conductance of the coating-film (Mujumdar, 2009). The reducing analyte (aq. NH_3) donates electrons to β-Fe_2O_3 NPs surface. Therefore, resistance is decreased, or conductance is increased. This is the reason why the analyte response (current response) increases with increasing potential.

On the other way, due to large surface area, the NPs of provide a supportive nanoenvironment for the chemical detection with fast and good capacity. The high sensitivity of NPs provides high electron communication features which progress the direct electron movement between the active sites of NPs and GE. The modified thin film had a good stability and reproducibility. As for the nanomaterials, β-Fe_2O_3 NPs provide a path to a new generation of chemical sensors, but a premeditate effort has to be expended for nanostructures to be taken critically for large scale applications, and to achieving high appliance density with accessibility to individual sensors. Reliable methods for fabricating, assembling and integrating building blocks onto sensors need to be explored.

CONCLUSION

Finally, the present work provides a hydrothermal method to synthesize low-dimensional iron oxide NPs, which was characterized using several conventional techniques like XRD, FE-SEM, EDS, HR-TEM, UV/visible, FT-IR, and Raman spectroscopy. The detailed morphological characterizations by FE-SEM and TEM revealed that the synthesized NPs possess almost spherical shape with typical diameters of ~ 60 ± 10 nm. The optical properties of as-grown β-Fe_2O_3 NPs were investigated by UV-visible absorption which shows the presence of characteristic β-Fe_2O_3 peak in the spectrum. It was also investigated the photo-catalytic degradation with Acridine orange under UV radiation sources and found the 50% degradation held with as-grown β-Fe_2O_3 NPs. The as-grown NPs were employed for the detection of aqueous ammonia in solution phase as chemical sensors. The performance of the proposed aqueous ammonia sensor using low-dimensional β-Fe_2O_3 NPs film was excellent in terms of sensitivity, lower detection limit, and large linear dynamic ranges in short response time. This new approach is also introduced a new route for efficient chemical sensor development to control the environmental toxicity as well as carcinogenicity, and can also play an important role in environmental and health care fields.

ACKNOWLEDGMENT

We are greatly acknowledged to Najran University for their financial supports and research facilities. Centre for Advanced Materials and Nano-Engineering (CAMNE), Najran University, Najran is also greatly acknowledged for their chemicals and nanomaterials analyses.

REFERENCES

1. Ansari, SG., Ansari, ZA., Wahab, R., Kim, YS., Khang, G., Shin, HS. (2008). Glucose sensor based on nano-baskets of tin oxide templated in porous alumina by plasma enhanced CVD. Biosens. Bioelectron. Vol. 23, pp. 1838.
2. Azároff, L. V.; R. Kaplow, N. Kato, R. J. Weiss, A. J. C. Wilson, R. A. Young (1974). X-ray diffraction. McGraw-Hill. Ballun, G., Hajdu, F., Harsanyi, G.

(2003). Highly sensitive ammonia sensor. IEEE 26th Int. Spring Sem. Elect. Tech. pp. 471.

3. Buzea, C., Pacheco, I., Robbie, K. (2007). "Nanomaterials and Nanoparticles: Sources and Toxicity". Biointerphases Vol. 2, pp. MR17.

4. Brown, ASC., Hargreaves, JSJ., Rijniersce, B. (1998) A study of the structural and catalytic effects of sulfation on iron oxide catalysts prepared from goethite and ferrihydrite precursors for methane oxidation. Catal. Lett. Vol. 53, pp. 7.

5. Chen, J., Xu, LN., Li, WY., Gou, Xl. (2005) β-Fe_2O_3 Nanotubes in Gas Sensor and Lithium-Ion Battery Applications. Adv. Mater. Vol. 17, pp. 582.

6. Cherepy, NJ., Liston, DB., Lovejoy, JA., Deng, H., Zhang, ZJ. (1998). Ultrafast Studies of Photoexcited Electron Dynamics in γ- and β-Fe_2O_3 Semiconductor Nanoparticles. J. Phys. Chem. B Vol. 102, pp. 770.

7. Compendium of Chemical Terminology, 2nd ed. (1997) the "Gold Book". Online corrected version: (2006).

8. Cornell, RM., Schwertmann, U. (2003) The Iron Oxides-Structure, Properties, Reactions, Occurrences and Uses. Darmstadt: Wiley–VCH GmbH & Co. KGaA.

9. Cornell, RM., Schwertmann, U. (2003). The iron oxides: structure, properties, reactions, occurrences and uses. Wiley VCH.

10. Christie, S., Scorsone, E., Persaud, K., Kvasnik, F. (2003). Remote detection of gaseous ammonia using the near infrared transmission properties of polyaniline. Sens. Actuators B Vol. 90, pp. 163.

11. Dale, L., Huber. (2009) Synthesis, Properties, and Applications of iron NPs. Small. Vol. 1, pp. 482. Fahlman, BD. (2007). Materials Chemistry. Springer. pp. 282. Kumar, A., Singhal, A. (2007) Synthesis of colloidal β-Fe_2O_3nanostructures—influence of addition of Co2+ on their morphology and magnetic behavior. Nanotechn. Vol. 18, pp. 475703.

12. Garcıa, KE., Morales, AL., Barrero, CE., Arroyave, CE., Greneche, JM. (2004) Magnetic and crystal structure in akaganeite nanoparticle. Physica B. Vol. 354, pp. 187.

13. Hunger, K., ed. (2003). Industrial Dyes. Chemistry, Properties, Applications. Weinheim: Wiley-VCH.

14. Garfield, S. (2000). Mauve: How One Man Invented a Color That Changed the World. Faber and Faber. ISBN 0-393-02005-3.

15. Jordan, A., Scholz, R., Maier-Hauff, K., Johannsen, M., Wust, P., Nadobny, J., Schirra, H., Schmidt, H., Galatsis, K., Cukrov, L., Wlodarski, W., McCormick, P., Kalantarzadeh,

16. K., Comini, E., Sberveglieri, G. (2003) p- and n-type Fe-doped SnO2 gas sensors fabricated by the mechanochemical processing technique. Sens. Actuators B. Vol. 93, pp. 562.

17. Kesavan, V., Sivanand, Chandrasekaran, PSS., Koltypin, Y., Gedanken, A. (1999) Catalytic aerobic oxidation of cycloalkanes with nanostructured

amorphous metals and alloys. Angew. Chem. Int. Ed. Vol. 38, pp. 3521.

18. Long, GL., Winefordner, JD. (1983). Linearization of electron capture detector response to strongly responding compounds. Anal. Chem. Vol. 55, pp. 713.
19. MacDougall, D., Crummett, WB. (1980). Guidelines for Data Acquisition and Data Quality Evaluation in Environmental Chemistry. Anal. Chem. Vol. 52, pp. 2242.
20. McMichael, RD., Shull, RD., Swartzendruber, LJ., Bennett, LH. (1992) Magnetocaloric effect in superparamagnets. J. Magn. Magn. Mater. Vol. 111, pp. 29.
21. Mujumdar, S. (2009). Synthesis and characterisation of SnO2 films obtained by a wet chemical process. Mat. Sci. Poland. Vol. 27, pp. 123.
22. Music, S., Saric, A., Popovic, S. (1997) Effect of urotropin on the formation of β-FeOOH. J. Mol. Struct. Vol. 410–411, pp. 153.
23. O'Donoghue, M. (1983). A guide to Man-made Gemstones. Great Britain: Van Nostrand Reinhold Company. pp. 40.
24. Patil, DR., Patil, LA., Amalnerkar, PP. (2007). Ethanol gas sensing properties of Al2O3- doped ZnO thick film resistors. Bull. Mat. Sci. Vol. 30, pp. 553.
25. Pare, B., Jonnalagadd, SB., Tomar, H., Singh, P., Bhagwat, BW. (2008). ZnO assisted photocatalytic degradation of acridine orange in aqueous solution using visible irradiation. Desalination Vol. 232, pp. 80.
26. Raj, VB., Nimal, AT., Parmar, Y., Sharma, MU., Sreenivas, K., Gupta, V. (2010). Crosssensitivity and selectivity studies on ZnO surface acoustic wave ammonia sensor. Sens. Actuators B Vol. 147, pp. 517.
27. Raj, K., Moskowitz, B., Casciari, R. (1995) Advances in ferrofluid technology. J. Magn. Magn. Mater. Vol. 149, pp. 174.
28. Rahman, MM., Jamal, A., Khan, SB., Faisal. M. (2011) Fabrication of Highly Sensitive Ethanol Chemical Sensor Based on Sm-Doped Co3O4 Nano-Kernel by Solution Method. J. Phys. Chem. C. Vol. 115 pp. 9503.
29. Rahman, MM., Jamal, A., Khan, SB., Faisal. M. (2011) Cu-doped ZnO Based Nanostructured Materials for Sensitive Chemical Sensor Applications. ACS App. Mater. Inter. Vol. 3 pp. 1346
30. Rahman, MM., Umar, A., Sawada, K. (2009) Development of Amperometric Glucose Biosensor Based on Glucose Oxidase Enzyme Immobilized with Multi-Walled Carbon Nanotubes at Low Potential Sens. Actuator: B Vol. 137 pp. 327.
31. Rahman, MM., Jeon, IC., Hasnat, MA., Samed, AJF. (2006) Xenon Beam Radiation Assisted Degradation of an Organic dye in Aqueous Medium. Environ. Sci. Technol. Vol. 1 pp. 210.
32. Saquib, M., Tariq, MA., Faisal, M., Muneer, M. (2008). Photocatalytic degradation of two selected dye derivatives in aqueous suspensions of titanium dioxide. Desalination. Desalination Vol. 219, pp.301.
33. Spence, JCH. (1988) [1980]. Experimental high-resolution electron microscopy. New York: Oxford U. Press. ISBN 0195054059.

34. Whitesides, GM., Boncheva, M. (2002) Beyond molecules: Self-assembly of mesoscopic and macroscopic components. Proc. Natl. Acad. Sci. USA. Vol. 99, pp. 4769.

35. Zeng, H., Li, J., Liu, JP., Wang, ZP., Sun, SH. (2002). Exchanged-coupled nanocomposite magnets via nanoparticle self-assembly. Nature Vol. 420, pp. 395.

36. Zollinger, H. 3rd ed. (2003). Color Chemistry. Synthesis, Properties and Applications of Organic Dyes and Pigments, Weinheim: Wiley-VCH.

Chapter 5

INSTRUMENTATION FOR FERROMAGNETIC RESONANCE SPECTROMETER

Chi-Kuen Lo

Department of Physics, National Taiwan Normal University, Taipei, Taiwan

INTRODUCTION

Even FMR is an antique technique, it is still regarded as a powerful probe for one of the modern sciences, the spintronics. Since materials used for spintronics are either ferromagnetic or spin correlated, and FMR is not only employed to study their magneto static behaviors, for instances, anisotropies [1,2], exchange coupling [3,4,5,6], but also the spin dynamics; such as the damping constant [7,8,9], g factor [8,9], spin relaxation [9], etc. In this chapter a brief description about the key components and techniques of FMR will be given. For those who have already owned a commercial FMR spectrometer could find very helpful and detail information of their system from the instruction and operation manuals.

The purpose of this text is for the one who want to understand a little more detail about commercial system, and for researchers who want to build their own spectrometers based on vector network analyzer (VNA) would gain useful information as well.

FMR spectrometer is a tool to record electromagnetic (EM) wave absorbed by sample of interest under the influence of external DC or Quasi DC magnetic field. Simply speaking, the spectrometer should consist of at least an EM wave excitation source, detector, and transmission line which bridges sample and EM source. The precession frequency of ferromagnetics lies at the regime of microwave (μ-wave) ranged from 0.1 to about 100 GHz, therefore, FMR absorption occurs at μ-wave range. The generation, detection and transmission at such this high frequency are not as simple as those for DC or low frequency electronics. According to transmission line and network theories [10,11], impedance of 50 μ between transmission line and load has to be matched for optimization of energy transfer.

FMR spectrometer also has a resonator and an electro magnet which produces magnetic field to vary the sample's magnetization during the measurement. Sample which is mounted inside the cavity absorbs energy from the μ-wave source. The detector electronics records the changes on either the reflectance or transmittance of theμ -wave while magnetic field is being swept [12]. Most commercial spectrometers, for examples, ELEXSYS-II E580, Bruker Co [13], and JES-FA100, JOEL Ltd. [14], character FMR by measuring the reflectance at fixed band frequency. μ-wave is generated by a Klystron, which goes to a metallic cavity via a wave guide. Signal reflected back from the cavity through the same waveguide to a detector, the Schottky barrier diode. Noted that a circulator is employed, such that the μ-wave does go to the detector and not return back to the generator. The integration of the source, detector, circulator, protected electronics, etc. in a single box, is named "Microwave Bridge". The basic configuration of most FMR spectrometer is shown in Fig.1. FMR absorption spectrum is obtained by comparing the incident and reflected signals. However, the signal to noise ratio (SNR) given by this kind of reflectometer is still not large enough to recognize the spectrum, and lock-in technique is needed to enhance SNR to acceptable value, for example, bigger than 5.

The operational frequency of the source, detector and cavity cannot be varied broadly, therefore, it is necessary to change the microwave bridge and cavity while working with different band frequency. Dart a glance at a FMR spectrometer, the key parts are the microwave bridge, cavity, gauss meter, electromagnet, and lock-in amplifier for signa process. Surely, automation and data acquisition are also crucial.

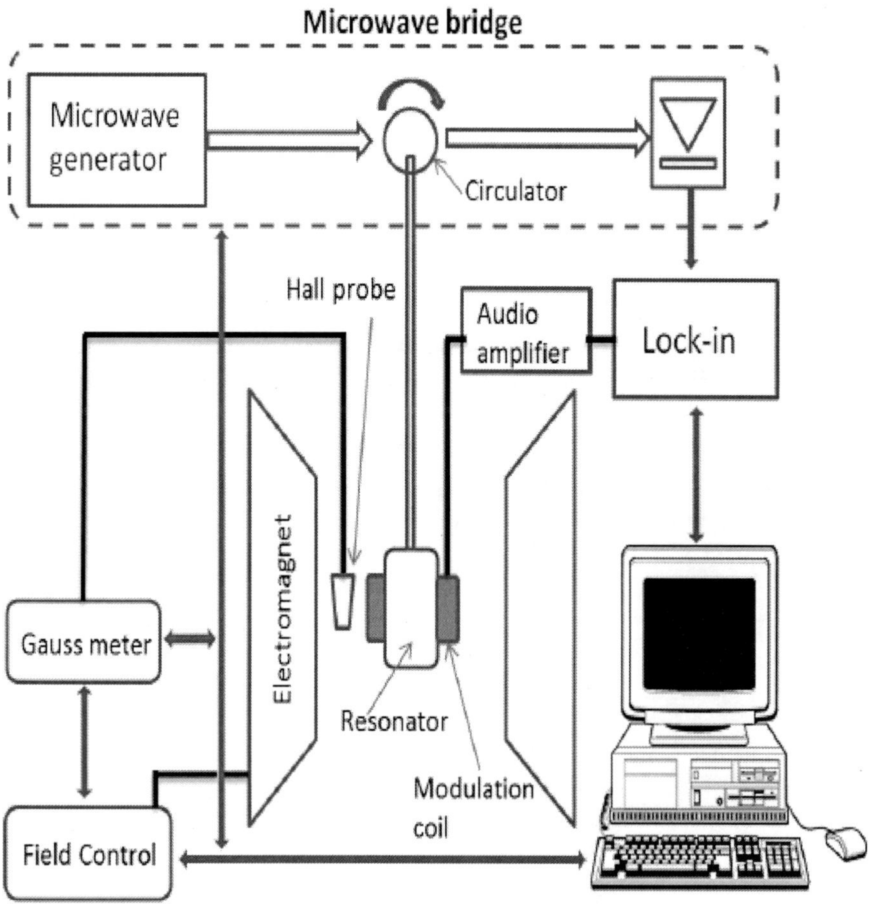

Figure 1 Basic configuration of FMR spectrometer. The microwave bridge mainly consists of the microwave generator, circulator, and detector.

KEY COMPONENTS OF FMR

Metallic cavity

A Metallic resonator (cavity) is a space enclosed by metallic walls which sustain electromagnetic standing wave or electromagnetic oscillation. The cavity has to be coupled with external circuit which provides excitation energy. On the other hand, the excited cavity supplies energy to the sample of interest (loading) through coupling [10,11]. The basic structure of a cavity set is sketched as in Fig.2(a), which consists of a wave guide, a coupler with iris control, and a metallic cavity. The working principle of coupled cavity can be understood by using LRC equivalent circuit which couples to a transmission line [10,11] as shown in Fig.2(b) and 2(c). The coupling structure is represented by an ideal transformer with transforming ratio 1:n. The parallel RLC circuit could be transformed to Ts from AB plane via the transformer, such that $C'=n^2C$, $R'=R/n^2$, and $L'=L/n^2$. The resonance frequency, fres, and Q-factor will not be affected by the transformation [10] as pointed out by eq.(1) and eq.(2), respectively:

$$\omega_0' = \frac{1}{\sqrt{L'C'}} = \frac{1}{\sqrt{\frac{L}{n^2}n^2C}} = \omega_0 \qquad (1)$$

$$Q_0' = \omega_0' C' R' = \omega_0 n^2 C \frac{R}{n^2} = \omega_0 CR = Q_0 \qquad (2)$$

A coupling coefficient, β, which defines the relation between energy dissipation (Ed) in a cavity and energy dissipation of external circuit (Ee), tells if the cavity is coupled with external circuit.

$$\beta \equiv \frac{E_d}{E_e} = \frac{\frac{U^2}{2Z_C}}{\frac{U^2}{2R/n^2}} = \frac{R}{Z_C n^2} = \frac{R'}{Z_C} \qquad (3)$$

U in eq.(3) stands for the voltage applied to the cavity, and 1:n is the transforming ratio for tuning to critical couple state $(Zc = R' \Rightarrow \beta = 1)$ before doing FMR measurement. The characterization impedance, ZC, is always designed to be $50\Omega(Z_0)$ for matching impedance as mentioned previously. Signal will be distorted and weakened if working at either undercoupling $(\beta < 1)$ or over-coupling states $(\beta > 1)$. The external Q factor (Qe) of a cavity, coupled to external circuit, states the energy dissipation in the external circuit, and also has the following relationship with β:

$$\beta = \frac{Q_o}{Q_e} \qquad (4)$$

In eq.(4), Q_0 is the intrinsic Q factor of cavity. Practically, the loaded Q-factor, Q_L, is used in experiments:

$$Q_L = Q_0 \frac{\beta}{1+\beta}\text{; or } \frac{1}{Q_L} = \frac{1}{Q_0} + \frac{1}{Q_e} \qquad (5)$$

Metallic cavities used for FMR measurement have to be TE mode, such that the cavity center which is also the sample location has maximum excitation magnetic field. Furthermore, this kind of cavity has very high Q-factor and plays key role in FMR detection. The signal output, VS, from the cavity can be expressed as [15]:

$$V_S = \chi'' \eta Q_L \sqrt{P Z_0} \qquad (6)$$

$x'', \eta, P \, and \, Z_0$ in eq.(6) stand for imaginary part of magnetic susceptibility of the sample, filling factor, microwave power, and characteristic impedance, respectively. The Q-factor of a cavity directly relates to the detecting sensitivity, which depends very much on the design and manufacturing technique. A good metallic cavity normally has unloaded Q-factor of more

than 5,000 which cannot be changed after being manufactured, and therefore is regarded as a constant in eq.(6). $Z_0 \, of \, 50\Omega$ is fixed for matching the network impedance. We cannot play too much on x″ and η as well, since the former is the intrinsic behaviour of the sample of interest to be tested, and the last parameter is the volume ratio of the sample and cavity. For thin film and multilayer samples, η is very small, hence this term cannot contribute too much to VS. The source power, P, at the first glance, is the only possible adjustable parameter for sensitivity.

Due to the occurrence of saturation, higher power may not be that helpful in increasing the FMR signal. Large excitation power drives magnetization precession in non-linear region which complicates the spectrum analysis. Besides, signal will be reduced and broadened while operating at saturation region. In order to determine the line shapes and widths precisely, this region should be avoided, hence low power is a good choice. The determination of saturation power is not difficult, since the signal intensity grows as the square root of power as indicated by eq.(6). Checking the signal intensity to see if eq.(6) is still validated as the power is increased. It is noted that the maximum excitation power of a common VNA is just a few mWs, and V_S is not that clear to follow $\sqrt{p}$.

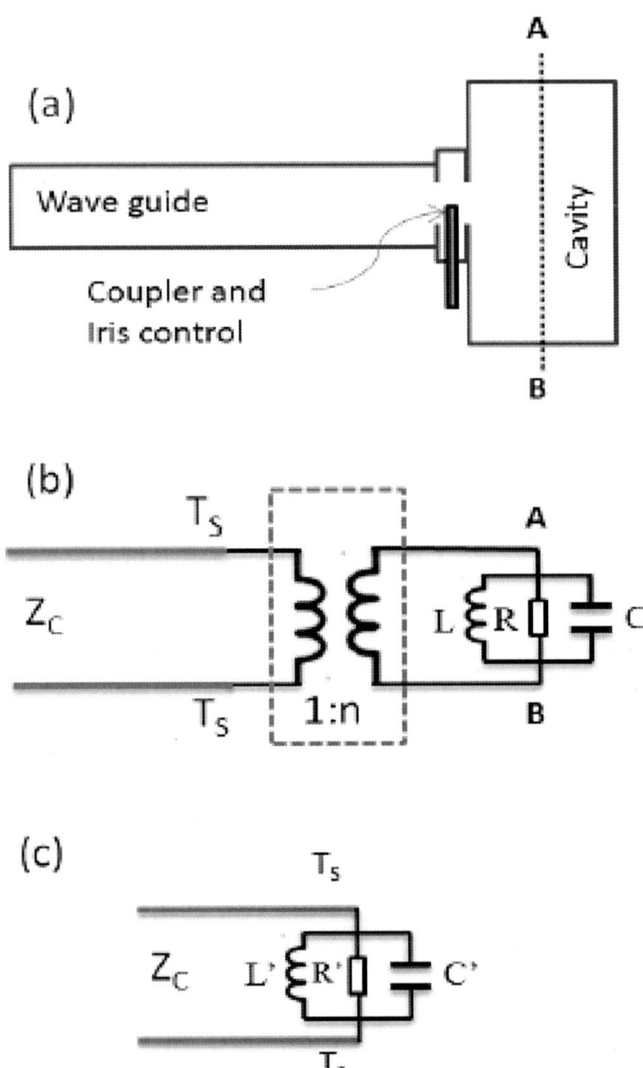

Figure 2 (a) The layout of a resonant cavity, (b) The equivalent LRC circuit. The iris control is regarded as a transformer, and (c) The equivalent circuit transformed to the Ts plane [10].

The Q-factor of commercial or home-made cavities can be determined easily by using a VNA, which can also be expressed as:

$$Q = \frac{f_{res}}{\Delta f} \tag{7}$$

f_{res} is the resonant frequency of a cavity, and Δf, is the full width at half maximum (FWHM) of the resonant peak. Eq.(7) can be simply found by a built-in function of VNA which measures directly the reflected power of the cavity in dB at fres. Besides with dB unit, FWHM (Δf) is also obtained at half power point, i.e. the power level at -3dB. As the values of f_{res} and FWHM are known, Q is determined as shown in Fig.3(d).

Q will be decreased once a sample is inserted. This is because the inserted sample and its holder absorb energy and change the coupling conditions resulting in Q reduction and resonance frequency shift. However, these deviations can be amended a little bit by adjusting the iris which controls the effective impedance by varying the aperture size between the cavity and wave guide, and by adjusting the position of metallic cap. This is equivalent to change the transformer ratio, that is, the equivalent inductance and capacitance. The iris control is a plastic screw (low μ-wave absorption material) which has a gold coated metal cap on one end, and the effective impedance of the whole (wave guide + iris + cavity) depends on the size of the aperture and location of the metal cap. The function of the iris acts as a device to tune the L and C, such that the cavity is critical couple again. As field is swept, the variations of sample's magnetization break the coupling condition so the cavity is deviated from critically coupled state. Consequently, wave is reflected back to the -wave bridge, and FMR signal is resulted.

For an ideal metallic cavity (infinite conductivity and perfectly smooth inside wall surface), there will be a unique resonant peak with infinite large Q-factor, However, this is not the case in practice. Since the cavity itself has finite conductivity, and further, the inside wall is not perfectly flat, these cause the existence of many peaks with very low Q values as extracted by a VNA shown in Fig.3(a). Fig.3(b) shows the simulation of a copper made cavity with a very rough inside wall. As the roughness is reduced, the spectrum is clearer as shown in Fig. 3(c). This could be due to the μ-wave which is multi-scattered by lumps on the wall surface, and these lumps could also enhance power dissipation further. In consequence, there exists many low Q peaks. Even there are many peaks other than the eigen frequency, these peaks do not response to the change of magnetization, and hence useless for FMR characterization. The manufacture of metallic cavity with Q-factor of higher than 3,000 is laboring. This is because the inside wall has to be mirror polished and coated with a few μm thick Au layer to reduce the imperfection.

Fig.3(a) and Fig.3(d) are the frequency response of a Bruker X-band cavity (TE102, ER 4104OR) exhibiting in larger and smaller frequency windows, respectively. The unloaded Qfactor of this cavity is ~9,000 as claimed by the manufacturer, however, our measurement tells that the Q factor is more than 14,000. This is because frequency resolution of the VNA is very high. As the resolution is decreased, Q is found to be decreased as well.

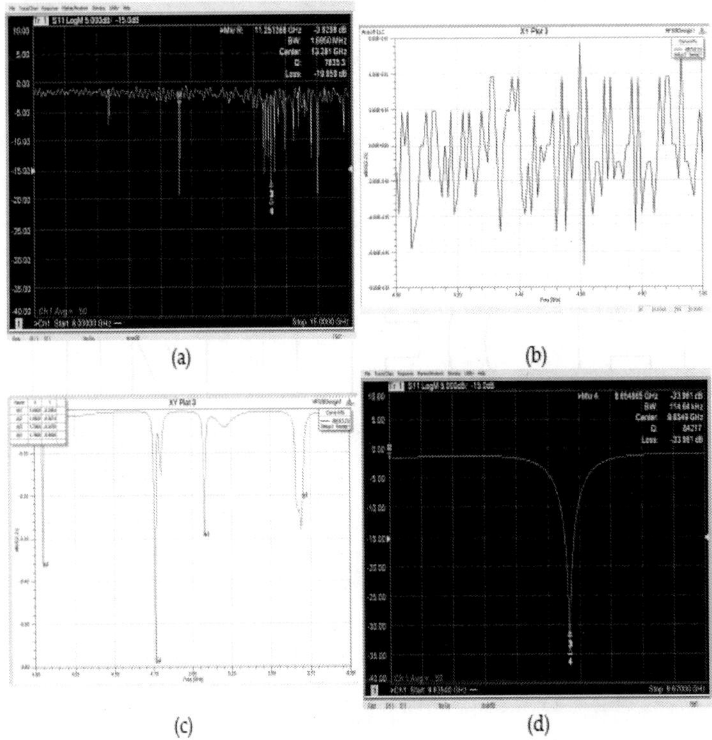

Figure 3 (a) the frequency response of a X-band cavity. Simulation result of a Cu cavity with (b) a very rough inside wall surface, and (c) a smooth inside wall surface. (d) VNA measurement of an X-band cavity with Q higher than 87, 000. Note that there are many side peaks but not appears in this frequency range.

Shorted Waveguide Cavity

The fabrication of metallic cavity with very high Q-factor is not easy as mentioned above. If the lossy of the sample is not very big, cavity with Q of a few hundred to thousand may be good enough to recognize FMR. If so, a shorted waveguide cavity (SWC) could be an alternative. This kind of cavity is just a section of waveguide sealed by a metal plate at one end [16] as show in Fig.4. Since metallic wave guide are commercially standard with wide range of frequency from L to W band, it is worth to obtain FMR information at different bands with this cheap, simple and effective method. If the length of the waveguide tube is equal to $n\lambda/2$, standing wave can be formed inside. In that n and λ represents for integer and wave length of the μ-wave, respectively. In order to excite FMR signal, sample has to be placed at the $n\lambda/4$ position away from the shorted plate at which maximum magnetic field is located. In Fig.4, n=0 and n=1 are the suitable locations, however, n =0 gives advantages of easier sample manipulation, and lesser interference from irrelevant insertions. Sample is mounted at the top of a plastic screw through the shorted plate, such that its position can be adjusted for maximum signal.

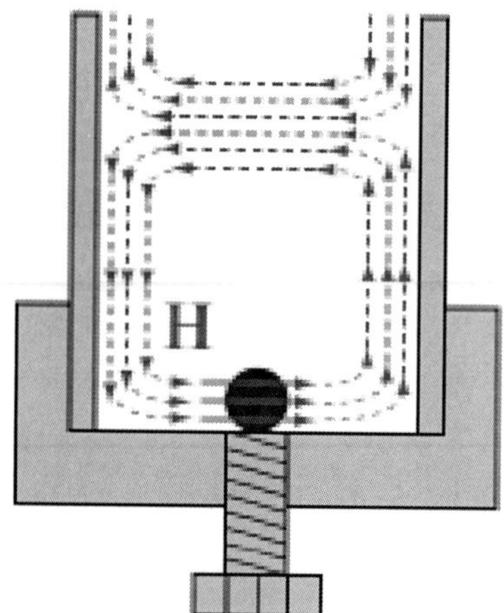

Figure 4 A shorted wave guide with possibility of sample manipulation. Magnetic field distribution originated from the μ-wave is also sketched [16].

However, SWC does not have an iris to tune the impedance, hence critical coupling may not be easy to obtain. Although the Q is somewhat lower (less than 2,000), SWC is easy to make from commercial waveguide tube. Furthermore, different band frequency cavity can be built in the same manner, and this is particular convenient to study FMR at different band frequency with network analyzer.

Microstrip and co-planar waveguide

Microstrip (MS) and co-planar waveguide (CPW) which are indeed transmission lines, are commonly used to extract FMR at broad frequency range [16,17,18,19]. Sample is mounted on top of their signal lines, and μ-wave is conducted into MS (CPW) via Port 1 of the VNA, and the differential change in either reflected or transmitted signal is analyzed. In this operation scheme, frequency is swept at fixed field, and once the FMR conditions are fulfilled μ-wave absorption occurs. Fig.5 and Fig.6 are the sketches of MS and CPW. Since $h << \lambda$, μ-wave propagates in these two lines is regarded as quasi-TEM mode. In order to match the impedance of 50Ω, the dimensions of these two planar transmission lines are restricted.

In case of MS, we have the followings [11]:

The effective dielectric constant, ε_{eff} has approximately the form:

$$\varepsilon_{eff} = \frac{\varepsilon_r+1}{2} + \frac{\varepsilon_r-1}{2}\frac{1}{\sqrt{1+12h/w}} \quad (8)$$

The dimension of the strip line and characteristic impedance, ZC, can be worked out as below:

$$Z_C = \begin{cases} \frac{60}{\sqrt{\varepsilon_{eff}}} ln\left(\frac{8h}{w} + \frac{w}{4h}\right) & ; \text{ for } w/d \leq 1 \\ \frac{120\pi}{\sqrt{\varepsilon_{eff}\left[\frac{w}{d}+1.393+0.667\ ln\left(\frac{w}{d}+1.444\right)\right]}} & ; \text{ for } w/d \geq 1 \end{cases} \quad (9)$$

In case of CPW if the thickness of signal line is ignorable, then [10]:

$$\varepsilon_{eff} = \frac{\varepsilon_r + 1}{2}$$

$$Z_C = \begin{cases} \dfrac{\eta_0}{\pi\sqrt{\varepsilon_{eff}}} ln\left(2\sqrt{\dfrac{b}{w}}\right)\ \Omega & : 0 < w/b < 0.173 \\ \dfrac{\pi\,\eta_0}{4\sqrt{\varepsilon_{eff}}}\left[ln\left(2\dfrac{1+\sqrt{w/b}}{1-\sqrt{w/b}}\right)\right]^{-1}\ \Omega & ; 0.173 < \dfrac{w}{b} < 1 \end{cases} \quad (10)$$

η_0 in eq.(11) stands for the wave impedance in free space. The ZC = Z0 is set to 50Ω , then b, d and ω can easily be determined from eq.(9) and eq.(11). The line width of the strip can be ranged from sub-millimeter to millimeters depending on the design and dielectric material used as implied by these equations. The size of this kind of line width can be fabricated simply by photo lithography together with life-off process. Due to the large loss, the characterization of nano size sample may not be easy and lock-in technique is needed for SNR enhancement as described in next section.

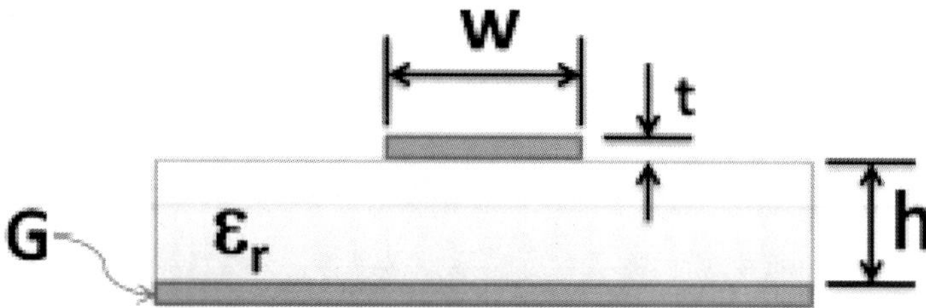

Figure 5 The geometry of a microstrip line, where ε_r is the relative dielectric constant of the substrate.

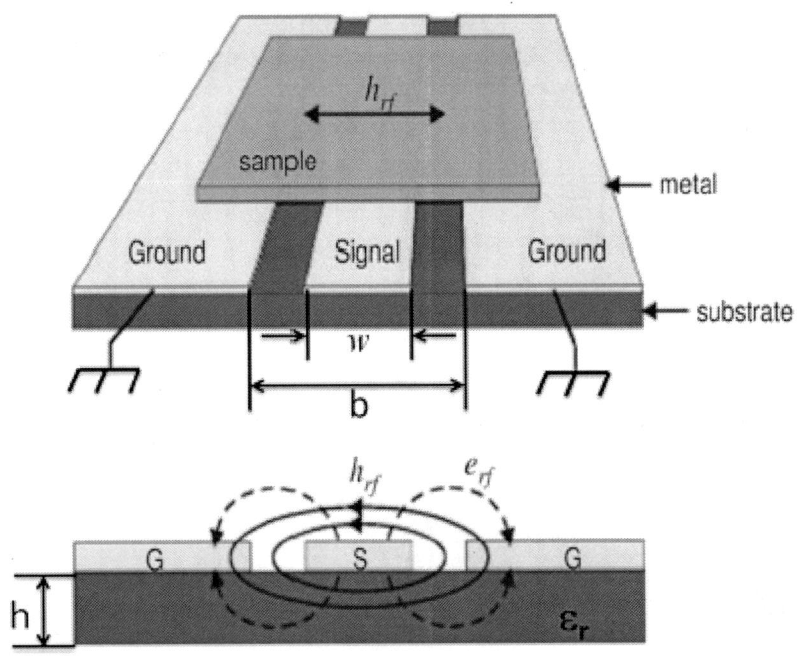

Figure 6 Schematical diagram of a coplanar waveguide along with the dimension. The lower diagram shows the magnetic (full lines) and electric (dashed lines) field lines winding around the signal conductor (S).

FMR can determined by characterizing the transmittance, ie. $S_{12.}$ However, if shorted MS, and CPW are used, S_{11}, the reflectance FMR can also be found. Fig.7 shows the FMR of permalloy films extracted by VNA and CPW [18].

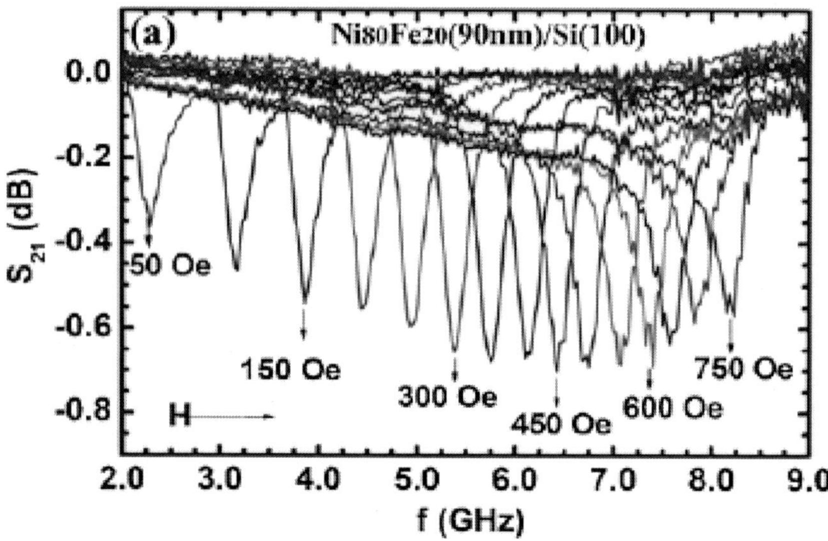

Figure 7 A typical FMR transmittance spectrum of permalloy films extracted by VNA with CPW [18].

TECHNIQUE FOR SIGNAL TO NOISE ENHANCEMENT

Lock-in phase detection

If signal is deeply buried inside noise floor, lock-in method is normally employed to enhance the signal to noise ratio (SNR) by narrowing the detecting band window [20].

Considering a DC Signal, S, which is heavily contaminated by white noise, n, and for every measurement comes out from a measurement equipment, a voltmeter for an example, the reading can be expressed as:

$$V=s+n \tag{12}$$

If S can be modulated sinusoidally at a certain frequency somehow, then:

$$V(t) = S\cos(\omega t) + n \tag{13}$$

Note in eq.(13) that the modulation of white noise is still a white noise.

There is also a reference signal with same frequency but could be different in modulating amplitude (Va) and phase (ϕ), i.e.,

$$V_{ref} = V_a \cos(\omega t + \phi) \tag{14}$$

The product of eq.(13) and (14) gives:

$$V(t)V_{ref} = SV_a \cos(\omega t)\cos(\omega t + \phi) + nV_a \cos(\omega t + \phi) = \frac{SV_a}{2}[\cos(\phi) + \cos(2\omega t + \phi)] + n\,V_a \cos(\omega t + \phi) \tag{15}$$

The second and third terms can be eliminated if these two parts are passed to a low pass filter with cutoff frequency setting at $\omega/2$ or lower. Finally, we have:

$$V \propto \frac{SV_a}{2}\cos(\phi) \tag{16}$$

That is, the signal is amplified by Va, and has a maximum while ϕ is "phase-locked" to zero, hence the name of lock-in amplifier.

Feld Modulation Lock-in Technique and Derivative Spectrum

It is easier to determine H_{res}, and ΔH_{pp} by differentiating the original signal. To do so, field modulation lock-in detection is employed and described below.

There is a small field, H_a, with modulation frequency of ω superposing on top of external DC magnetic field, H_0:

$$H(t) = H_0 + H_a \cos(\omega t) \tag{17}$$

Assuming we have FMR signal, VFMR, whose Taylor expansion at H0 is given below:

$$V_{FMR}(H) = V_{FMR}(H_0) + \left.\frac{dV_{FMR}}{dH}\right|_{H=H_0} H_a \cos(\omega t) + \cdots \tag{18}$$

Meanwhile, we also have another signal, the reference, which has the same frequency as that of the modulation field, but with a phase angle, ϕ:

$$V_{ref} = \cos(\omega t + \phi) \tag{19}$$

The product of eq.(18) and (19) gives eq.(20) below:

$$V_{FMR}(H) \times V_{ref} = V_{FMR}(H_0)\cos(\omega t + \phi) \left.\frac{dV_{FMR}}{dH}\right|_{H=H_0} + H_a \cos(\omega t)\cos(\omega t + \phi) + \cdots :$$
$$V_{FMR}(H_0)\cos(\omega t + \phi) + \frac{1}{2}\frac{dV_{FMR}}{dH} H_a \cos(\phi) + \frac{1}{2}\frac{dV_{FMR}}{dH} H_a \cos(2\omega t + \phi) + \cdots \tag{20}$$

The first and third terms in eq.(20) can again be removed by using a low pass filter with cutoff frequency setting at ω/2 or even lower. The second term is time independent, proportional to derivative of the input signal and magnify by Ha, which has a maximum if the phase, ϕ, is locked to 0. Another advantage of use this method is that 1/f noise and drift problem can be excluded by setting the sampling window at high frequency.

The way to turn the DC or quasi DC signal from to AC is an important subject. The simple and normal way to do this is to insert two coils which sandwich the cavity. These coils are driven by a power amplify at certain high frequency, such that an AC magnetic field of a few mT at a frequency up to 200 kHz can be produced. Therefore, the output signal consists of the modulation and DC components. All these together with the reference are sent to the lock-in amplify for SNR enhancement and derivative spectrum as mentioned previously.

Modulation amplitude (MA), modulation frequency (MF), and time constant (TC) which is the reciprocal of the low pass filter's cutoff frequency, have large influence on the spectra.

Signal strength is linearly proportional to the MA while MA is small. Simply speaking, MA should be small enough for sampling in linear region, but needs to be large enough for gaining good sensitivity. However, too large the MA results in signal distortion as shown in

Fig.8(a). This can be understood from eq.(20) that high order term cannot be ignored for large MA which causes distortion of derivative signal. The choice of MF is critical as well that small MF cannot get rid of 1/f noise completely, and elongates the acquisition time.

Large MA and MF also cause passage effect that the rate of "passage" through the absorption line is faster than the relaxation rates, which results in distorted spectrum, or even inversion of signals upon reversal of the field scan direction. Therefore, it is important to check the FMR signal with either a standard or a well-known sample with different lockin conditions for the best settings. In order to obtain correct line shape spectrum, the rule of thumb is to set the MA about 1/10 of the FWHM and increases to about 1/3 if necessary. TC also needs to coordinate with MF. According to Nyquist sampling theorem, the sampling rate should be at least twice the highest frequency contained in the signal [21]. For example, if the MF is 100 kHz, the sampling frequency would be at least 200 kHz at which TC is about 5μS. If this restriction does not fulfill, sampling points cannot not be captured immediately.

Consequently, information loss and line shape distortion are always resulted. The best way to avoid this is to set the scan time for a FMR signal 10 times longer than the time constant.

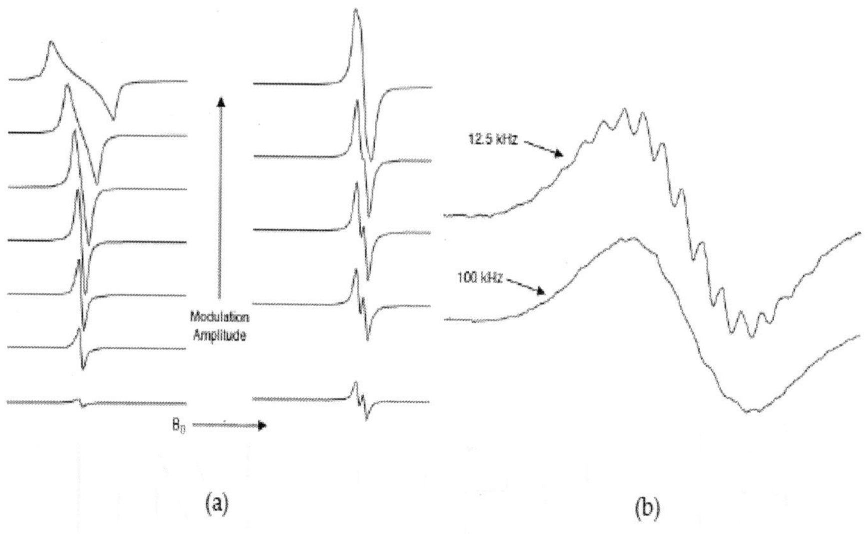

Figure 8 The influence of modulation amplitude (a) and modulation frequency (b) on signal. Distortion will be resulted if MA and MF are not appropriate [15].

VECTOR ANALYZER BASED FMR SPECTROMETER

Basics of vector network analyzer (VNA)

VNA is indeed an instrument developed for characterization of electrical devices (device under test, DUT) by sending an electromagnetic wave. As an analogue of that in optics, the incident wave (either optical or micro wave) will be reflected or/and transmit after interacting with the DUT. By examining the reflectance and transmittance, that is, the ratios of the powers of reflected, transmitted to that of the incident waves, scattering parameters of the DUT can be found as depicted in Fig.9. VNA cannot only find out the reflectance and transmittance, but also the impedance, phase lag, insertion loss, return loss, voltage standing ratio, etc., can be worked out. However for FMR experiment, only the first two functions are employed. VNA has at least two ports, and each port can produce and measure μ-wave. The scattering parameter Sij stands for scattering power ratio of incident wave produced by port i, and measured by port j. That is, for a two-port VNA, S11 and S22 characterize wave reflectance, while S12 and S21 determine the transmittance. If FMR is determined by reflected spectrum, one port is enough by analyzing either the S11 or S22. For transmittance spectrum, two ports are required for either the S12, or S21 determination. VNA has capability of microwave generation and detection, and furthermore, nowadays model has frequency ranged from about 0.1 to 100 GHz, or even higher at excitation power of tens dBm, Thus, it can serve as a microwave bridge to extract FMR parameters at a very broad frequency band.

VNA measures not only the scattering amplitudes, but also their correlated phases. This is in contrast to its counterpart, the scalar network analyzer (SNA) which cannot tell phase information. FMR signal is extracted from power absorption which contains no phase information, and SNA should be good enough for the purpose. However, this kind of machine has been obsoleted for many years.

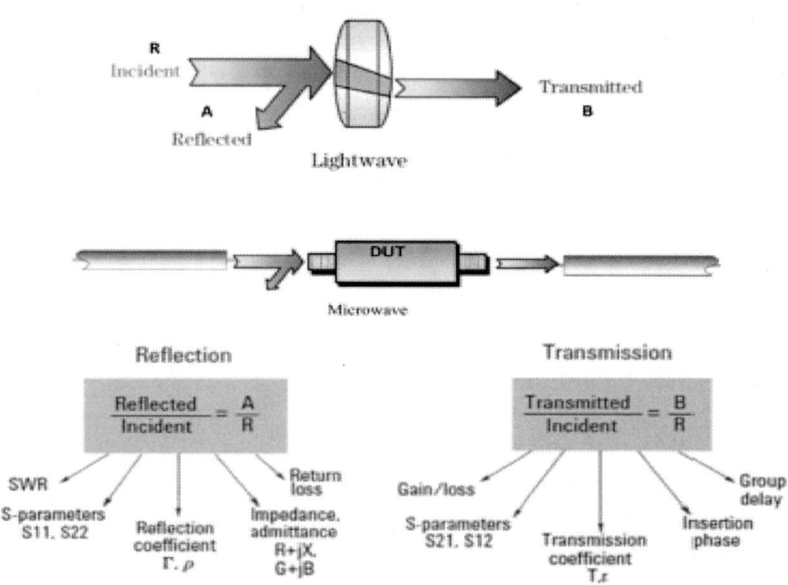

Figure 9 The fundamentals of microwave network analysis are analogue to that of the optics. The lower part of the panel indicates the definitions of scattering parameters [24].

VNA is commonly used with MS and CPW for FMR determination with frequency swept at fixed field. Since signal output from VNA cannot be plugged into lock-in amplifier directly, the characterization of nano scale sample is thus difficult without using lock-in. The application lock-in with VNA will be discussed in next section.

Field Swept VNA-FMR

Metallic cavity usually has rather higher Q-factor and could be simply used with a VNA for field swept FMR detection. Taking the advantages of high Q cavity and VNA, C.K. Lo, et. al. [23] demonstrated FMR measurement without employing field modulation lock-in technique. They employed a TE102 cavity with unload Q of ~9,000 at X-band, and FMR signal was recorded through the measurement of reflectance, and therefore one VNA port for S11 was used. This powerful method extracts signal easily and directly as shown in Fig.10(a). One could if necessary, differentiate the original data for derivative spectrum which is used to determine the peak position and line wide as those in commercial FMR spectrometer. Further, this combination has quite good sensitivity that 1.6 nm CoFeB can be detected with SNR better than 5 as shown in Fig.10(b).

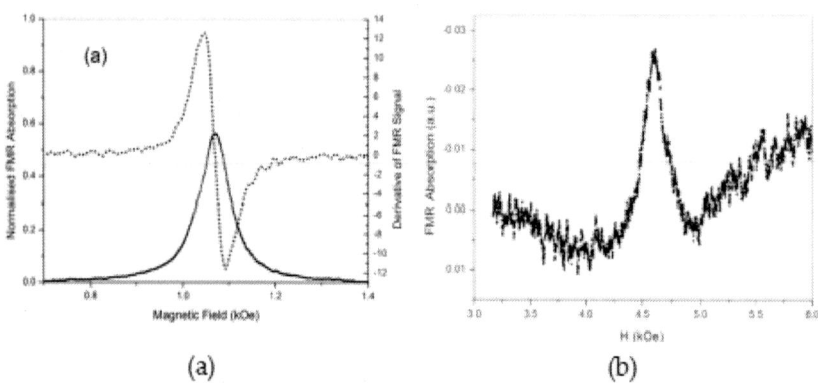

Figure 10 (a) FMR of 5 nm Py exacted at P = 10 dBm. The dotted line is the derivative of experimental data. (b) Signal extracted from a 1.6nm CoFeB

A new built spectrometer should be characterized before properly used and well known samples, such as permalloy, Fe, Co, etc., ferromagnetics are normally employed due to their ΔH and peak positions can be found widely in literatures. Also, these samples are easy to prepare with different thicknesses, and results should be comparable to those in literatures reported.

There are many build-in useful functions with nowadays VNA, and only some of them are used, for instances, the traces of valley (dip) position, Q factor, band width, average, etc. Dip frequency and Q-factor will be varied as the sample's magnetization state is changed by external field. The changes of these quantities were also recorded simultaneously for a Fe/Ag multilayer as shown in Fig.11 in which (a) is the FMR absorption, and (b), the post derivative of (a). The shifts of resonant frequency and Q-factor are shown in (c) and (d), respectively. Despite the variation of center frequency is just a few hundreds MHz, it allows us to determine ferromagnetic parameters precisely. The change of Q-factor is upside-down to that of the absorption. This is because any absorption of microwave inside the cavity will reduce the Q value.

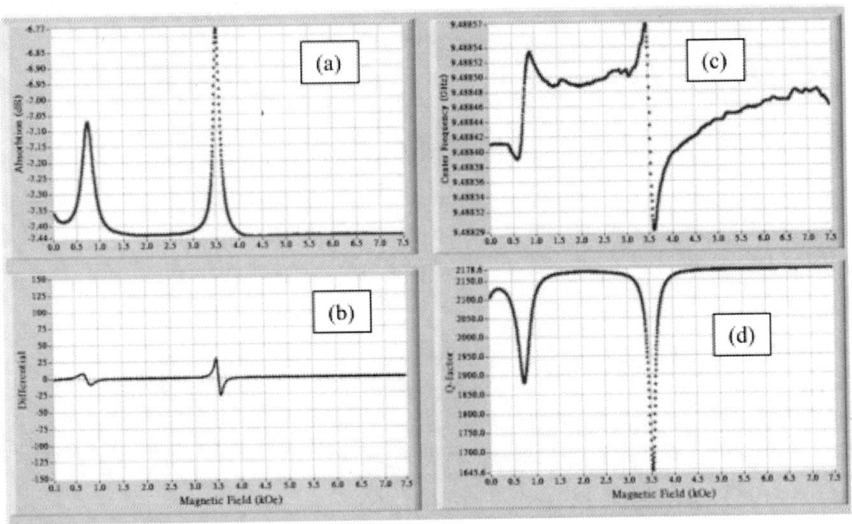

Figure 11 The changes of FMR (a), resonance frequency (c), and Q-factor are recorded. (b) is the derivative spectrum taken from (a) mathematically.

The excitation power of VNA is only tens dBm, and it is unlikely to drive the sample into non-linear regime. Signal intensity is proportional to $\sqrt{P}$ as indicated by eq. (6), however, SNR is found to be not much different if P is bigger than -20 dBm as demonstrated in Fig.12.

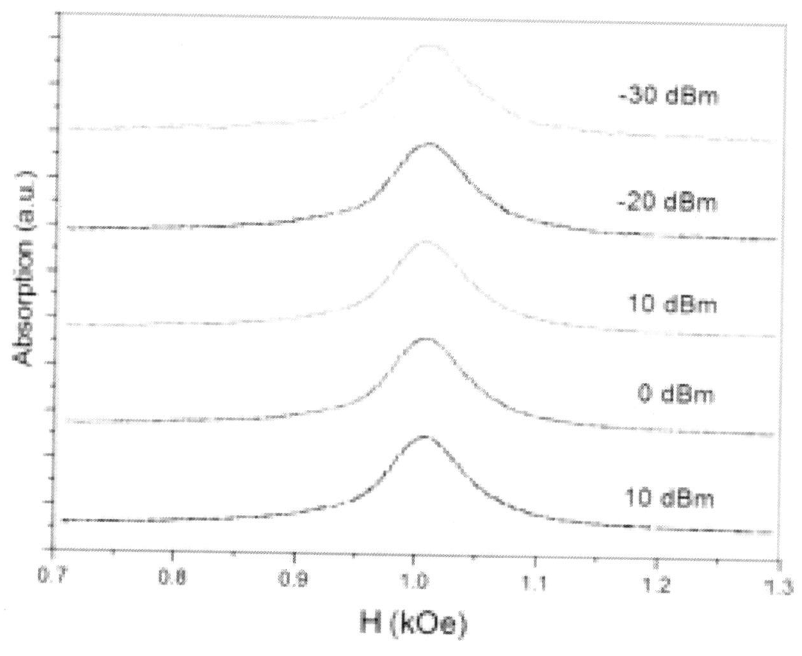

Figure 12 FMR signal of 10nm Py was recorded as function of excitation power. Note that these spectra have the same scale, but different offset for clear comparison.

As mentioned previously that VNA has very high frequency resolution, apparently, Q could be tuned as high as 140,000 and more by simply adjusting the iris. However, distorted spectrum is resulted at very high Q, and two peaks appear while Q is adjusted close to maximum Q as seen in Fig.13(a) and (b), respectively. It is also found that the SNR does not varied significantly if Q lies between 2,000 to 13,000 (the unloaded Q is about 9,000 as claimed). Surely, this finding is just for referral, and would depend very much on cavity used.

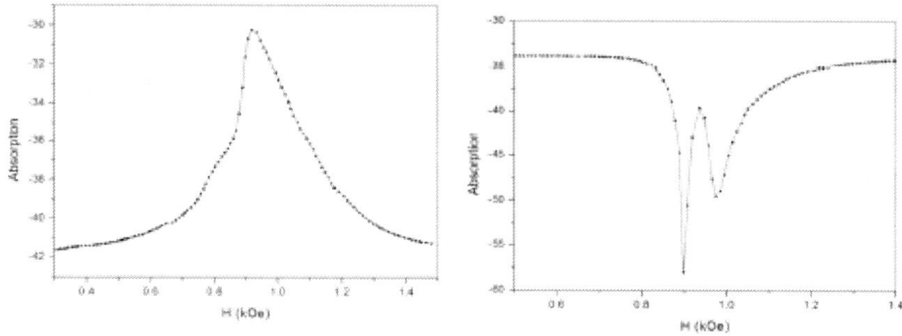

Figure 13 (a) data taken at Q = 50,000, and (b) at ~110,000. Clearly, distorted signals are obtained.

The advantages of the VNA-FMR with high Q cavity are: (1) the sensitivity which is comparable to that of commercials, but a lot cheaper, (2) multi frequencies can be done in the same manner by just changing to another band cavity only within the frequency range of the VNA. Nevertheless, some points needed to be well aware. Firstly, VNA has built-in sweep average function which increases the SNR for every sweeping. The penalty of using high average count is the very long acquisition time. For example, a commercial FMR spectrometer takes about a few seconds to sweep a spectrum over 2 kOe field at 1 Oe resolution with time constant of a 1mS. However, the settled time for a VNA with average of 1 is found to be about 0.1S a point. Therefore, it takes at least 3 min to obtain the spectrum with the same conditions. Average of 10, and sometime 50 are often used for better SNR, then the acquisition time is very long which could overheat the electromagnet. Secondary, VNA cannot work with lock-in amplifier directly, and a circulator is also needed to force the reflected wave to a new port. Before the signal is conducted to the lock-in amplifier, a microwave transducer (Schottky diode, for example) is required. This is because the output from VNA is power which has to be transduced into voltage before plugging into the lock-in amplifier. Besides, the output of VNA has already been averaged over a range of frequencies. Thus the Schottky detector cannot recognize this tiny average change. This problem can be amended by narrowing the starting and ending frequencies of the VNA. The two frequencies indeed can be set to the same value as the dip frequency. While working with lock-in, the VNA-FMR acquires data as fast as the commercial one, since the SNR enhancement is processed by the lock-in and not by the VNA. Further, MS-, and CPW- FMR can also work together with lock-in method in this manner. If one wants to study high power VNA-FMR, a high power microwave amplifier could be used. However, if the VNA is still used as the analyzer, a suitable attenuator has to be inserted between the returned line and VNA, otherwise damage is resulted.

AUTHOR DETAILS

Chi-Kuen Lo

Department of Physics, National Taiwan Normal University, Taipei, Taiwan

REFERENCES

1. M. Dı´az de Sihues, C. A. Durante-Rincon, and J. R. Fermin, J. Magn. Magn. Mater. 316, e462 (2007).
2. V. G. Gavriljuk, A. Dobrinsky, B. D. Shanina, and S. P. Kolesnik, J. Phys. Condens. Matter 18, 7613 (2006).
3. Z. Zhang, L. Zhou, P. E. Wigen, and K. Ounadjela, Phys. Rev. B 50(9), 6094 (1994).
4. B. Heinrich, Y. Tserkovnyak, G. Woltersdorf, A. Brataas, R. Urban, and G. E. W. Bauer, Phys. Rev. Lett. 90(18), 187601 (2003).
5. V. P. Nascimento and E. Baggio Saitovitch, F. Pelegrinia, L. C. Figueiredo, A. Biondo, and E. C. Passamani, J. Appl. Phys. 99, 08C108 (2006).
6. K. Lenz, E. Kosubek, T. Tolinski, J. Lindner, and K. Baberschke, J. Phys.: Condens. Matter 15, 7175 (2003).
7. M. Oogane, T. Wakitani, S. Yakata, R. Yilgin, Y. Ando, A. Sakuma, T. Miyazaki, J. J. Appl. Phys. 50 (5A) 3889 (2006)
8. R. Urban, G. Woltersdorf, and B. Heinrich, Phys. Rev. Lett. 87(21), 271204-1 (2009).
9. T. Kato, K. Nakazawa, R. Komiya, N. Nishizawa, S. Tsunashima, and S. Iwata, IEEE Trans. Magn., VOL. 44, NO. 11, NOVEMBER 2008
10. "Microwave Electronics: Measurement and Materials Characterization", L. F. Chen, C. K. Ong, C. P. Neo, John Wileys & Sons Ltd. 2004
11. "Microwave Engineering", D.M. Pozar, John Wileys & Sons Ltd. (2012)
12. "High-Frequency EPR Instrumentation", E.J. Reijerse, Appl Magn Reson (2010) 37:795- 818

13. See http://www.bruker-biospin.com/epr-products.html.
14. See http://www.jeol.com/PRODUCTS/AnalyticalInstruments/ElectronSpinResonance/tabid/98/Default.aspx.
15. "Quantitative EPR", G. R. Eaton, S. S. Eaton, D. P. Barr, and R. T. Weber, Springer Wien, New York, (2010)
16. Marina Vroubel, Yan Zhuang, Behzad Rejaei, and Joachim N. Burghartz, J. Appl. Phys. 99, 08P506 (2006)
17. C. Nistor, K. Sun, Z. Wang, M. Wu, C. Mathieu, and Matthew Hadley Appl. Phys. Lett. 95, 012504 (2009)
18. Y. Chen, D.S. Hung, Y.D. Yao, S.F. Lee, H.O. Ji, C. Yu, J. Appl. Phys. 101, 09C104 (2007)
19. H. G LOWI'NSKI, J. DUBOWIK, Acta Physicae Superficierum , Vol. XII, 2012
20. "The Art of Electronics", P. Horowitz, W. Hill, Cambridge Univ. Press, 2nd Ed. 1989
21. Advanced Digital Signal Processing and Noise Reduction", S.V. Vaseghi, John Wiley & Sons Ltd., 2008
22. Agilent VNA user manual
23. C.K. Lo, W.C. Lai, J.C. Cheng, Rev. Sci. Instruments, 82, 086114 (2011)
24. R.W. Damon, Rev. Mod. Phys., 25, No.1, 239 (1953)
25. N. Bloembergen, S. Wang, Phys. Rev., 93 No.1 72 (1953)

Chapter 6

FERROMAGNETIC RESONANCE

Orhan Yalçın

Niğde University, Niğde, Turkey

INTRODUCTION

Ferromagnetism is used to characterize magnetic behavior of a material, such as the strong attraction to a permanent magnet. The origin of this strong magnetism is the presence of a spontaneous magnetization which is produced by a parallel alignment of spins. Instead of a parallel alignment of all the spins, there can be an anti-parallel alignment of unequal spins. This results in a spontaneous magnetization which is called ferrimagnetism.

The resonance arises when the energy levels of a quantized system of electronic or nuclear moments are Zeeman split by a uniform magnetic field and the system absorbs energy from an oscillating magnetic field at sharply defined frequencies corresponding to the transitions between the levels. Classically, the resonance event occurs when a transverse ac field is applied at the Larmor frequency.

The resonance behaviour usually called magnetic resonance (MR) and nuclear magnetic resonance (NlV:IR). Main types of resonance phenomenon can be listed as nuclear magnetic resonance (NlV:IR), nuclear quadrupole resonance (NQR), electron paramagnetic/spin resonance (EPR, ESR), spin wave resonance (SWR), ferromagnetic resonance (FlV:IR), antiferromagnetic resonance (AFMR) and conductor electron spin resonance (CESR). The resonant may be an isolated ionic spin as in electron paramagnetic resonance (EPR) or a nuclear magnetic resonance (NMR). Also, resonance effects are associated with the spin waves and the domain walls. The resonance methods are important for investigating the structure and magnetic properties of solids and other materials.These methods are used for imaging and other applications.

The following information can be accessed with the help of such resonance experiments. (i) Electrical structure of point defects by looking at the absorption in a thin structure. (ii) The line width with the movement of spin or surroundings isn't changed. (iii) The distribution of the magnetic field in solid by looking at the of the resonance line position (chemical shift and etc.). (iv) Collective spin excitations.

The atoms of ferromagnetic coupling originate from the spins of d-electrons. The size of p permanent atomic dipoles create spontaneously magnetized. According to the shape of dipoles materials can be ferromagnetic, antiferromagnetic, diamagnetic, paramagnetic and etc.

Ferromagnetic resonance (FMR) technique was initially applied to ferromagnetic materials, all magnetic materials and unpaired electron systems. Basically, it is analogous to the electron paramagnetic resonance (EPR). The EPR technique gives better results at unpaired electron systems. The FMR technique depends on the geometry of the sample at hand. The demagnetization field is observed where the sample geometry is active. The resonance area of the sample depends on the properties of material. The FJviR technique is advantageous because it does not cause damage to materials. Also, it allows a three dimensional analysis of samples. The FJviR occurs at high field values while EPR occurs at low magnetic field values. Also, line-width of ferromagnetic materials is large according to paramagnetic materials. Exchange interaction energy between unpaired electron spins that contribute to the ferromagnetism causes the line narrowing. So, ferromagnetic resonance lines appear sharper than expected.

The FJviR studies have been increased since the EPR was discovered in 1945 (Zavosky, 1945; Kittel, 1946, 1947, 1949, 1953, 1958; Kip, 1949; Bloembergen, 1950, 1954; Crittenden, 1953; Van Vleck, 1950; Herring, 1950; Anderson, 1953; Damon, 1953; Young, 1953; Ament, 1955; Ruderman, 1954; Reich, 1955; Kasuya, 1956; White, 1956; Macdonald, 1956; Mercereau, 1956; Walker, 1957; Yosida, 1957; Tannenwald, 1957; Jarrett, 1958; Rado, 1958; Brown, 1962; Frait, 1965; Sparks, 1969). The beginnings of theoretical and experimental studies of spectroscopic investigations of basic sciences are used such as physics, chemistry, especially nanosciences and nanostrudures (Rodbell, 1964; Kooi, 1964; Bhagat, 1967, 1974; Sparks, 1970(a), 1970(b),1970(c), 1970(d); Rachford, 1981; Dillon, 1981; Schultz, 1983; Artman, 1957, 1979; Ramesh,1988(a), 1988(b); Fraitova, 1983(a), 1983(b), 1984; Teale, 1986; Speriosu, 1987; Vounyuk, 1991; Roy, 1992; Puszkarski, 1992; Weiss, 1955). The F:MR technique can provide information on the magnetization, magnetic anisotropy, dynamic exchange/dipolar energies and relaxation times, as well as the damping in the magnetization dynamics (Wigen, 1962, 1984, 1998; De Warnes, 1970; Wolfram, 1971; Yu, 1975; Frait, 1985, 1998; Rook, 1991; Bland, 1994; Patton,1995, 1996; Skomski, 2008; Coey, 2009). This spectroscopic method/FJvlR have been used to magnetic properties (Celinski, 1991; Farle, 1998, 2000; Fermin, 1999; Buschow, 2004; Heinrich, 2005(a), 2005(b)), films (Ozdemir, 1996, 1997), monolayers (Zakeri, 2006), ultrathin and multilayers films (Layadi, 1990(a), 1990(b), 2002, 2004; Wigen, 1993; Zhang, 1994(a),1994(b); Farle, 2000; Platow, 1998; Anisimov, 1999; Y1ld1z, 2004; Heinrich, 2005(a);Lacheisserie, 2005; de Cos, 2006; Liua, 2012; Schafer, 2012), the angular, the frequency (Celinski, 1997; Farle, 1998), the temperature dependence (Platow, 1998), interlayer exchange coupling (Frait, 1965, 1998; Parkin, 1990, 1991(a), 1991(b), 1994; Schreiber, 1996; Rook, 1991; Wigen, 1993; Layadi, 1990(a); Heinrich, 2005; Paul, 2005), Brillouin light scattering (BLS) (Griinberg, 1982; Cochran, 1995; Hillebrands, 2000) and sample inhomogeneities (Artman,1957, 1979; Damon, 1963; McMichael, 1990; Arias, 1999; Wigen, 1998; Chappert, 1986;Gnatzig, 1987; Fermin, 1999) of samples. Besides using FMR to characterize magnetic properties, it also allows one to study the fundamental excitations and technological applications of a magnetic system (Sclunool, 1998; Voges, 1998; Zianni, 1998; Grunberg, 2000, 2001; Vlasko-Vlasov, 2001; Zhai, 2003; Alda§, 2004;

Birkhauser Verlag, 2007; Seib, 2009). The various thickness, disk array, half-metallic ferromagnetic electrodes, magnon scattering and other of some properties of samples have been studied using the FMR tehniques (lv:lazur, 1982; da Silva, 1993; O,ikazumi, 1997; Song, 2003; Mills, 2003; Rameev, 2003(a), 2003(b), 2004(a), 2004(b); An, 2004; Ramprasad, 2004; Xu, 2004; Wojtowicz, 2005; Zakeri, 2007; Tsai, 2009; O,en, 2009). The magnetic properties of single-crystalline (Kambe, 2005; Brustolon, 2009), polycrystalline (Singh, 2006; Fan, 2010), alloy fihns (Sihues, 2007), temperature dependence and similar qualities have been studied electromagnetic spectroscopy techniques (Ozdemir, 1998; Birlikseven, 1999(a), 1999(b); Fermin, 1999; Rameev, 2000; Akta§, 2001; Budak, 2003; Khaibullin, 2004). The magnetic resonance techniques (EPR, FMR) have been applied to the iron oxides, permalloy nanostructure (Kuanr, 2005), clustered, thermocouple connected to the ferromagnet, thin permalloy layer and et al. (Guimariies, 1998; Spoddig, 2005; Can, 2012; Rousseau, 2012; Valenzuela, 2012; Ba.kker, 2012; Macia, 2012; Dreher, 2012; Kind, 2012; Li, 2012; Estevez, 2012; Sun, 2012(a), 2012(b), 2012(c); Richard, 2012). Magneto-optic (Paz, 2012), dipolar energy contributions (Bose, 2012), nanocrystalline (lv:laklakov, 2012; Raita, 2012), Lao.1Sro. aMnOa fihns (Golosovsky, 2012), L<ll.67Bao.aaN:Im-yAy0a, A - Fe, Cr (Osthover, 1998), voltage-controlled magnetic anisotropy (VCMA) and spin transfer torque (Zhu, 2012) and the typical properties of the inertial resonance are investigated (Olive, 2012).The exchange bias (Backes, 2012), Q cavities for magnetic material (Beguhn, 2012), MgO/CoFeB(fa structure (O,en, 2012), the interfacial origin of the giant magnetoresistive effect (GMR) phenomenon (Prieto, 2012), self demagnetization field (Hinata, 2012), Fea04/InAs(100) hybrid spintronic structures (Huang, 2012), granular fihns (Kakazei, 1999, 2001; Sarmiento, 2007; Krone, 2011; Kobayashi, 2012), nano-sized powdered barium (BaFet20t9) and strontium (Sr Fet20t9) hexaferrites (Korolev, 2012), Nio.1Jvfm.a.x CoxFe2C¢ ferrites (NiMnCo: x = 0.00, 0.04, 0.06, and 0.10) (Lee, 2012), thin fihns (Demokritov, 1996,1997; Nakai, 2002; Lindner, 2004; Aswal, 2005; Jalali-Roudsar, 2005; Cochran, 2006; Mizukami, 2007; Seemann, 2010), Ni2MnGa films (Huang, 2004), magnetic/electronic order of fihns (Shames, 2012), Fet.xGd(Tb)x films (Sun, 2012), in i AloD6Fet.940a (Yoshikiyo, 2012). 10 nm thick Fe/GaAs(llO) fihn (R&ner, 2012), triangular shaped permalloy rings (Ding, 2012) and C02-Y hexagonal ferrite

single rod (Bai, 2012) structures and properties have been studied by FMR tecniques (Spaldin, 2010). Biological applications (Berliner, 1981; Wallis, 2005; Gattesci:U, 2006; Kopp, 2006; Fischer, 2008; Mastrogiacomo, 2010), giant magneto-impedance (Valenzuela, 2007; Park, 2007), dynamics of feromagnets (Vilasi, 2001; Rusek, 2004; Limmer, 2006; Selhnyer, 2006; Spinu, 2006; Azzerboni, 2006; I<rivoruchko, 2012), magneto-optic kerr effect (Suzuki, 1997; Neudecker, 2006), Heusler alloy (HA) films (Kudryavtsev, 2007), ferrites (Kohmoto, 2007), spin polarized electrons (Rahman, 2008) and quantum mechanics (Weil, 2007) have been studied by FMR technique in generally (Hillebrands, 2002, 2003, 2006). In additional, electric and magnetic properties of pure, Cu2• ions doped hydrogels have been studied by ESR techniques (Co§kun, 2012).

The FMR measurements were performed in single crystals of silicon- iron, nickel-iron, nickel and hcp cobalt (Frait, 1965), thin films (Knorr, 1959; Davis, 1965; Hsia, 1981; Krebs, 1982; Maksymowich, 1983, 1985, 1992; Platow, 1998; Durusoy, 2000; Baek, 2002; Kuanr, 2004), CoCr magnetic thin films (Cofield, 1987), NiFe/FeN:rn thin films (Layadi, 1988), single-crystal Fe/Cr/Fe(100) sandwiches (Krebs, 1989), polyc:rystalline single films (Hathaway, 1981; Rezende, 1993) and ultrathin multilayers of the system Au/Fe/Au/Pd/Fe (001) prepared on GaAs(001) (Woltersdorf, 2004). The FMR techniques have been succesfully applied peak--to peak linewidth (Yeh, 2009; Sun, 2012), superconducting and ferromagnetic coupled structures (Richard, 2012) and thin Co films of 50 nm thick (Maklakov, 2012). The garnet materials (Ramesh, 1988 (a), 1988 (b)), polar magneto-optic kerr effect and brillouin light scattering measurements (Riedling, 1999), giant-magnetoresistive (GMR) multilayers (Griinberg, 1991; Borchers, 1998) and insulated multilayer film (de Cos, 2006; Lacheisserie, 2005) are the most intensely studied systems. The technique of FMR can be applied to nano-systems (Poole, 2003; Parvatheeswara, 2006; Mills, 2006; Schmool, 2007; Vargas, 2007; Seemann, 2009; Wang, 2011; Patel, 2012; De Biasi, 2013). The FMR measurement on a square array of permalloy nanodots have been comparion a numerical simulation based on the eigenvalues of the linearized Landau Lifshitz equation (Rivkin, 2007). The dynamic .fluctuations of the nanoparticles and their anisotropic behaviour have been recorded with FMR signal (Owens, 2009). Ferromagnetic resonance (FMR) modes for FeroCoao magnetic nanodots of 100 nm in diameter in a mono domain state are studied under different in-plane and out-of-plane magnetic fields (Miyake, 2012). The FMR techniques have been accomplished applied to magnetic microwires and nanowire arrays (Adeyeye, 1997; Wegrowe, 1999, 2000; Garda-Miquel, 2001; Jung, 2002; Arias, 2003; Raposo, 2011; Boulle, 2011; Kraus, 2012; Klein, 2012). In additional, FMR measurements have been performed for nanocomposite samples of varying particles packing fractions with demagnetization field (Song, 2012). The ferromagnetic resonance of magnetic .fluids were theoretically investigated on thermal and particles size distribution effects (Marin, 2006). The FMR applied to nanoparticles, superparamagnetic particles and catalyst particles (de Biasi, 2006; Vargas, 2007; Duraia, 2009).

In the scope of this chapter, we firstly give a detailed account of both magnetic order and their origin. The origin of magnetic orders are explained and the equations are obtained using Fig.1 which shows rotating one electron on the table plane. Then, the dynamic equation of motion for magnetization was derived. We mentioned MR and damping terms which have consisted three terms as the Bloch-Bloembergen, the Landau-Lifshitz and the Gilbert form. We indicated electron EPR/ESR and their historical development. The information of spin Hamiltonian and g.-tensor is given. The dispersion relations of monolayer, trilayers, five-layers and multilayer/ n-layers have regularly been calculated forferromagnetic exchange-couple systems (Grunberg, 1992; Nagamine, 2005, Schmool, 1998).The theoretical FMR spectra were obtained by using the dynamic equation of motion for magnetization with the Bloch-Bloembergen type damping term. The exchange-spring (hard/soft) system which is the best of the sample for multilayer structure has been explained by using the FMR technique and equilibrium condition of energy of system. The FMR spectra originated from the iron/soft layers as shown in the exchange spring magnets in Fig.9. Finally, superparamagnetic/single-domain nanoparticles and their resonance are described in detail.

MAGNETIC ORDER

Magnetic materials are classified as paramagnetic, ferromagnetic, ferromagnetic, antiferromagnetic and diamagnetic to their electronic order. Magnetic orders are divided in two groups as (i) paramagnetic, ferromagnetic, ferrimagnetic, antiferromagnetic and (ii) diamagnetic.

The magnetic moments in diamagnetic materials are opposite to each other as well as the moments associated with the orbiting electrons so that a: zero magnetic moment (11) is produced on macroscopic scale.In the paramagnetic materials, each atom possesses a small magnetic moment. The orientation of magnetic moment of each atom is random, the net magnetic moment of a large sample (macroscopic scale) of dipole and the magnetization vector are zero when there is no applied field.

Nanoscience, nanotechnology and nanomaterials have become a central field of scientific and technical activity. Over the last years the interest in magnetic nanostructures and their applications in various electronic devices, effective opto-electronic devices, bio-sensors, photo-detectors, solar cells, nanodevices and plasmonic structures have been increasing tremendously. This is caused by the unique properties of magnetic nanostructures and the outstanding performance of nanoscale devices. Dimension in the range of one to hundred nanometers, is called the nano regime. In recent years, nanorods, nanoparticles, quantum dots, nanocrystals etc. are in a class of nanostructures (Yal m, 2012; Kartopu & Yal m, 2010; Alda§, 2006) studied extensively. As the dimensions of nano materials decrease down to the nanometer scale, the surface of nanostructures starts to exhibit new and interesting properties mainly due to quantum size effects.

ORIGIN OF MAGNETIC MOMENT

The magnetization of a matter is derived by electrons moving around the nucleus of an atom. Total magnetic moment occurs when the electrons such as a disc returns around its axis consist of spin angular momentum and returns around the nucleus consist orbital angular momentum. The most of matters which have unpaired electrons have a little magnetic moment.This natural angular momentum consists of the result of charged particle return around its own axis and is called spin of the particle.The origin of spin is not known exactly, although electron is point particle the movement of an electron in an external magnetic field is similar to the movement of the disc. In other words, the origin of the spin is quantum field theoretical considerations and comes from the representations of the Poincare algebra for the elementary particles. The magnetism related to spin angular momentum, orbital angular momentum and spin-orbit interactions angular momentum. The movement of the electron around the nucleus can be considered as a current loop while electron spin is considered very small current loop which generate magnetic field. Here, orbital angular momentum was obtained by the result of an electron current loop around the nucleus. Thus, both it is exceeded the difficulty of understanding the magnetic moment and the magnetic moment for an electron orbiting around the nucleus is used easily. The result of orbital angular momentum (I) adapted for spin-angular momentum (S) (Cullity, 1990).

One electron is rotating from left to right on the table plane as shown in Fig.l. The rotating electron creates a current (i) on the circle with radius of r.

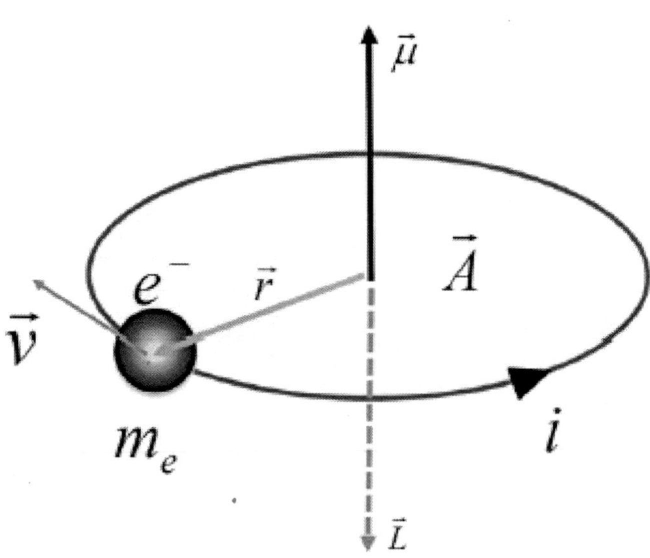

Figure 1 Schematic representation of the precession of a single electron on the table plane.

The magnetic moment of a single electron is defined as below

$$\vec{\mu} = i \cdot \vec{A} \qquad (1)$$

Where, $\vec{A}$ is the circle area.The magnetic moment is written as follows by using the current -e=i.t one cycle 2p r=v.t and angular momentum L=me.v.r definition

$$\vec{\mu} = -\frac{e}{2m_e}\vec{L} \qquad (2)$$

Where r =e /2me. and Iis the gyromagnetic (magneto-mechanical or magneto-gyric) ratio and the orbital-angular momentum, respectively. Therefore, the magnetic moment fi is obtained from Eq. (2) as below

$$\vec{\mu} = -\gamma \vec{L} \qquad (3)$$

The following expression is obtained when derivative of Eq.(3)

$$d\vec{\mu} + \gamma d\vec{L} + \vec{L} d\gamma = 0. \qquad (4)$$

For our purpose, we only need to know that r is a constant and dy = 0 . From this results, dfi +r df.= 0 . The derivative of time of this equation, the equation of motion for magnetic moments of an electron is found as below

$$\frac{1}{\gamma}\frac{d\vec{\mu}}{dt} = \frac{d\vec{L}}{dt} = \vec{\tau} \qquad (5)$$

This equation is related to $\vec{\tau} = d\vec{L}/dt$ in two dimensional motions on the plane and $\vec{F} = d\vec{P}/dt$ in one dimensional motion.

This motion corresponds to Newton's dynamic equations. When an electron is placed in an applied magnetic field $\vec{H}$,the magnetic field will produce a torque $(\vec{\tau})$ on the magnetic moment $(\vec{\mu})$ of amount $\vec{\mu}\times\vec{H}$ .The equation of motion for magnetic moment $(\vec{\mu})$ is found by equating the torque as below

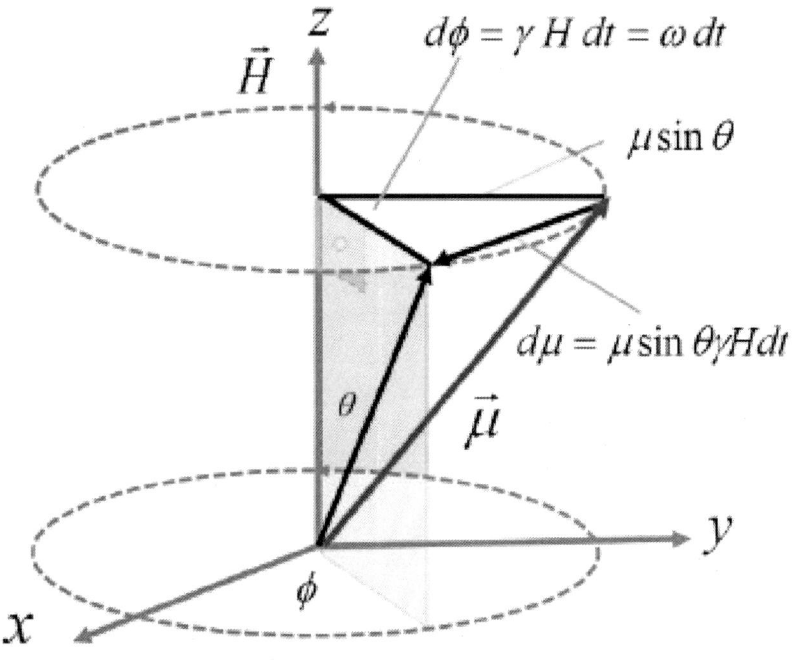

Figure 2. Schematic representation of precession of a single magnetic moment ,.U in the external magnetic field around the z-axis.

$$\frac{1}{\gamma}\frac{d\vec{\mu}}{dt} = \vec{\mu}\times\vec{H} \qquad (6)$$

This expression is called the equation of motion for magnetic moment $(\vec{\mu})$. The motion of magnetic moment $(\vec{\mu})$ forms a cone related to $\vec{H}$ when the angle e of magnetic moment and external magnetic field does not change.Therefore, in time (dt) the tip of the vector $(\vec{\mu})$ moves an angle (g H)dt. The magnetic moment vector make precession movement about $\vec{H}$ at a frequency of g H/2p. This frequency V=w/2p=gH/2p, is called the Larmor frequency. In general this Larmor frequency is used this form w= γH in literature.

MAGNETIC RESONANCE

Magnetic Resonance (MR) is a research branch which examines magnetic properties of matters. The magnetic properties of atom originate from electrons and nucleus. So, it is studied in two groups such as electron paramagnetic resonance (EPR)/electron spin resonance (ESR) and nuclear magnetic resonance (NMR). At ESR and NMR all of them are the sample is placed in a strong static magnetic field and subjected to an orthogonally amplitude-frequency. While EPR uses a radiation of microwave frequency in general, NMR is observed at low radio frequency range. The energy absorption occurs when radio frequency is equal with energy difference between electrons two levels. But, the transition must obey the selection rules. The splitting between the energy levels occurs when total angular moment of electron is different from zero. On the other hand, the splitting of energy levels has not been observed in the filled orbit. The precession motion o£ a paramagnetic sample in magnetic field is seen schematically in Fig. 2. I£ microwave field with v frequency at perpendicular is applied to the static field, it comes out power absorption when precession (ω_0) is same with v -frequency. The power increases when these frequencies come near to each other and it occurs maximum occurs at point when they are equal. This behaviour is called magnetic resonance (lV:IR). The magnetic materials contain a large number of atomic magnetic moment in generally. Net atomic magnetic moment can be calculated by $\vec{M} = N\vec{\mu}$. Where, N is the number of atomic magnetic moment in materials. td.M - -

$$\frac{1}{\gamma}\frac{d\vec{M}}{dt} = \vec{M} \times \vec{H}_{eff} \quad (7)$$

This precession movement continue indefinitely would take forever when there is no damping force. The damping term may be introduced in different ways. Indeed, since the details of the damping mechanism in a ferromagnet have not been completely resolved, different mathematical forms for the damping have been suggested. The three most common damping terms used to augment the right-hand side of Eq. (7) are as follows:

(i) The Bloch-Bloembergen form: $-\frac{\vec{M}_{\theta,\varphi}}{T_2}-\frac{\vec{M}_z-M_0}{T_1}$

(ii) The Landau-Lifshitz form: $\frac{-\lambda}{|\vec{M}|^2}\vec{M}\times\vec{M}\times\vec{H}$

(iii) The Gilbert form: $\frac{\alpha}{|\vec{M}|}\vec{M}\times\frac{d\vec{M}}{dt}$

Bloch-Bloembergen type damping does not converse M so it is equivalent to the type of Landau-Lifshitz and the Gilbert only when ais small and for small excursion ofM. For large excursion o£ M , the magnitude of M is certainly not protected, as the damping torque is in the direction of the magnetization component in this formularization. Hence, the observation of M in the switching experiments in thin films should be provide a sensitive test on the appropriate form o£ the damping term for ferromagnetism since M which is conserved during switching. This would suggest that the form of the Bloch-Bloembergen damping term would not be applicable for this type of experiment. The Gilbert type (Gilbert, 1955) is essentially a modification of the original form which is proposed firstly by Landau and Lifshitz (Landau & Lifshitz, 1935).It is very important to note that the Landau Lifshitz and Gilbert type of damping conserve while the Bloch-Bloembergen (Bloembergen, 1950) type does not. Landau and Lifshitz observed that the ferromagnetic exchange forces between spins are much greater than the Zeeman forces between the spins and the magnetic fields in their formulation of the damping term. Therefore, the exchange will conserve the magnitude of M.Inthis formulation, since the approach of M towards His due completely to the relatively weak interaction between M and H , we must require that A. « y M . In this small damping limit, the Landau-Lifshitz and the Gilbert forms are equivalence so that whether one uses one or the other is simply a matter of convenience or familiarity. However, Callen has obtained a dynamic equation by quantizing the spin waves into magnons and treating the problem quantum-mechanically (Callen, 1958). Subsequently, Fletcher, Le Craw, and Spencer have reproduced the same equation using energy consideration (Fletcher, 1960).In their reproduction, they found the mean the rate of energy transfer between the uniform precession, the spin waves (Grunberg, 1979, 1980) and the lattice.

ELECTRON PARAMAGNETIC RESONANCE

Stem and Gerlach (Gerlach, 1922) proved that the electron-magnetic moment of an atom in an external magnetic field originates only in certain directions in the experiment in 1922. Uhlenbek and Goudsmit found that the connection between the magnetic moment and spin angular momentum of electron (Uhlenbek, 1925), Rabi and Breit found the transition between the energy levels in oscillating magnetic field (Rabi, 1938). This also proved to be observed in the event of the first magnetic resonance. The EPR technique is said to be important of Stern-Gerlach experiment. Zavoisky observed the first peak in the electron paramagnetic resonance for CuCI22H20 sample and recorded (Zavoisky, 1945). The most of EPR experiments were made by scientists in the United Kingdom and the United States. Important people mentioned in the experimental EPR studies; Abragam, Bleaney and Van Vleck. The historical developments of :MR have been summarized by Ramsey (Ramsey, 1985). NMR experiments had been done by Purcell eta!. (Purcell, 1946). Today it has been used as a tool for clinical medicine.:MRI was considered as a basic tool of cr scan in 1970s.

The behaviors of spin system under the external magnetic field with the gradient of spin system are known NMR tomography. This technique is used too much for medicine, clinics, diagnostic and therapeutic purposes.General structure of the EPR spectrometer consist four basic parts in general. (i) Source system (generally used in the microwave 1-100 GHz), (ii) cavity-grid system, (iii) Magnet system and (iv) detector and modulation system. EPR/ESR is subject of the MR. An atom which has free electron when it is put in magnetic field the electron's energy levels separate (Yal m, 2003, 2007(a), 2007(b)). This separation originates from the interaction of the electrons magnetic moment with external magnetic field. Energy separating has been calculated by the following Hamiltonian.

$$\hat{H} = g\,\mu_B\,\vec{H}\cdot\hat{S} \tag{8}$$

It is called Zeeman Effect.If the applied magnetic field oriented z-axis energy levels are;

$$E_{Ms} = g\,\mu_B\,H\cdot M_s \tag{9}$$

Here, g is the g-value (or Lande g-value) (for free electron g.= 2.0023193and proton g_N = 2.7896), ,μ_B is Bohr magneton (, μ_B = (eh/4πm_{e}.) =9.2740x10^{-24} J/T) and M_s is the number of magnetic spin quantum. If the orbital angular momentum of electron is large of zero (L > 0) g -value for free atoms is following

$$g = 1 + \frac{S(S+1) - L(L+1) + J(J+1)}{2J(J+1)} \tag{10}$$

The anisotropy of the g-factor is described by taking into account the spin-orbit interaction combined (Yal9Jl, 2004(c)).The total magnetic moment can be written at below;

$$\mu_{eff} = g\,\mu_B\sqrt{J(J+1)}$$

The values of orbital angular momentum of unpaired electrons for most of the radicals and radical ions are zero or nearly zero. Hence, the number of total electron angular momentum J equals only the number of spin quantum 5 .So, these values are nearly 2.For free electron

(M.= ±1/2) and for this electron;

$$\Delta E = E_{+1/2} - E_{-1/2} = g\,\mu_B\,H \qquad (11)$$

When the electromagnetic radiation which frequency v is applied to such an electron system;

$$h\upsilon = g\,\mu_B\,H \qquad (12)$$

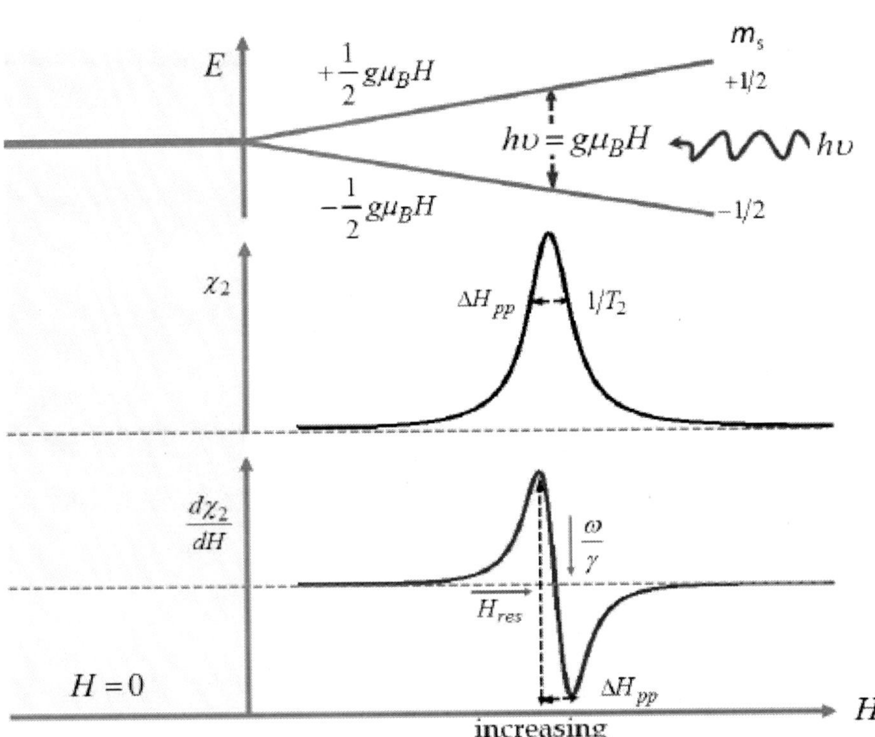

Figure 3 The energy levels and resonance of free electron at :zero field and increasing applied magnetic field.In this figure,while the value of magnetic field increases, the separating between energy levels increase. Arbitraryunitsusedin vertical axes for ;x_2 and dx_2/dH.

If this equation is provided the system absorbs energy from applied electromagnetic wave (see Fig.3). It is called resonance effect. Material absorbs energy in two different ways from applied electromagnetic wave by according to the Eq.(12). Firstly, in Eq.(12), frequency of electromagnetic wave doesn't change whlle the external magnetic field changes. Secondly, its opposite can be provided.

In thls Fig.(3) it has been seen that magnetic susceptibility ;x_2 versus magnetic field. At the same time it is said the absorption curve. The magnetic field derivative beneath of this figure is FMR absorption spectrum (dx_2/dH). Here, Δ_{PP} and $1/T_2$ are linewidth, H_{res} is resonance field, aJ/Y is resonance frequency.

Spin Hamiltonian

The spin Hamiltonian is total electronic spins and nucleon spin 1 which have crystal lattice under the static magnetic field following;

$$\hat{H} = \mu_B \vec{H} \cdot \vec{\vec{g}} \cdot \hat{S} + \hat{S} \cdot \vec{\vec{D}} \cdot \hat{S} + \vec{S} \cdot \vec{\vec{A}} \cdot \hat{I} - \mu_N \cdot \vec{H} \cdot \vec{g}_N \cdot \hat{I} + \hat{I} \cdot \vec{\vec{P}} \cdot \hat{I} \qquad (13)$$

$\hat{S}$ and $\hat{I}$ operators of electronic and nucleus, respectively. In this equation, first term is Zeeman effect, second one is thin layer effects, the thlrd one is the effect of between electronic spin and nucleus-spin of ion and it is known that thin layer effects. The fourth term is the effect of nucleus with the magnetic field. The last one is quadrupole effect of nucleus.It can be added different terms in Eq.13 (Slichter, 1963).

$\vec{\vec{g}}$ Tensor

The total magnetic moment of ion is $\vec{\mu} = g\mu_B \vec{J}$. $\vec{J}$ is the ratio of total angular momentum to planck constant. Lande factor g is depend on $\vec{S}, \vec{L}\, \vec{J}$ For the base energy level if $\vec{L}$. is :zero, g factor is equal free electron's g -factor.But, g -factor in the exited energy levels separated from the g -factor of free electron. The hamiltonian for an ion which is in the magnetic field is following (Weil, 1994).

$$\hat{H} = \mu_B \vec{H} \cdot (\hat{L} + g_e \hat{S}) + \lambda \hat{L} \cdot \hat{S} \tag{14}$$

In this equation, first term is Zeeman effects, second one is spin-orbit interaction. The first order energy of ion whlch shows IJ,M} and it is excepted not degenerate is seen at below.

$$E_J = \langle J,M | g_e \mu_B H_z \hat{S}_z | J,M \rangle + \langle J,M | (\mu_B H_Z + \lambda \hat{S}_z) | J,M \rangle \tag{15}$$

We can write the hamiltonian equation whlch uses energy equations. There is two terms in the hamiltonian equations. The first term is the independent temperature coefficient for paramagnetic for paramagnetic, the last terms are only for spin variables. If the angular moment of ion occurs because of spin, $\vec{\vec{g}}$ tensor is to be isotropic.

FERROMAGNETIC RESONANCE

The most important parameters for ferromagnet can be deduced by the ferromagnetic resonance method. FMR absorption curves may be obtained from Eq.(12) by chancing frequency or magnetic field. FlV:IR signal can be detected by the external magnetic field and frequency such as EPR signal. The field derivative FlV:IR absorption spectra are greater than in EPR as a generally. The linear dependence of frequency of resonance field may be calculated from 1GHz to 100 GHz range in frequency spectra (L-,S-, C-, X-, K-, Q-, V-, E-, W-, F-, and D-band). The resonance frequency, relaxation, linewidth, Lande g-factor (spectroscopic g-factor), the coercive force, the anisotropy field, shape of the specimen, symmetry axes of the crystal and temperature characterized FMR spectra. The broadening of the FlV:IR absorption line depend on the line width (so called $1/T_2$ on the Bloch -Bloembergen type damping form). The nonuniform modes are seen in the EPR signal. The nonlinear effects for FMR are shown by the relationship between the uniform precessions of magnetic moments. The paramagnetic excitation of unstable oscillation of the phonons displays magneto-elastic interaction in ferromagnetic systems. This behaviour so called magnetostriction. The FIV:IR studies have led to the development of many micro-wave devices. These phenomenon are microwave tubes, circulators, oscillators, amplifiers, parametric frequency converters, and limiters. The resonance absorption curve of electromagnetic waves at centimeter scale by ferromagnet was first observed by Arkad'ev in 1913 (Arkad'ev, 1913) .

The sample geometry, relative orientation of the equilibrium magnetization M,the applied de magnetic field H and experimental coordinate systems are shown in Fig.4.

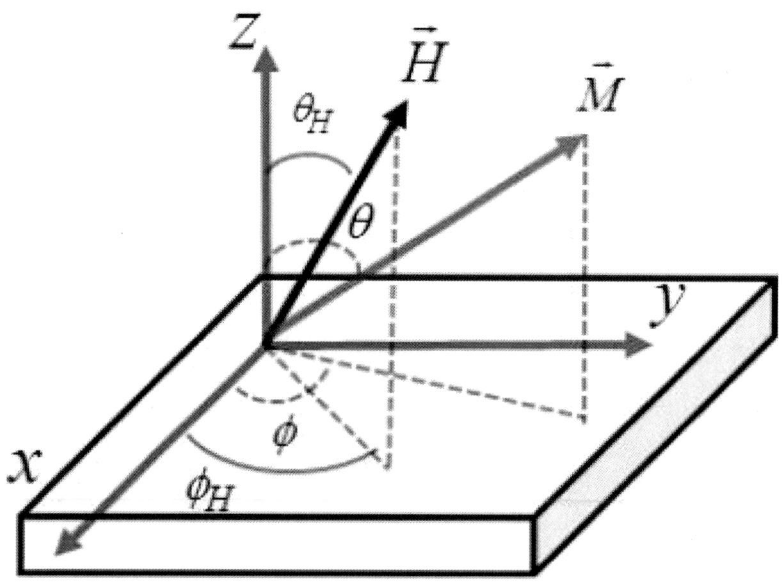

Figure 4 Sample geometries and relative orientations of equilibrium magnetization $\vec{M}$ and the de components of external magnetic field, $\vec{H}$ far thin films.

The ferromagnetic resonance data analyzed using the free energy expansion similar to that employed

$$\begin{aligned}
E_T &= E_Z + E_a + E_d + E_{ex} + \ldots \\
E_Z &= -M \cdot H\left(\sin\theta \sin\theta_H \cos(\phi - \phi_H) + \cos\theta\cos\theta_H\right) \\
E_a &= K_{u1}\sin^2\theta + K_{u2}\sin^4\theta + K_{u3}\sin^6\theta \ldots \\
E_d &= -2\pi M^2 \sin^2\theta \\
E_{ex} &= -JS_iS_j
\end{aligned} \tag{16}$$

Where, E_Z ,E_a.,E_d,E_{ex} are Zeeman, magnetocrystalline anisotropy, demagnetization and ferromagnetic exchange energy. (θ, ϕ) and (θ_H, ϕ_H) are the angles for magnetization and applied magnetic field vector in the spherical coordinates, respectively. Magnetic anisotropy energy arises from either the interaction of electron spin magnetic moments with the lattice via spin-orbit coupling. On the other hand, anisotropy energy induced due to local atomic ordering. The θ in anisotropy energy is the angle between magnetization orientation and local easy axis of the magnetic anisotropy. K_{u1} and K_{u2} are energy density constants. The demagnetization field is proportional to the magnetic free pole density. The exchange energy for thin magnetic film may be neglected in generally. Because associated energies is small. But, this exchange energy are not neglected for multilayer structures. This energy occurs between the magnetic layers, so that this energy called interlayer exchange energy. This expression is seen at the end of this subject in details. $K_{eff} = \pi M^2 + K_u$ is the effective uniaxial anisotropy term and Kv takes into account some additional second-order uniaxial anisotropy and $H_{eff} = 2\pi M_s + (2K_u/M_s)$ is the effective field for a single magnetic films. The equilibrium values of polar angles e for the magnetization vector $\vec{M}$ are obtained from static equilibrium conditions. E_θ, E_ϕ,$E_{\theta\theta}$ and $E_{\phi\phi}$ can be easily calculated using the Eq. (16).

Neglecting the damping term one can write the equation of motion for the magnetization vector $\vec{M}$ as

$$\frac{1}{\gamma}\frac{d\vec{M}}{dt} = \vec{M} \times \vec{H}_{eff} \tag{17}$$

Here the $\vec{H}_{eff}$ is the effective magnetic field that includes the applied magnetic field and the internal field due to the anisotropy energy. The dynamic equation of motion for magnetization with the Bloch-Bloembergen type damping term is given as Eq.(18).

$$\frac{1}{\gamma}\frac{d\vec{M}}{dt} = \vec{M} \times \vec{H}_{eff} - \frac{\vec{M} - \delta_{iz} M_0}{T} \tag{18}$$

Here, T = (T_2 , T_2 , T_1) represents both transverse (for M_x and M_y components)and the longitudinal (for M_z, components) relaxation times of the magnetization. That is, T_1 is the spin-lattice relaxation time, T_2 is the spin-spin relaxation time, and δ_{iz}= (0,0,1) for (x,y,z) projections of the magnetization. In the spherical coordinates the Bloch-Bloembergen equation can be written as below;

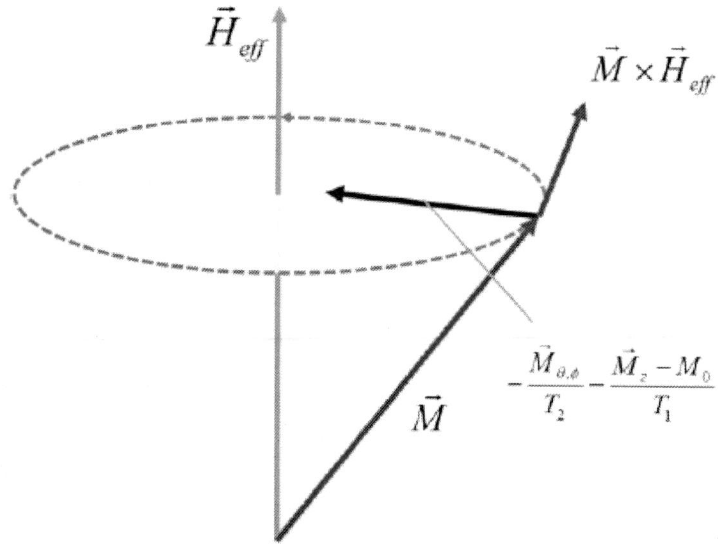

Figure 5 Damped precession of a magnetic moment $\vec{M}$ toward the effective magnetic field $\vec{H}_{eff}$ according to the Bloch-Bloembergen type equation (Akta,5,1993,1994; Yal m,2008(a)),

$$\frac{1}{\gamma}\frac{d\vec{M}}{dt}=\frac{\vec{M}}{|\vec{M}|}\times\vec{\nabla}E-\frac{\vec{M}_{\theta,\phi}}{\gamma T_2}-\frac{\vec{M}_z-M_0}{\gamma T_1} \tag{19}$$

Where, the torque is obtained from the energy density through the expression.

$$\vec{\nabla}E=-\left(\frac{\partial E}{\partial\theta}\right)\hat{e}_\phi+\frac{1}{\sin\theta}\left(\frac{\partial E}{\partial\phi}\right)\hat{e}_\theta \tag{20}$$

For a small deviation from the equilibrium orientation, the magnetization vector $\vec{M}$ can be approximated by

$$\vec{M}=M_s\hat{e}_r+m_\theta\hat{e}_\theta+m_\phi\hat{e}_\phi \tag{21}$$

Where the dynamic transverse components are assumed to be sufficiently small and can be given as

$$\begin{aligned}m_\theta(z,t)&=m_\theta^0\exp i(\omega t\pm kz)\\ m_\phi(z,t)&=m_\phi^0\exp i(\omega t\pm kz)\end{aligned} \tag{22}$$

Dispersion relation for films can be derived by using these solutions (Eq. (22)) in Eqs, (19) and (20), On the other hand, the eigen frequency of thin £hns mode is determined by the static effective field and can be derived directly from the total free energy for magnetic system/ ferromagnet, It is given by the second derivatives of the total energy with respect to the θ and ϕ (Smit, 1955; Artman, 1957; Wigen, 1984, 1988, 1992; Baseglia, 1988; Layadi, 1990; Farle, 1998), The matrices form for meandmis calculated using the Eq.(19) with Eq.(20, 21, 22),

$$\begin{pmatrix} i\frac{\omega}{\gamma}+\frac{1}{\gamma T_2}+\frac{E_{\theta\phi}}{M_s \sin\theta} & \frac{E_{\phi\phi}}{M_s \sin^2\theta} \\ -\frac{E_{\theta\theta}}{M_s} & i\frac{\omega}{\gamma}+\frac{1}{\gamma T_2}-\frac{E_{\theta\phi}}{M_s \sin\theta} \end{pmatrix} \cdot \begin{pmatrix} m_\theta \\ m_\phi \end{pmatrix} = \tag{23a}$$

$$\left(\frac{\omega}{\gamma}\right)^2 = \frac{1}{M^2 \sin^2\theta}\left(E_{\theta\theta} E_{\phi\phi} - E_{\theta\phi}^2\right) + \left(\frac{1}{\gamma T_2}\right)^2 \tag{23b}$$

Here (ω/γ) = $g\mu_B H$ is the Larmour frequency of the magnetization in the external de effective magnetic field. This dispersion relation can be related as the angular momentum analogue to be linear momentum oscillator described $\omega = \sqrt{k / \mu_i}$. Here, restoring force constant k is the second derivative of the potential part in the energy of systemi k= E_{xx}.The inverse mass $\mu^{-1}$is given by the second derivative of the kinetic part in the energy with respect to linear momentum μ_i^{-1} = E_{PP}.The restoring constant in this chapter corresponds to $E_{\theta\theta}$. The inverse mass is proportional to$E_{\phi\phi}$. The $E_{\theta\phi}$; arises when the coordinate system is not parallel to the symmetry and last term originated from relaxation term in Eq.(23) (Sparks, 1964; Morrish, 1965; Vittoria, 1993;Gurevich, 1996;Chikazumi, 1997)

The power absorption from radio frequency (if) field in a unit volume of sample is given by

$$P = \frac{1}{2}\omega\chi_2 h_1^2 \tag{24}$$

where ω is the microwave .frequency, h_1 is the amplitude of the magnetic field component and x_2 is the imaginary part of the high-frequency susceptibility. The field derivative FJv:IR absorption spectrum is proportional to dx_2 / dH and the magnetic susceptibility x is given as

$$\chi = 4\pi\left(\frac{m_\phi}{h_\phi}\right)_{\phi\to 0} \tag{25}$$

The theoretical absorption curves are obtained by using the imaginary part of thc high frcquency magnctic susccptibility as a function of applied field (Oner, 1997; Min, 2006; Cullity, 2009)

$$\chi = \chi_1 + i\chi_2 = \frac{4\pi M_s\left(\frac{E_{\theta\theta}}{M_s}\right)\left(\left(\frac{\omega_0}{\gamma}\right)^2 - \left(\frac{\omega}{\gamma}\right)^2 + i\left(\frac{2\omega}{\gamma^2 T_2}\right)\right)}{\left(\left(\frac{\omega_0}{\gamma}\right)^2 - \left(\frac{\omega}{\gamma}\right)^2\right)^2 + \left(\frac{2\omega}{\gamma^2 T_2}\right)^2} \tag{26}$$

The dispersion relation can be derived by substituting Eq.(16) into Eq. (23) (Ak:ta§, 1997; Yal m, 2004(a), 2004(b), 2008(a);Giiner, 2006; Kharmouche, 2007;Stashkevich, 2009)

$$\left(\frac{\omega}{\gamma}\right)^2 = \left[H\cos(\theta-\theta_H) + H_{eff}\cos(2\theta)\right]\cdot\left[H\cos(\theta-\theta_H) + H_{eff}\cos^2\theta\right] + \left(\frac{1}{\gamma T_2}\right)^2 \tag{27}$$

here, $\omega_0\pi$vis the circular frequency of the EPR spectrometer. Fitting Eq.(27) with experimental results of the FMR measurement at different out-of-plane-angle (θ_H), the values for the effective magnetization can be obtained.

Figure 6 uses of both experimental and theoretical coordinate systems for the nanowire sample geometry. Equilibrium magnetization $\vec{M}$ and de-magnetic field $\vec{H}$ are shown in this figure and also the geometric factor and hexagonal nanowire array presentation of nanowire are displayed. The ferromagnetic resonance theory has been developed for thin films applied to nanowires with the help of the following Fig.6. The effective uniaxial anisotropy term for nanowire arrays films $k_{eff} = \pi M^2(1-3p) + K_u$ is written in this manner for arrayed nanowires. The first term in the K_{eff} is due to the magnetostatic energy of perpendicularly-arrayed

NWs (Dubowik, 1996; Encinas-Oropesa, 2001; Demand, 2002; Yal m, 2004(a); Kartopu, 2009, 2010, 2011(a)) and constant with the symmetry axis along wire direction. The second term in the K_{eff} is packing factor for a perfectly ordered hcp NW arrays. The packing factor is defined as $p = (\pi / 2\sqrt{3})(d/r)^2$. The packing factor (P) of nanowires increases, nanowire diameter increases, the preferential orientation o£ the easy direction of magnetization changes from the parallel to the perpendicular direction to the wire axis (Kartopu, 2011(a)). As further, the effective uniaxial anisotropy (K_{eff}) for a perfectly ordered hcp NWs should decrease linearly with increasing packing factor. $H_{eff} = 2\pi M_s(1-3p) + (2K_u / M_s)$, which is the effective anisotropy field derived from the total magnetic anisotropy energy of NWs Eq. (16). The values for total magnetization have been obtained by fitting H_{eff} with experimental results of EMR measurements at different angles (θ_H) of external field $\vec{H}$ .The experimental spectra are proportional to the derivative of the absorbed power with respect to the applied field which is also proportional to the imaginary part of the magnetic susceptibility.

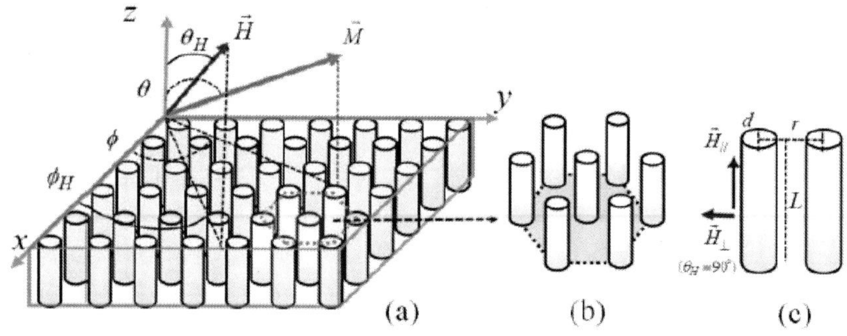

Figure 6 (a) Schematic representation of the cobalt nanowires and the relative orientation of the equilibrium magnetization M and the de component of the external magnetic field H,for the FMR experiments and their theoretical calculations.(b) Hexagonal NW attay exhibiting a total of seven wires and the dashed lines bottom of the seven wires indicate the sixfold symmet:ry. (c) Sample parameters used in the packing factors P calculation.

The experimental data were analyzed by using magnetic energy density for a system consisting of n magnetic layers with saturation magnetization M_s, and layer thickness t_i;.The magnetic energy density for the nanoscale multilayer structures the energy per unit surface area can be written as below

$$E=-\begin{pmatrix} \sum_{i=1}^{n} t_i M_s H(\sin\theta_i \sin\theta_H \cos(\phi_i-\phi_H)+\cos\theta_i\cos\theta_H)+ \\ \sum_{i=1}^{n} t_i K_{a,i}\sin^2\phi_i\cos^2\theta_i+\sum_{i=1}^{n} t_i K_{b,i}\sin^4\phi_i\cos^4\theta_i+ \\ \sum_{i=1}^{n-1} A_{i,i\mp1}(\sin\theta_i\sin\theta_{i\mp1}\cos(\phi_{i\mp1}-\phi_i)+\cos\theta_i\cos\theta_{i\mp1})+ \\ \sum_{i=1}^{n-1} B_{i,i\mp1}(\sin\theta_i\sin\theta_{i\mp1}\cos(\phi_{i\mp1}-\phi_i)+\cos\theta_i\cos\theta_{i\mp1})^2 \end{pmatrix} \quad (28)$$

Where, θ_i is the polar angle of the magnetization M_s to the z-axis and ϕ_i is the azimuth angle to the x-axis in the film plane. The first term is the Zeeman energy. The second and third terms correspond to first and second order magnetocrystalline energy with respectively. These energies due to the demagnetization field and any induced perpendicular anisotropy energy. On the other hand, these energies qualitatively have the same angular dependence with respect to the film normal. The second order magnetocrystalline energy term can be neglected for most of the ferromagnetic systems.The last two terms corresponds to bilinear and biquadratic interactions of ferromagnetic layers through nonmagnetic spacer via conduction energies. $A_{i,i\pm1}$ and $B_{i,i\pm1}$ are bilinear and biquadratic coupling constants, respectively. The bilinear exchange interaction can be written from Eq.16. $A_{i,i\pm1}$ can be either negative and positive depending on antiferromagnetic and ferromagnetic interactions, respectively.The antiparallel/perpendicular and parallel alignments of magnetization of nearest neighboring layers are energetically favorable for a negative/positive value of $B_{i,i\pm1}$. Biquadratic interaction for spin systems have been analysed for Ising system in detail (ellen, 1973; Erdem, 2001). The biquadratic term is smaller than the bilinear interaction term. Therefore, it can be neglected for most of the ferromagnetic systems. The indirect exchange energy depends on spacer thickness and even shows oscillatory behavior with spacer thickness (Ruderman, 1954;

Yosida, 1957; Parkin, 1990, 1991(a), 19901(b), 1994). The current literature on single ultrathin films and multilayers is given in below at table (Layadi, 1990(a), 1990(b); Wigen, 1992; Zhang, 1994(a), 1994 (b); Goryunov, 1995; Ando, 1997; Farle; 1998; Platow, 1998;Schmool, 1998; Lindner, 2003;Sklyuyev, 2009; Topkaya, 2010; Erkovan, 2011).

This type exchange-coupling system is located in an external magnetic field, the magnetic moment in each layer.The suitable theoretical expression may be derived in order to deduce magnetic parameter for ac susceptibility. The equation of precession motion for magnetization of the ifll layer in the spherical coordinates with the Bloch-Bloembergen type relaxation term can be written as

$$\frac{1}{\gamma}\frac{d\bar{M}}{dt} = \frac{1}{t_i}\frac{\bar{M}}{M_{i,s}} \times \bar{\nabla}_{M_i} E - \frac{\bar{M}_{\theta i,\varphi i}}{\gamma T_2} - \frac{\bar{M}_{z,i} - M_{i,s}}{\gamma T_1}. \quad (i = 1,\ n) \tag{29}$$

The matrices form for $m_{\theta,i-1}, m_{\phi,i-1}, m_{\phi,i}, m_{\theta,i+1}\ and\, m_{\phi,i+1}$ of each magnetic layers calculated using the Eq.(29) with Eq.(20, 21,28).

$$\begin{pmatrix}
\Omega + \frac{E_{\theta_1\phi_1}}{t_1 M_s \sin\theta_1} & \frac{E_{\phi_1\phi_1}}{t_1 M_s \sin^2\theta_1} & \frac{E_{\theta_2\phi_1}}{t_1 M_s \sin\theta_1} & \frac{E_{\phi_1\phi_2}}{t_1 M_s \sin\theta_1 \sin\theta_2} & \frac{E_{\theta_3\phi_1}}{t_1 M_s \sin\theta_1} & \frac{E_{\phi_1\phi_3}}{t_1 M_s \sin\theta_1 \sin\theta_3} \\
-\frac{E_{\theta_1\theta_1}}{t_1 M_s} & \Omega - \frac{E_{\theta_1\phi_1}}{t_1 M_s \sin\theta_1} & -\frac{E_{\theta_1\theta_2}}{t_1 M_s} & -\frac{E_{\theta_1\phi_2}}{t_1 M_s \sin\theta_2} & -\frac{E_{\theta_1\theta_3}}{t_1 M_s} & -\frac{E_{\theta_1\phi_3}}{t_3 M_s \sin\theta_3} \\
\frac{E_{\theta_1\phi_2}}{t_2 M_s \sin\theta_2} & \frac{E_{\phi_1\phi_2}}{t_2 M_s \sin\theta_1 \sin\theta_2} & \Omega + \frac{E_{\theta_2\phi_2}}{t_2 M_s \sin\theta_2} & \frac{E_{\phi_2\phi_2}}{t_2 M_s \sin^2\theta_2} & \frac{E_{\theta_3\phi_2}}{t_2 M_s \sin\theta_3} & \frac{E_{\phi_2\phi_3}}{t_2 M_s \sin\theta_2 \sin\theta_3} \\
-\frac{E_{\theta_1\theta_2}}{t_2 M_s} & -\frac{E_{\theta_2\phi_1}}{t_2 M_s \sin\theta_1} & -\frac{E_{\theta_2\theta_2}}{t_2 M_s} & \Omega - \frac{E_{\theta_2\phi_2}}{t_2 M_s \sin\theta_2} & -\frac{E_{\theta_2\theta_3}}{t_3 M_s} & -\frac{E_{\theta_2\phi_3}}{t_3 M_s \sin\theta_3} \\
\frac{E_{\theta_1\phi_3}}{t_3 M_s \sin\theta_3} & \frac{E_{\phi_1\phi_3}}{t_3 M_s \sin\theta_1 \sin\theta_3} & \frac{E_{\theta_2\phi_3}}{t_3 M_s \sin\theta_2} & \frac{E_{\phi_2\phi_3}}{t_3 M_s \sin\theta_2 \sin\theta_3} & \Omega + \frac{E_{\theta_3\phi_3}}{t_3 M_s \sin\theta_3} & \frac{E_{\phi_3\phi_3}}{t_3 M_s \sin^2\theta_3} \\
-\frac{E_{\theta_1\theta_3}}{t_3 M_s} & -\frac{E_{\theta_3\phi_1}}{t_3 M_s \sin\theta_3} & -\frac{E_{\theta_2\theta_3}}{t_3 M_s} & -\frac{E_{\theta_3\phi_2}}{t_3 M_s \sin\theta_2} & -\frac{E_{\theta_3\theta_3}}{t_3 M_s} & \Omega - \frac{E_{\theta_3\phi_3}}{t_3 M_s \sin\theta_3}
\end{pmatrix}
\begin{pmatrix} m_{\theta_1} \\ m_{\phi_1} \\ m_{\theta_2} \\ m_{\phi_2} \\ m_{\theta_3} \\ m_{\phi_3} \end{pmatrix} = 0 \tag{30}$$

1, 2, 3 number in this Eq.(30) corresponds to i -1, i and i +1 , respectively. Here

$\Omega = i\left(\frac{\omega}{\gamma}\right) + \frac{1}{\gamma T_2}$ Then dispersion relation for ferromagnetic exchange-coupled n-layers has been calculated using the (2nx2n) matrix on the left-hand side of Eq. (30) in below in detail.

$$\left(\frac{\omega}{\gamma}\right)^{2n} + C_{(2n-2)/2}\left(\frac{\omega}{\gamma}\right)^{2n-2} + \ldots + C_1\left(\frac{\omega}{\gamma}\right)^{2} + C_0 = 0 \tag{31}$$

Here, n is the number of ferromagnetic layer. $C_{0,}$ C_1 ⋯ etc. are constant related to $t_i, M_s, E_{\theta i\theta i}, E_{\varphi i\varphi i}, E_{\theta i\varphi i}$ · $\sin\theta_i$ and $\sin^2\theta_i$ .The dispersion relations for monolayer, trilayers, five-layers obtained from the Eq.(31). For tri-layers detail information are seen in ref. (Zhang, 1994(a); Sclunool, 1998; Lindner, 2003). It is given that the dispersion relation for monolayer, trilayers, five-layers and multilayers/ n-layers in Fig. 7.

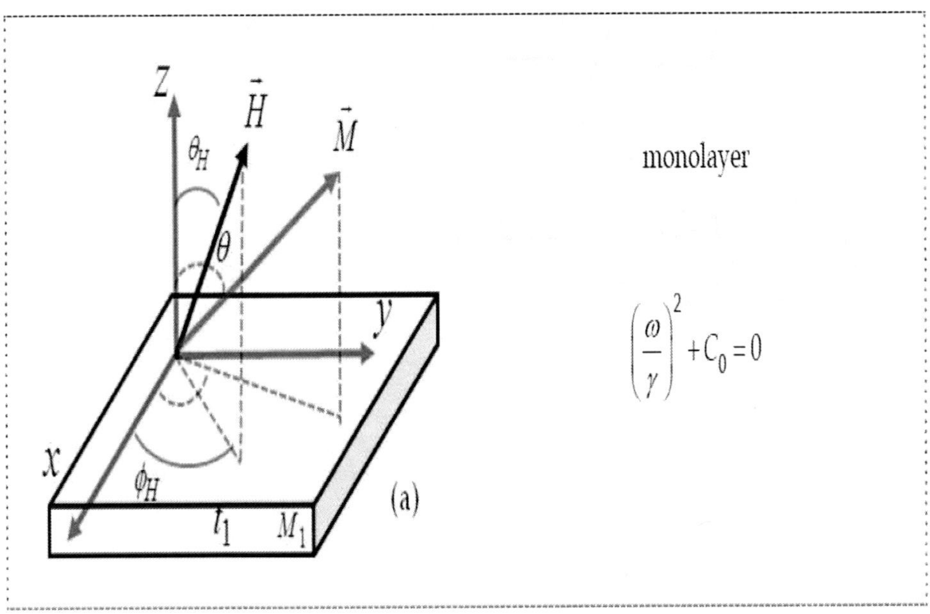

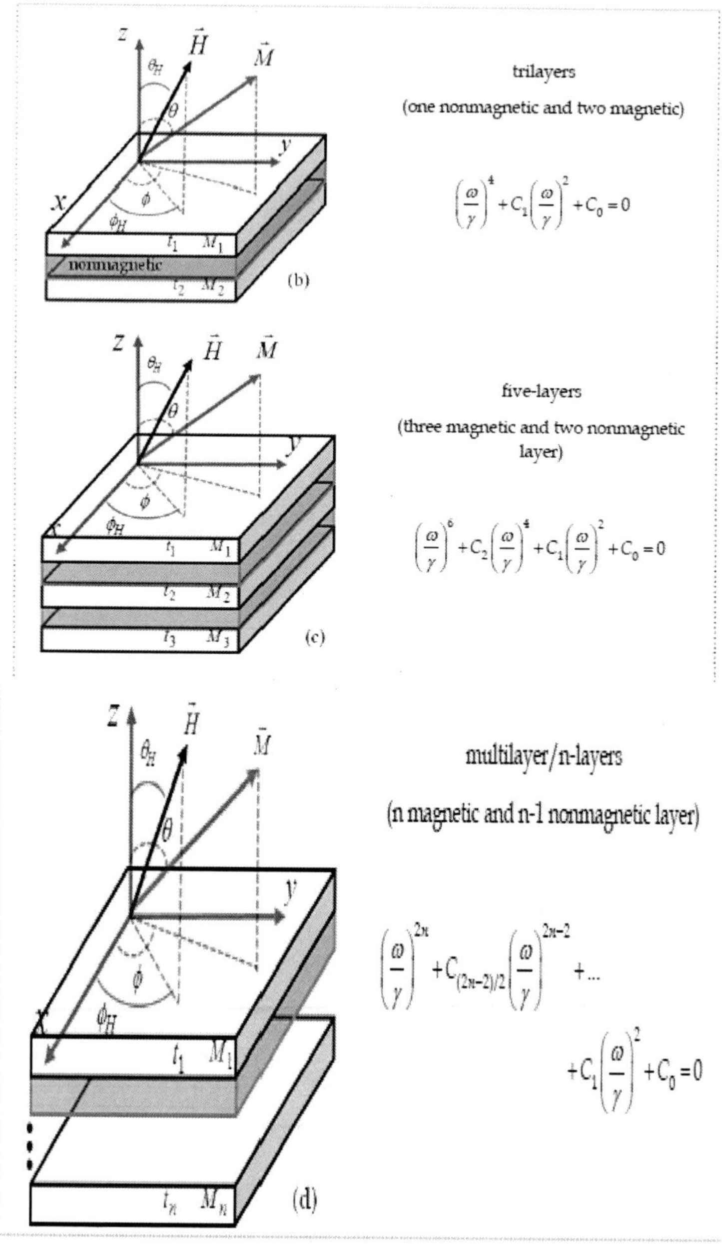

Figure 7. Schematic representation of the (a) one layer, (b) three layer, (c) five layer and (d) n magnetic layer and their relative orientation of the equilibrium magnetization $\vec{M}$ and the de component of the external magnetic field $\vec{H}$ for the FMR experiments and their theoretical calculations.

Example:exchange Spring (Hard/Soft) Behaviour

The Bloch wall, Neelline and magnetization vortex are well known properties for magnetic domain in magnetic systems. The multilayer structures are ordered layer by layer. The best of the sample for multilayer structure are exchange-spring systems. The equilibrium magnetic properties of nano-structured exchange-spring magnets may be studied in detail for some selected magnetic systems. The exchange systems are oriented from the exchange coupling between ferromagnetic and antiferromagnetic fims or between two ferromagnetic fims. This type structure has been extensively studied since the phenomenon was discovered (Meiklejohn, 1956, 1957). Kneller and Hawing have been used firstly the "exchange-spring" expression (Kneller, 1991). Spring magnet £lms consist of hard and soft layers that are coupled at the interfaces due to strong exchange coupling between relatively soft and hard layers. The soft magnet provides a high magnetic saturation, whereas the magnetically hard material provides a high coercive £eld. Skomski and Coey explored the theory of exchanged coupled £lms and predicted that a huge energy about three times of commercially available permanent magnets (120 MGOe) can be induced (Skomski, 1993; Coey, 1997). The magnetic reversal proceeds via a twisting of the magnetization only in the soft layer after saturating hard layers, if a reverse magnetic £eld that is higher than exchange £eld is applied. The spins are sufficiently closed to the interface are pinned by the hard layer, while those in deep region of soft layer rotate up to some extent to follow the applied £eld (Szlaferek, 2004).To be more specific, the angle of the rotation depends on the distance to the hard layer. That is the angle of rotating in a spiral spin structure similar to that of a Bloch domain wall. If the applied field is removed, the soft spins rotate back into alignment with the hard layer.

The general expression of the free energy for exchange interaction spring materials at film $\left(\theta i_{,i\pm1} = \pi / 2 \, and \, \theta_H = \pi / 2\right)$ plane in spherical coordinate system as below.

$$E=-\sum_{i=1}^{N}\bar{H}\cdot\bar{M}_i-\sum_{i=1}^{N}K_{a,i}\cos^2\phi_i-\sum_{i=1}^{N}K_{b,i}\cos^4\phi_i$$
$$-\sum_{i=1}^{N-1}\frac{A_{i,i\mp1}}{t^2}\cos(\phi_{i\mp1}-\phi_i)-\sum_{i=1}^{N-1}\frac{B_{i,i\mp1}}{t^2}\cos(\phi_{i\mp1}-\phi_i)^2 \tag{32}$$

The expression is obtained as following using $\phi_i \rightarrow \phi i$ for magnetization's equilibrium orientations of each layer at a state of equilibrium under the external magnetic field.

$$\tan\phi_i' = \frac{HM_i t^2\sin\phi_H + A_{i,i\mp1}\sin\phi_{i\mp1} + 2B_{i,i\mp1}\sin\phi_{i\mp1}\cos(\phi_i-\phi_{i\mp1})}{HM_i t^2\cos\phi_H + 2t^2K_{a,i}\cos\phi_i + 4t^2K_{b,i}\cos^3\phi_i + A_{i,i\mp1}\cos\phi_{i\mp1} + 2B_{i,i\mp1}\cos\phi_{i\mp1}\cos(\phi_i-\phi_{i\mp1})} \tag{33}$$

In this example, second-order anisotropy term $\left(K_{b,i}=0\right)$ and biquadratic interaction constant $\left(B_{i,\pm1}=0\right)$ considered and the result obtained show as following as adapted with spring magnets SmCo(hard)/Fe(soft). For theoretical analysis, the exchange-spring magnet SmCo/Fe is divided into subatomic multi-layers $\left(d=2\mathring{A}\right)$, and the spins in each layer are characterized by the average magnetization M_i, and the uniaxial anisotropy constant K_i (Fig.8).

Sublayers are coupled by an exchange constant $A_{i,\pm1}$ (Astalos, 1998; Fullerton, 1998, 1999; Jiang, 1999, 2002, 2005; Grimsditch, 1999; Scholz, 2000; Hellwig, 2000; Pollmann, 2001; Dumesnil, 2002). is the angle formed by the magnetization of the ith plane with the in plane (where the external field is always perpendicular to the £1m normal) easy axis of the hard layer (Y1lchz, 2004(a), 2004(b)).

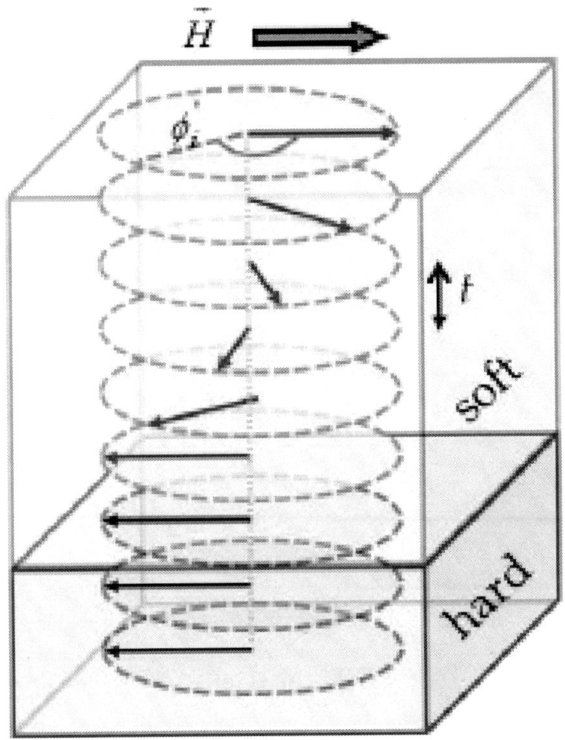

Figure 8 Schematic illustration of phases of exchange spring magnets.

The FMR spectra for exchange-spring magnet of SmCo/Fe have been analyzed using the Eqs. (26, 27, and 33) in Fig.9. Sm-$(200\,\mathring{A})/Fe\left(200\,\mathring{A}\; and\, 100\,\mathring{A}\right)$ bilayers have been grown on epitaxial 200 A Cr(211) buffer layer on single crystal MgO(llO) substrates by magnetron sputtering technique (Wiichner, 1997). To prevent oxidation Sm-Co/Fe film was coated with a 100 A thick Cr layer. The FMR spectra for exchange-spring magnets of 200 A and 100 A Fe samples for different angles of the applied magnetic £eld in the film plane are presented in Fig.9.

There are three peaks that are one of them corresponds to the bulk mode and the remaining to the surface modes for 200 A Fe sample. For more information about the FMR studies exchange spring magnets look at the ref. (Yildiz, 2004(a), 2004(b)). Exchange-spring coupled magnets are promising systems for applications in perpendicular magnetic data recording storage devices and permanent magnet (Schre£1, 1993(a), 1993(b),1998, 2002; Mibu, 1997, 1998).

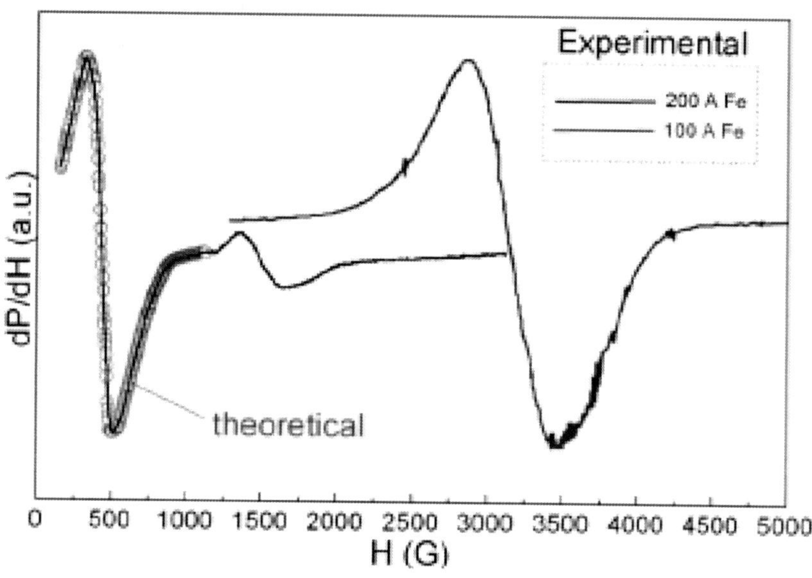

Figure 9 FMR spectra for SmCo $(200\overset{\circ}{A})/Fe\,200\overset{\circ}{A}$ (blackline) and SmCo $(200\overset{\circ}{A})/Fe\,200\overset{\circ}{A}$ (blue line) samples.These FMR spectra origmated from the iron,!soft layers.

SUPERPARAMAGNETIC RESONANCE

Magnetic nanoparticles have been steadily interested in science and nanotechnology. As the dimensions of magnetic nanoparticles decrease to the nanometer scale, these nanoparticles start to exhibit new and interesting physical properties mainly due to quantum size effects (Yal9Jl, 2004(a), 2008(b), 2012). A single domain particle is commonly referred to as superparamagnetic (Held, 2001; Diaz, 2002; Fonseca, 2002). The superparamagnetic/single domain nanoparticles are important for non surgical interfere of human body. Even the intrinsic physical characteristics of nanoparticles are observed to change drastically compared to their macroscopic counterparts. Stoner-Wohlfarth (Stoner, 1948) and Heisenberg model (Heisenberg, 1928) to describe the fine structure were firstly used in detail. A simple (Bakuzis, 2004) and the first atomic-scale models of the ferrimagnetic and heterogeneous systems in which the exchange energy plays a central role in determining the magnetization of the NPs, were studied (Kodama, 1996, 1999; Kodama & Berkowitz, 1999). Superparamagnetic resonance (SPR) studies of fine magnetic nanoparticles is calculated a correlation between the line-width and the resonance field for superparamagnetic structures (Berger, 1997, 1998, 2000(a), 2000(b), 2001; Kliava, 1999). The correlation of the line-width and the resonance field is calculated from Bloch-Bloembergen equation of motion for magnetization. The SPR spectra, line width and resonance field may be analyzed by using the Eq.(34) in below. The equation of motion for magnetization with Bloch-Bloembergen type relaxation term for FMR adapted for superparamagnetic structures from Eqs.(18) and (19) in below.

$$\chi_2(H) = \frac{1}{\pi} \frac{\Delta_H(\Delta_H^2 + H^2 + H_r^2)}{\left(\Delta_H^2 + (H - H_r)^2\right)\cdot\left(\Delta_H^2 + (H + H_r)^2\right)} \tag{34}$$

Here, $\Delta_H = 1/\gamma T_2$, $H_r = -(\omega/\gamma)$. This equation for SPR system so called modified Bloch for fine particle magnets. The SPR microwave absorption is proportional to the imaginary part of the dynamic susceptibility. The line shape and resonance field for superparamagnet is obtained. The temperature evolution for the SPR line-width for nanoparticles can be calculated by $\Delta_H = \Delta_T L(x)$. In this expression Δ_T is a saturation line-width at a temperature T $L(x) = \coth(x) - (1/x)$ is the Langevin function with $x = MVH_{eff}/k_B T$ V is the particle volume. The superparamagnetic (Chastellain, 2004; Dormer, 2005; Hamoudeh, 2007), core shell nanoparticles and nanocrystalline nanoparticles (Woods, 2001; Wiekhorst, 2003; Tartaj, 2004) have been performed for possible biological applications (Sun,2005; Zhang,

2008). In additional, superparamagnetic nanoparticles have been used for hydrogels, memory effects and electronic devices (Raikher, 2003;Sasaki, 2005; Heim, 2007).

RESULT AND DISCUSSIONS

The EPR, FMR and SPR signals have been observed in Fig.lO. The EPR signal has reached approaching peak level about 3000 Gas seeing at Fig.lO. It's symmetric and line width are narrower than resonance field, in generally.If EPR samples show crystallization, resonance field value starts to change. The EPR signal can be observed at lower temperature about 3000 G and the signal can show crystalline property.The signal is observed in two different areas at FMR spectra as the magnetic field is parallel and perpendicular to the film. The FMR spectra are observed at low field when the magnetic field is parallel to the film, in generally. On the other hand, the FMR spectra are observed at highest field when the magnetic field is perpendicular to the film. For other conditions FMR signals are observed between these two conditions for thin films. FMR spectra can be seen a wide range of field so as to the thin films are full.

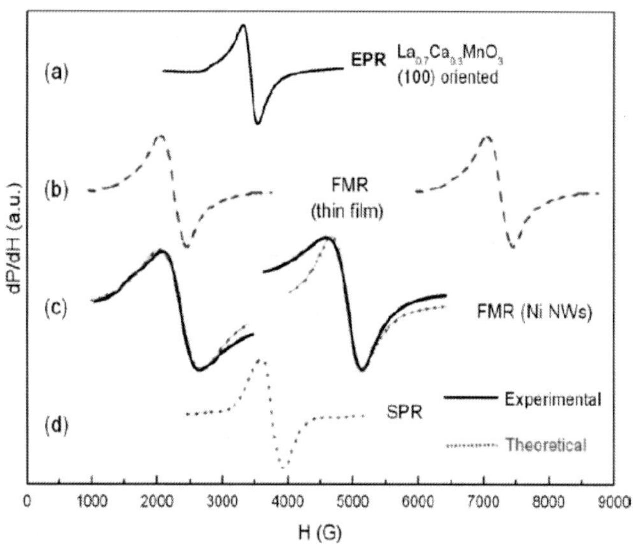

Figure 10 (a) The EPR/ESR experimental signal fot $L_{ao.7}C_{ao.3}$.MnO. samples at room tempetatute (see, Kattopu,2011 (b)).(b) Theoretical FMR spectra calculated from Eq. (26) with Eq.(27) at parallel (θ= 90°

;- 2000G) and perpendicular (θ = 0°;-7000G) position of OPG case.(c) The theoretical (ted dot line in online) and experimental FMR spectra for NiNWs $(P = 29,6; L = 0,8\,\mu m, \tau = 13)$ (see fot detail, Kattopu,2011 (a)).(d) The theotetical SPR signal for supetpatamagnet by Eq.(34) at room tempetatute.

FMR spectra are similar to thin films at nanowire samples.In case of occupancy rate is that as the theoretical P <33% for Nickel (Ni) it behaviors like thin £1m. But, in case of occupancy rate is that as the theoretical P < 33% it behaviors different from thin film. This situation is clearly visible from $H_{eff} = 2\pi M_s(1-3P)+(2K_u/M_s)$. If the occupancy rate is P <33% sample's signals show the opposite behavior according to thin £1m FJviR signals. Look at for more information (Kartopu, 2011(a)). This is perceived as changes the direction of the easy axis. The changes of easy axes depend on magnetization (Terry, 1917) and porosity (Kartopu, 2011 (a)) for magnetic materials/transition elements. The SPR signal is similar to EPR signal.SPR peak may show symmetrical properties both at room temperature and low temperatures.The SPR signal is in the form of Lorentzian and Gaussian line shapes at all temperature range. Specially prepared nanoparticles SPR peak exhibit shift in symmetry.The line width of SPR peak expands at low temperature.

AUTHOR DETAILS

OrhanYal m

Nigde University, Nigde,Turkey

ACKNOWLEDGEMENT

I would like to thank MuhiHin Oztiirk and SongiilOziim of Nigde University for valuable discussions an the critical reading of the chapter. This study was supported by Research found (Grant No.FEB2012/12) of Nigde University.

REFERENCES

1. Adeyeye, A.O.; Bland, J.A.C.; Daboo, C.; Hasko, D.G. (1997). Magnetostatic interactions and magnetization reversal in ferromagnetic wires. Phys. Rev. B. Vol. 56, pp.3265.
2. Akta§, B. (1993). Clear evidence for field induced unidirectional exchange surface anisotropy in NiMn alloys. Solid State Commun.Vol.87, pp. 1067.
3. Akta§, B.; Ozdemir, M. (1994). Simulated spin wave resonance absorption curves for ferromagnetic thin £lms and application to NiMn films. Physica B. Vol. 119, pp. 125. Akta§, B. (1997). FJviR properties of epitaxial F€204 films on Mg0(100). Thin Solid Films. Vol. 307, pp.250.
4. Akta§, B.; Ozdemir, M.; Yilgin, R.; Oner, Y.; Sato, T.; Ando, T. (2001). Thickness and temperature dependence of magnetic anisotropies of Ni77 Mn22 £1ms. Physica B. Vol.305, pp.298.
5. Akta§, B.; Yildiz, F.; Rameev, B.; I<haibullin, R; Tagirov, L.;Ozdemir, M. (2004). Giant room temperature ferromagnetism in rutile Ti02 implanted by Co. Phys. stat. sol. (c). Vol. 12, pp.3319.
6. Akta§, B.; Tagirov, L. & Mikailov, F. (October, 2006). Magnetic Nanostructures,Springer Series in materials science, Vol.94, ISBN 978-3-540-49334-1.
7. Ament, W. S.; Rado, G. T. (1955). Electromagnetic effects of spin wave resonance in ferromagnetic metals. Phys. Reu.Vol.97, pp. 1558.
8. Ando, Y.; Koizumi, H.; Miyazaki, T. (1997). Exchange coupling energy determined by ferromagnetic resonance in 80 Ni-Fe/Cu multilayer films. J. Magn. Magn. Mater. Vol. 166, pp. 75.
9. Anisimov, A. N.; Farle, M.; Poulopoulos, P.; Platow, W.; Baberschke, K; Isberg, P.; Wappling, R;Niklasson, A.M. N.; Eriksson, 0.(1999).Orbital magnetism and magnetic anisotropy probed with ferromagnetic resonance. Phys. Rev. Lett.Vol.82, pp.2390.
10. An, S. Y.; Krivosik, P.; Kraemer, M.A.; Olson, H. M.; Nazarov, A. V.; Patton, C. E. (2004).
11. High power ferromagnetic resonance and spin wave instability processes in permalloy thin films.f. Appl. Phys. Vol.96, pp. 1572.
12. Anderson, P. W. (1953). Exchange narrowing in paramagnetic resonance. Reu. Mod. Phy. Vol. 25, pp.269.
13. Arias, R.; Mills, D.L. (1999). Extrinsic contributions to the ferromagnetic resonance response in ultrathin films, Phys. Reu. B. Vol. 60, pp. 7395.
14. Arias, R.; Mills, D. L. (2003). Theory of collective spin waves and microwave response of ferromagnetic nanowire arrays. Phys. Rev. B. Vol. 67, pp.094423.
15. Arkad'ev, V.K. (1913). The Reflection o£ Electric Waves from a Wire, Sro. Phys.-JETP, Vol. 45A, issue 45, pp.312.
16. Artman, J. 0. (1957). Ferromagnetic resonance in metal single crystals. Phys. Rev. Vol. 105, pp.74.

17. Artman, J. 0. (1979). Domain mode FMR in materials with Kt and Ku J. Appl. Phys.Vol. 50, pp.2024.

18. Astalos, R J.; Carnley, R E. (1998). Magnetic permeability for exchange-spring magnets: application to Fe/Sm-Co. Phys. Reu. B. Vol. 58, pp.8646.

19. Aswal, D.K.;Singh, A.; Kadam, R.M.; Bhide, M.K.; Page, A.G.; Bhattacharya, S.; Gupta,S.K; Yakhmi, J.V.; Sahni, V.C. (2005). Ferromagnetic resonance studies of nanocrystalline Lao.6Pbo.4Mn0athlnfilms. Mater. Lett. Vol.59, pp. 728.

20. Azzerboni,B.;Asti,G.;Pareti,L.;Ghldini,M.(2006).Magneticnanostructuresin modem technology, spintronics, magnetic MEMS and recording. Proceedings of the NATO advanced study institute on magnetic nanostructures for micro-electromechanical systems and spintronic applications catona, Published by Springer. Italy. ISBN 978-1- 4020-6337-4.

21. Backes, D.; Bedau, D., Liu, H., Langer, J.; Kent, A.D. (2012). Characterization of interlayer interactions in magnetic random access memory layer stacks using ferromagnetic resonance.f. Appl. Phys.Vol. 111, pp.07C721.

22. Baek, J.S.; Min, S.G.; Yu, S.C.; Lim, W.Y. (2002). Ferromagnetic resonance of Fe--Sm-0 thin fihns.J. App. Phys.Vol.93, pp. 7604.

23. Bai, Y .;Xu, F.;Qiao, L. (2012). The twice ferromagnetic resonance in hexagonal ferrite single rod and paired rods. Phys. Lett. A.Vol.376, pp.563.

24. Bakker, F.L.; Flipse, J.; Slachter, A.; Wagenaar, D.; van Wees, B.J. (2012). Thermoelectric detection of ferromagnetic resonance o£ a nanoscale ferromagnet. Phys. Reu. Lett. Vol. 108, pp. 167602.

25. Bakuzis, A.F.; Morais, P.C. (2004). Magnetic nanoparticle systems: an Ising model approximation.f. Magn. Magn. Mater.Vol. 272-276, pp.e1161.

26. Baseglia, L.; Warden, M.; Waldner, F.; Hutton, S.L.; Drumheller, J.E.; He, Y.Q.; Wigen, P.E.; Marysko M. (1988). Derivation o£ the resonance frequency from the free energy of ferromagnets.Phys.Rev.B. Vol.38, pp.2237.

27. Beguhn, S.; Zhou, Z.; Rand, S.; Yang, X.; Lou, J.; Sun, N.X. (2012). A new highly sensitive broadband ferromagnetic resonance measurement system with lock-in detection. f. Appl. Phys.Vol. 111, pp.07A503.

28. Berger, R; Bissey, J.; Kliava, J. (2000(a)). Lineshapes in magnetic resonance spectra.f. Phys.: Ccndens. Matter. Vol. 12, pp.9347.

29. Berger, R.; Kliava, J.; Bissey, J.C. (2000(b)). Magnetic resonance o£ superparamagnetic iron containing nanoparticles in annealed glass.f. Appl. Phys.Vol.87, pp. 7389.

30. Berger, R.; Bissey, J.C.; Kliava, J.; Soulard, B. (1997). Superparamagnetic resonance of ferric ions in devitri£eld borate glass.f. Magn. Magn. Mater.Vol. 167, pp. 129.

31. Berger, R; Bissey, J. C.; Kliava, J.; Daubric, H.; Estournes, C. (2001). Temperature dependence of superparamagnetic resonance o£ iron oxide nanoparticles. f. Magn. Magn. Mater.Vol. 234, pp.535.

32. Berger, R; Kliava, J.; Bissey, J. C.; Ba1ettoz, V. (1998). Superparamagnetic

resonance of annealed iron containing borate glass.f. Phys.: Condens. Matter. Vol. 10, pp.8559.

33. Birkhauser Verlag, A.G. (2007).Spin glasses statics and dynamics. Basel, Boston,Berlin.ISBN 978-3-7643-8999-4.

34. Birlikseven, C.; Topacli, C.; Durusoy, H.Z.; Tagirov, L.R; Koymen, A.R.; Akta§, B. (1999 (a)). Magnetoresistance, magnetization and FMR study of Fe/Ag/Co multilayer fihn.f. Magn. Magn. Mater. Vol. 192, pp.258.

35. Birlikseven, C.; Topacli, C.; Durusoy, H.Z.; Tagirov, L.R.; Koymen, A.R.; Akta§, B. (1999 (b)). Layer-sensitive magnetization, magnetoresistance and ferromagnetic resonance (FMR) study of NiFe/Ag/CoNi trilayer £hn.f. Magn. Magn. Mater.Vol.202, pp.342.

36. Bhagat, S.M.; Anderson, J.R.; Wu, N. (1967). Influence o£ the anomalous skin effect on the ferromagnetic resonance linewidth in iron. Phys. Rev.Vol. 155, pp.510.

37. Bhagat, S.M.; Lubitz, P. (1974). Temperature variation of ferromagnetic relaxation in the 3d transition metals. Phys. Rev.B. Vol.10, pp. 179.

38. Bland, J.A.C.; Heinrich, B. (1994). Ultrathin magnetic structures I: An introduction to the electronic magnetic and structural properties. Springer-Verlag Berlin Heidelberg.ISBN 3- 540-57407-7.

39. Bloembergen, B. (1950). On the ferromagnetic resonance in nickel and supermalloy. Phys. Rev.Vol. 78, pp.572.

40. Bloembergen, N.; Wang, S. (1954). Relaxation effects in para- and ferromagnetic resonace. Phys. Rev.Vol.93, pp. 72.

41. Borchers, J.A.; Dura, J.A.; Unguris, J.; Tulchinsky, D.; Kelley, M.H.; Majkrzak, C.F. (1998). Observation of antiparallel magnetic order in weakly coupled CoyCu multilayers. Phys. Rev. Lett. Vol.82, pp.2796.

42. Bose, T.; Trimper,S. (2012). Nonlocal feedback in ferromagnetic resonance. Arxiv: 1204-5342. Ccnd-mat. Mes-hall. pp. 1.

43. Boulle, 0.; Malinowski, G.; I<laui, M. (2011). Current-induced domain wall motion in nanoscale ferromagnetic elements. Materials Science and Engineering R. Vol. 72, pp. 159.

44. Brown, F.M. (1962).Magnetostatic principles in ferromagnetism. North Holland Publishing Ccmpany.

45. Brustolon, M.; Giamello, E. (2009). Electron paramagnetic resonance. john Wiley & Sons, Inc. LSBN 978..()-470-25882.

46. Budak,S.; Yildiz, F.;Ozdemir, M.; Akta§, B. (2003). Electron spin resonance studies on single czystalline F€204 fihns.J. Magn. Magn. Mater. Vol. 258-259, pp.423.

47. Buschow, K.H.J.; de Boer, F.R (2004).Physics of magnetism and magnetic materials. Kluwer Academic Publishers.ISBN:0-306-47421-2.

48. Callen, H.B. (1958). A ferromagnetic dynamical equation. f. Phys. Chem. Solids. Vol. 4, pp. 256.

49. Can, M.M.; Co§kun, M.; Fuat, T. (2012). A comparative study of nanosized iron oxide particles; magnetite (Fea04), maghemite (y-Fe20a) and hematite (a-Fe20a), using ferromagnetic resonance.f. Alloy. Compd. Vol.542, pp.241.

50. Celinski, Z.; Urquhart, B.; Heinrich, B. (1997). Using ferromagmetic resonance to measure the magnetic moments of ultrathin films, f. Magn. Magn. Mater. Vol. 166, pp. 6.

51. Celinski, Z.; Heinrich, B. (1991). Ferromagnetic resonance line width of Fe ultrathin fihns grown on a bee substrate.f. Appl. Phys.Vol. 70, pp.5936.

52. Chappert, C.; Le Dang, K; Beauvillain, P.; Hurdequint, H.; Renard, D. (1986). Ferromagnetic resonance studies of very thin cobalt films on a gold substrate. Phys. Reu. B. Vol.34, pp. 3192.

53. Chastellain, M.; Petri, A.; Hofmann, H. (2004). Particle size investigations of a multi-step synthesis of PVA coated superparamagnetic nanoparticles.J. Colloid. interf Sci. Vol.278, pp.353.

54. Chen, Y.S.; Cheng, C.W.; Chern, G.; Wu, W.F.; Lin, J.G. (2012). Ferromagnetic resonance probed annealing effects on magnetic anisotropy of perpendicular CoFeB/MgO bilayer. J. Appl. Phys.Vol. 111, pp.07C101.

55. Chen, S.H.; Chang, C.R.; Xiao, J.Q.; Nikolic, KB. (2009). Spin and charge pumping in magnetic tunnel junctions with precessing magnetization: A nonequilibrium green function approach. Phys. Rev. B. Vol. 79, pp.054424.

56. Chen, H.H.; Levy, P.M. (1973). Dipole and quadrupole phase transitions in spin-1 models. Phys. Rev. B. Vol. 7, 4267.

57. Chikazumi,S. (1997). Physics of ferromagnetism. Oxford University Press. ISBN 0-19-851776- 9.

58. Cochran, J.F. (1995).Light scattering from ultrathin magnetic layers and bilayers in magnetic ultrathin films. Heinrich, B.; Bland, J.A.C. (Eds.) Springer, Berlin, Heidelberg Vol. II, pp. 222.B.Hillebrands:Brillouin light scattering in magnetic superlattices.ibid pp.258.

59. Cochran, J.F.; Kambersky, V. (2006). Ferromagnetic resonance in very thin films. f. Magn. Magn. Mater. Vol. 302, pp.348.

60. Coey, J.M.D. (1997).Permanent magnetism. Solid State Ccmm.Vol. 102, pp. 101.

61. Coey, J.M.D. (2009).Magnetism and magnetic materials. Cambridge University Press.LSBN-13 978052181614-4.

62. Cofield, M.L.; Glocker, D.; Gau, J.S. (1987). Spin-wave resonance in CoCr magnetic thin films. J. Appl. Phys. Vol. 61, pp.3810.

63. Co§kun, R.; Okutan, M.; Yal m, 0.; Kosemen, A. (2012). Electric and magnetic properties of hydrogels doped with Cu ions. Acta.Phys.Pol. A Vol. 122, pp. 683.

64. Crittenden, E.C.Jr; Hoffman, R.W. (1953). Thin films of ferromagnetic materials. Reu. Mod. Phys.Vol. 25, pp.310.

65. Cullity, B.D.;Graham, C.D. (1990).Introduction to Magnetic Materials.Wiley. New York. pp. 199.

66. Cullity, B.D.; Graham, C.D. (2009). Introduction to magnetic materials 2M Edition.

67. Damon, R.W. (1953). Relaxation effect in the ferromagnetic resonance. Rev. Mod. Phys. Vol. 25, pp.239.

68. Damon, R.W. (1963). Ferromagneticre sonance at high power in Rado, G.T.; Suhl, H. (Eds.). Magnetism. Vol.I da Silva, E.C.; Meckenstock, R; von Geisau, 0.; Kordecki, R.; Pelzl, J.; Wolf, J.A; Grunberg, P. (1993). Ferromagnetic resonance investigations of anisotropy fields of Fe(OOl) epitaxial layers.f. Magn. Magn. Mater. Vol. 121, pp.528.

69. Davis, J.A (1965). Effect of surface pinning on the magnetization of thin films. f. Appl. Phys. Vol. 36, pp.3520.

70. Demokritov, S.; Rucker, U.; Griinberg, P. (1996). Enhancement of the Curie temperature of epitaxial EuS(100) films caused by growth dislocations.f. Magn. Magn. Mater. Vol. 163, PP ·21.

71. Demokritov,S.;Rucker,U.;Arons,R.R.;Grunberg,P.(1997).Antiferromagnetic interlayer coupling in epitaxial Fe/EuS (100) bilayers.J. Appl. Phys. Vol. 81, pp.5348.

72. De Warnes, RE.; Wolfram, T. (1970). Dipole-exchange spinwaves in ferromagnetic films. f. Appl. Phys. Vol. 41, pp.87.

73. Demand, M.; Encinas-Oropesa, A; Kenane, S.; Ebels, U.; Huynen, I.; Pirax, L. (2002). Ferromagnetic resonance studies of nickel and permalloy nanowire arrays. f. Magn. Magn. Mater.Vol. 249, pp.228.

74. De Biasi, E.; Lima, Jr. E.; Ramos, C.A; Butera, A.; Zysler, R.D. (2013). Effect of thermal .fluctuations in FMR experiments in uniaxial magnetic nanoparticles: Blocked vs. superparamagnetic regimes.f. Magn. Magn. Mater. Vol.326, pp. 138.

75. De Biasi, R.S.; Gondim, E.C. (2006). Use of ferromagnetic resonance to determine the size distribution of y-Fe20a nanoparticles.Solid State Commun. Vol. 138, pp.271.

76. De Cos, D.; Arribas, AG.; Barandiaran, J.M. (2006). Ferromagnetic resonance in gigahertz magneto-impedance of multilayer systems.f. Magn. Magn. Mater. Vol. 304, pp.218.

77. Diaz, L.L.; Torres, L.; Mora, E. (2002). Transition from ferromagnetism to superparamagnetism on the nanosecond time scale. Phys. Reu. B. Vol. 65, pp.224406.

78. Dillon, J.F.Jr.; Gyorgy, E.M.; Rupp, L.W.Jr.; Yafet, Y.; Testardi,L.R. (1981). Ferromagnetic resonance in compositionally modulated CuNi films.f. Appl. Phys. Vol.52, pp.2256. Ding, J.; Kostylev, M.; Adeyeye, AO. (2012). Broadband ferromagnetic resonance spectroscopy of permalloy triangular nanorings.f. Appl. Phys.Vol. 100, pp.062401.

79. Dormer, K;Seeney, C.; Lewelling, K.; Lian, G.; Gibson, D.; Johnson, M. (2005). Epithelial internalization of superparamagnetic nanoparticles and response to external magnetic field. Biomaterials. Vol. 26, pp.2061.

80. Dreher, L.; Weiler, M.; Pempeintner, M.; Huebl, H.; Gross, R.; Brandt, M.S.;

Goennenwein, S.T.B. (2012). Surface acoustic wave-driven ferromagnetic resonance in nickel thin films: Theory and experiment. Arxiv:Cond-mat. Mes-hall. Vol. 1208, pp.1.

81. Dubowik, J. (1996). Shape anisotropy of magnetic heterostructures. Phys. Rev. B. Vol. 54, pp.1088.

82. Dumesnil, K.; Dufour, C.; Mangin, Ph.; Rogalev, A. (2002). Magnetic springs in exchange coupled DyFe2/YFe2 superlattices: An element-selective x-ray magnetic circular dichroism study. Phys. Rev. B. Vol. 65, pp.094401.

83. Duraia, E.M.; Abdullin, I<h.A. (2009). Ferromagnetic resonance of cobalt nanoparticles used as a catalyst for the carbon nanotubes synthesis.f. Magn. Magn. Mater. Vol. 321, pp. 69.

84. Durusoy, H.Z.; Akta§, B.; Yilgin, R.; Terada, N.; Ichikawa, M.; Kaneda, T.; Tagirov, L.R (2000). New technique for measuring the microwave penetration depth in high- Tc superconducting thin films. Phys. B. Vol.284-288, pp.953.

85. Encinas-Oropesa, A.; Demand, M.; Piraux, L.; Huynen, I.; Ebels, U. (2001). Dipolar interactions in arrays of nickel nanowires studied by ferromagnetic resonance. Phys. Rev. B. Vol. 63, pp. 104415.

86. Erdem, R.; I<eskin, M. (2001). Dynamics of a spin-1 Ising system in the neighborhood of equilibrium states, Phys. Rev. E.Vol. 64, pp.026102.

87. Erkovan, M.; Oztiirk, S.T.; Topkaya, R; Ozdemir, M.; Akta§, B.; Oztiirk, 0. (2011).

88. Ferromagnetic resonance investigation of Py/Cr multilayer system. f. Appl. Phys. Vol. 110, pp.023908.

89. Estevez, D.C.; Betancourt, I.;Montiel, H. (2012). Magnetization dynamics and ferromagnetic resonance behavior of melt spun FeBSiGe amorphous alloys.f. Appl. Phys.Vol. 112, pp. 053923.

90. Fan, W.J.; Qiu, X.P.; Shi, Z.; Zhou, S.M.; Cheng, Z.H. (2010). Correlation between isotropic ferromagnetic resonance field shift and rotatable anisotropy in polycrystalline NiFe/FeMn bilayers. Thin Solid Films. Vol. 518, pp.2175.

91. Farle, M. (1998). Ferromagnetic resonance of ultrathin metallic layers. Rep. Frog. Phys. Vol. 61, pp. 755.

92. Farle, M.; Lindner, J.; Baberschke, K. (2000).Ferromagnetic resonance of Ni(111) on Re(0001). f. Magn. Magn. Mater. Vol.212, pp. 301.

93. Fermin, J.R.; Azevedo, A.; Aguiar, F.M.; Li, B.; Rezende, S.M. (1999). Ferromagnetic resonance linewidth and anisotropy dispersion in thin Fe films. f. Appl. Phys. Vol. 85, pp. 7316.

94. Fischer, H.; Mastrogiacomo, G.; Lo££ler, J.F.; Warthmann, R.J.; Weidler, P.G.; Gehring, A.U. (2008). Ferromagnetic resonance and magnetic characteristics of intact magnetosome chains in Magnetospirillum gryphiswaldense. Earth and Planetary Science Letters. Vol. 270, pp.200.

95. Fletcher, RC.; Le Craw, R.C.; Spencer, E.G. (1960). Electron spin relaxation in ferromagnetic insulators. Phys. Rev. Vol. 117, pp.955.

96. Fonseca, F.C.; Goya, G.F.; Jardim, RF.; Muccillo, R.; Carreno, N.L.V.; Longo,

E.; Leite, E.R (2002). Superparamagnetism and magnetic properties of Ni nanoparticles embedded in SiOz. Phys. Reu. B. Vol. 66, pp.104406.

97. Frait, Z.; Macfaden, H. (1965). Ferromagnetic resonance in metals frequency dependence. Phys. Rev. Vol.139, pp.A1173.

98. Frait, Z.; Fraitova, D.; Zarubova, N. (1985). Observation of FMR surface spin wave modes in bulk amorphous ferromagnets.Phys. Stat. Sol. (b). Vol. 128, pp.219.

99. Frait, Z.; Fraitova, D. (1998).Low energy spinwave excitation in highly conducting thin films and surfaces, in Frontiers in Magnetism of Reduced Dimension Systems. Nato AS!Series, Bar'yakhtar, V.G.; Wigen, P.E.; Lesnik, N.A. (Eds.) (Kluwer, Dordrecht) pp. 121.

100. Frait, Z.; Macfaden, H. (1965). Ferromagnetic resonance in metals frequency dependence. Phys. Rev.Vol. 139, pp. 1173.

101. Fraitova, D. (1983(a)). An analytical theory of FMR in bulk metals, I. Dispersion relations. Phys. Stat. Sol. Vol. 120, pp.341.

102. Fraitova, D. (1983(b)). An analytical theory of FMR in bulk metals, II. Penetration depths. Phys. Stat. Sol. Vol. 120, pp. 659.

103. Fraitova, D. (1984). An analytical theory of FMR in bulk metals,III.Surface impedance. Phys. Stat. Sol. Vol. 124, pp.587.

104. Fullerton, E.E.; Jiang, J.S.; Bader, S.D. (1999). Hard/soft magnetic heterostructures: model exchange-spring magnets.f. Magn. Magn. Mater. Vol.200, pp.392.

105. Fullerton, E.E.; Jiang, J.S.;Grimsditch, M.; Sowers, C.H.; Bader,S.D. (1998). Exchange-spring behavior in epitaxial hard/soft magnetic bilayers.Phys. Reu. B. Vol.58, pp.12193.

106. Gatteschi, D.;Sessoli, R.; Villain, J. (2006).Molecular nanomagnets. Oxford University Press.

107. Garcia-Miquel, H.;Garcia, J.M.; Garcia-Beneytez, J.M.; Vazquez, M. (2001). Surface magnetic anisotropy in glass-coated amorphous microwires as determined from ferromagnetic resonance measurements.f. Magn. Magn. Mater.Vol.231, pp.38.

108. Gerlach, W.; Stem, 0. (1922). Dasmagnetische moment dessilber atoms. Zeitschrift for Physik.Vol.9, pp.353.

109. Gilbert, T.A. (1955). Armour research foundation rept. Armour Research Foundation.Chicago. No.ll.

110. Gnatzig, K.;. DOtsch, H.; Ye, M.; Brockmeyer, A. (1987). Ferrimagnetic resonance in garnet films at large precession angles. f. Appl. Phys. Vol. 62, pp.4839. Golosovsky, M.; Monad, P.; Muduli, P.K; Budhani, R.C. (2012). Low-field microwave absorption in epitaxial Lao.1Sro.aMnOa films resulting from the angle-tuned ferromagnetic resonance in the multidomain state. Arxiv: 1206-3041. Cont-mat. mtrl-sci. pp. 1.

111. Goryunov, Yu. V.; Garifyanov, N.N.; Khaliullin, G.G.; Garifullin, LA.; Tagirov, L.R; Schreiber, F.; Miihge, Th.; Zabel, H. (1995). Magnetican isotropies of

sputtered Fe films on MgO substrates. Phys. Reu. B. Vol.52, pp. 13450.

112. Grimsditch, M.; Carnley, R; Fullerton, E.E.; Jiang, S.; Bader, S.D.; Sowers, H. (1999).

113. Exchange-spring systems: Coupling of hard and soft ferromagnets as measured by magnetization and Brillouin light scattering (invited).f. Appl. Phys. Vol.85, pp. 5901.

114. Grunberg, P.; Schwarz, B.; Vach, W.; Zinn, W.; Dabkowski, D. (1979). Light scattering from spin waves in bubble films.f. Magn. Magn. Mater. Vol. 13, pp. 181.

115. Grunberg, P. (1980). Brillouin scattering from spin waves in thin ferromagnetic fihns. f. Magn. Magn. Mater. Vol. 15-18, pp. 766.

116. Grunberg, P.; Mayr, C.M.; Vach, W. (1982).Determination of magnetic parameters by means of brillouin scattering. Examples: Fe, Ni, NioaFe>.2. f. Magn. Magn. Mater. Vol. 28, pp. 319.

117. Grunberg, P.; Barnas, J.; Saurenbach, F.; Fu!3, J.A.; Wolf, A.; Vohl, M. (1991). Layered magnetic structures: antiferromagnetic type interlayer coupling and magnetoresistance due to antiparallel alignment. J. Magn. Magn. Mater. Vol. 93, pp.58.

118. Grunberg, P.; Demokritov, S.; Fuss, A.; Schreiber, R.; Wolf, J.A.; Purcell, S.T. (1992). Interlayer exchange, magnetotransport and magnetic domains in Fe/ Cr layered structures. f. Magn. Magn. Mater. Vol. 104-107, pp.1734.

119. Grunberg, P. (2000). Layered magnetic structures in research and application. Acta mllter. Vol. 48, pp. 239. Grunberg, P. (2001). Layered magnetic structures: I:Ustory, facts and figures.f. Magn. Magn. Mater. Vol.226-230, pp. 1688.

120. Guimariies, A.P. (2009).Principles of nanomagnetism. Springer-Verlag Berlin Heidelberg.ISBN 978-3-642-01481-9. Guimariies, A.P. (1998). Magnetism and magnet1c resonance in solids. A Wiley-Interscience Publication.Canada.

121. Gurevich, A.G.; Melkov, G.A. (1996). Magnetization Oscillations and Waves (CRC, Boca Raton).

122. Giiner, S.; Yal m, 0.; Kazan, S.; Y1ld1z, F.; ai:Ungoz, R. (2006). FMR studies of bilayer Co90Feto/NiatFet9, NiatFet9/Co<nFeto and monolayer Ni&Fet9 ti:Un fihns. Phys. Stat. Solid (a). Vol. 203, pp.1539.

123. Hamoudeh, M.; A1 Faraj, A.; Canet-Soulas, E.; Bessueille, F.; Leonard, D.; Fessi, H. (2007). Elaboration of PLLA-based superparamagnetic nanoparticles: Characterization, magnetic behaviour study and in vitro elaxivity evaluation. Int. f. Phaftnllceut. Vol. 338, pp.248.

124. Hathaway, K.; Cullen, J. (1981). Magnetoelastic softening of moduli and determination of magnetic anisotropy in RE-TM compounds. f. Appl. Phys. Vol. 52, pp.2282.

125. Heim, E.; Harling, S.; Pohlig, K.; Ludwig, F.; Menzel, H.; Sci:Ulling, M. (2007). Fluxgate magnetorelaxometry of superparamagnetic nanoparticles for hydrogel characterization. f. Magn. Magn. Mater.Vol.311, pp. 150.

126. Heinrich, B. (2005(a)). Ferromagnetic resonance in ultrati:Un film structures.

In magnetic ultrati:Un films. Heinrich, B.; Bland, J.A.C. (Eds.) Springer, Berlin, Heidelberg.Vol. II,pp. 195.

127. Heinrich, B.; Bland, J.A.C. (2005(b)). Ultrati:Un magnetic structures N, applications of nanomagnetism. Springer Berlin Heidelberg New York .ISBN 3-540-21954-4.

128. Heisenberg, W. (1928).Theory of ferromagnetism. ZeitschriftforPhysik.Vol. 49, pp. 619.

129. Held, G.A.; Grinstein, G.; Doyle, H.; Sun, S.; Murray, C.B. (2001). Competing interactions in dispersions of superparamagnetic nanoparticles. Phys. Rev. B. Vol. 64, pp.012408.

130. Hellwig, 0.; Kortrigh, J.B.; Takano, K.; Fullerton, E.E. (2000).Switci:Ung behavior of Fe-Pt/Ni Fe exchange- spring fihns studied by resonant soft-x-ray magneto-optical Kerr effect. Phys. Rev. B. Vol. 62, pp.11694.

131. Herring, C.; Kittel, C. (1950).On the theory of spin waves in ferromagnetic media. Phys. Reu. Vol. 81, pp.869.

132. Hillebrands, B. (2000).Light scattering in solids VII. Topics. Appl. Phys. Vol. 75, M. Cardona, Gilntherodt, G. (Eds.). Springer, Berlin, Heidelberg.

133. Hillebrands, B.; Thiaville, A. (2006). Spin dynamics in confined magnetic structures III. Springer-Verlag Berlin Heidelberg.ISBN-10 3-540-20108-4.

134. Hillebrands, B.; Ounadjela, K. (2002). Spin dynamics in confined magnetic structures I. Springer-Verlag Berlin Heidelberg.

135. Hillebrands, B.; Ounadjela, K (2003). Spin dynamics in confined magnetic structures II. Springer-Verlag Berlin Heidelberg. Hinata, S.; Saito, S.; Takahashi, M. (2012). Ferromagnetic resonance analysis of internal effective field of classified grains by switching field for granular perpendicular recording media. f. Appl. Phys.. Vol. 111, pp. 07B722.

136. Hsia, L.C.; Wigen, P.E. (1981). Enhancement of uniaxial anisotropy constant by introducing oxygen vacancies inCa-doped YIG.f. Appl. Phys. Vol.52, pp. 1261.

137. Huang, Z.C.; Hu, X.F.; Xu, Y.X.; Zhai, Y.; Xu, Y.B.; Wu, J.; Zhai, H. R. (2012). Magnetic properties of ultrathin single crystal Fea04 film on InAs(100) by ferromagnetic resonance.f. Appl. Phys.Vol. 111, pp.07C108.

138. Huang, M.D.; Lee, N.N.; Hyun, Y.H.; Dubowik, J.; Lee, Y.P. (2004). Ferromagnetic resonance study of magnetic-shape-memory Ni21v:InGa films.f. Magn. Magn. Mater. Vol. 272-276, pp.2031.

139. Jalali-Roudsar, A.A.; Denysenkov, V.P.; I<hartsev, S.I. (2005). Determination of magnetic anisotropy constants for magnetic garnet epitaxial fihns using ferromagnetic resonance. f. Magn. Magn. Mater. Vol.288, pp. 15.

140. Jarrett, H.S.; Waring, R.K. (1958). Ferrimagnetic resonance in NiMnOa. Phys. Rev. Vol. 111, pp.1223.

141. Jiang, J.S.; Fullerton, E.E.; Sowers, C.H.; Inomata, A.; Bader, S.D.; Shapiro, A.J.; Shull, R.D.; Gomakov, V. S.; Nikitenko, V.I. (1999). Spring magnet films. IEEE Trans. Magn. Vol. 35, pp.3229.

142. Jiang, J.S.; Pearson, J.E.; Liu, Z.Y.; Kabius, B.; Trasobares,S.; Miller, D.J.; Bader,S.D. (2005). A new approach for improving exchange-spring magnets. J. Appl. Phys. Vol. 97, pp. 10!<311.

143. Jiang, J.S.; Bader, S.D.; Kaper, H.; Leaf, G.K; Shull, R.D.; Shapiro, A.J.; Gomakov, V.S.; Nikitenko, V.I.; Platt, C.L.; Berkowitz, A.E.; David, S.; Fullerton, E.E. (2002). Rotational hysteresis of exchange-spring Magnets. f. Phys. D: Appl. Phys. Vol. 35, pp.2339.

144. Jung, S.; Watkins, B.; DeLong, L.; Ketterson, J.B.; Chandrasekhar, V. (2002). Ferromagnetic resonance in periodic particle arrays. Phys. Reu. B. Vol. 66, pp.132401.

145. Kakazei, G.N.; Kravets, A.F.; Lesnik, N.A.; de Azevedo, M.M.P.; Pogorelov, Y.G.;Sousa, J.B. (1999).Ferromagnetic resonance in granular thin fihns.J. Appl. Phys. Vol. 85, pp.5654. Kakazei, G.N.; Pogorelov, Yu.G.; Sousa, J.B.; Golub, V.O.; Lesnik, N.A.; Cardoso,S.; Freitas, P.P. (2001). FMR in CoFe(t)/ AhOa multilayers: from continuous to discontinuous regime.f. Magn. Magn. Mater. Vol. 226-230, pp. 1828.

146. Kambe, T.; Kajiyoshi, K; Oshima, K.; Tamura, M.; Kinoshita, M. (2005). Ferromagnetic resonance in !3-p-NPNN at radio-frequency region. Polyhedron. Vol.24, pp.2468.

147. Kartopu, G.; Yal :m, 0.; Kazan, S.; Akta§, B. (2009). Preparation and F1v:IR analysis of Co nanowires in alumina templates.f. Magn. Magn. Mater. Vol.321, pp. 1142.

148. Kartopu, G.; Yal m, 0. (2010). Electrodeposited nanowires and their applications. edited by N. Lupu. INTECH. available from: http://sciyo.com/articles/show/title/fabrication-and applications-of-metalnanowire-arrays-electrodeposited-in-ordered-porous.-templates.

149. Kartopu, G.; Yal m, 0.; Choy, K.-L.; Topkaya, R; Kazan, S.; Akta§, B. (2011 (a)). Size effects and origin of easy-axis in nickel nanowire arrays.f. Appl. Phys. Vol. 109, pp.033909. Kartopu, G; Yal m, 0; Demiray, A.S. (2011 (b)). Magnetic and transport properties of chemical solution deposited (100).-textured Lao.1Sro.aMnOa and L< >.7Cao.aMnOa nanocrystalline thin £lms. Phys. Scr. Vol.83, pp.015701.

150. Kasuya, T. (1956). A theory of metallic ferro- and antiferromagnetism on Zener's model. Frog. Theor. Phys. Vol. 16, pp.45.

151. Kind, J.; van Raden, U.J.; Garda-Rubio, I.; Gehring, A.U. (2012). Rock magnetic techniques complemented by ferromagnetic resonance spectroscopy to analyse a sediment record. Geophys. f. Int. Vol. 191, pp.51.

152. Kip, A.F.; Arnold, R.D. (1949). Ferromagnetic resonance at microwave frequencies in iron single crystal. Phys. Rev. Vol. 75, pp. 1556.

153. Kittel, C. (1947).On the theory of ferromagnetic resonance absorption. Phys. Rev. Vol. 73, pp. 155.

154. Kittel, C. (1946).Theory of the structure of ferromagnetic domains in £lms and small Particles. Phys. Rev. Vol. 70, pp.965.

155. Kittel, C. (1949). On the gyromagnetic ratio and spectroscopic splitting factor

of ferromagnetic substances. Phys. Rev. Vol. 76, pp. 743.

156. Kittel, C.; Abrahams, E. (1953).Relaxation processes in ferromagnetism. Rev. Mod. Phys. Vol. 25, pp.233.

157. Kip, A.F.; Arnold, RD. (1949). Ferromagnetic Resonance at Microwave Frequencies in Iron Single Crystal.Phys. Rev.Vol. 75, pp. 1556.

158. Kittel, C. (1958). Interaction of spin waves and ultrasonic waves in ferromagnetic crystals. Phys. Rev.Vol.110, pp.836. I<harmouche, A.; Ben Youssef, J.; Layadi, A.; Cheri£,S.M. (2007). Ferromagnetic resonance in evaporated Co/Si(100) and Co/glass thin £hns.f. App. Phys.Vol. 101, pp. 13910. I<haibullin, R.I.; Tagirov, L.R; Rameev, B.Z.; Ibragimov, S.Z.; Y:lldJZ, F.; Akta§, B. (2004). High curie--temperature ferromagnetism in cobalt-implanted single-crystalline rutile. f. Phys.: Condens. Matter. Vol.16, pp. 1.

159. Klein, P.; Varga, R.; Infante, G.; Vazquez, M. (2012). Ferromagnetic resonance study of FeCoMoB microwires during devitri£cation process.f. Appl. Phys. Vol. 111, pp.053920.

160. Kliava, J.; Berger, R. (1999). Size and shape distribution of magnetic nanoparticles in disordered systems: computer simulations of superparamagnetic resonance spectra. f. Magn. Magn. Mater.Vol. 205, pp.328.

161. Kneller, E.F.; Hawing, R. (1991). The exchange-spring magnet: a new material principle for permanent magnets. IEEE Trans. Magn.Vol.27, pp.3588.

162. Kobayashi, T.; Ishida, N.; Sekiguchi, K; Nozaki, Y. (2012). Ferromagnetic resonance properties of granular Co-Cr-Pt films measured by micro-fabricated coplanar waveguides.f. Appl. Phys.Vol. 111, pp.07B919.

163. Kodama, R.H. (1999).Magnetic Nanoparticles.f. Magn. Magn. Mater. Vol. 200, pp.359.

164. Kodama, R.H.; Berkowitz, A.E. (1999). Atomic-scale magnetic modeling of oxide nanoparticles. Phys. Reu. B. Vol. 59, pp. 6321.

165. Kodama, R.H.; Berkowitz, A.E.; McNiff Jr. E.J.; Foner S. (1996). Surface spin disorder in NiFe20manoparticles. Phys. Reu. Lett. Vol. 77, pp.394.

166. Kooi, C.F.; Wigen, P.E.; Shanabarger, M.R.; Kerrigan J.V. (1964). Spin-wave resonance in magnetic films on the basis of surface -spin -pinning model and the volume inhomogeneity model. f. Appl. Phys. Vol.35, pp. 791.

167. Kopp, R.E.; Weiss, B.P.; Maloof, A.C.; Vali, H.; Nash, C.Z.; Kirschvink, J.L. (2006). Chains, clumps, and strings: magneto fossil taphonomy with ferromagnetic resonance spectroscopy. Earth Planet Sc. Lett. Vol. 247, pp. 10.

168. Korolev, K.A.; McCloy, J.S.; Afsar, M.N. (2012). Ferromagnetic resonance of micro- and nano-sized hexagonal ferrite powders at millimeter waves. f. Appl. Phys. Vol. 111, pp. 07E113.

169. Kohmoto, 0. (2007). Ferromagnetic resonance equation of hexagonal ferrite in c-plane. f. Magn. Magn. Mater. Vol. 310, pp.2561.

170. Knorr, T.G.; Hoffman, R.W. (1959). Dependence o£ geometric magnetic anisotropy in thin iron films. Phys. Reu. Vol.113, pp. 1039.

171. Kraus, L.; Frait, Z.; Ababei, G.; Chayka, 0.; Chiriac, H. (2012). Ferromagnetic resonance in submicron amorphous wires.f. Appl. Phys.Vol. 111, pp.053924.

172. Krebs, J.J.; Rachford, F.J.; Lubitz, P.; Prinz, G.A. (1982). Ferromagnetic resonance studies of very thin epitaxial single crystals o£ iron. f. Appl. Phys. Vol.53, pp.8058.

173. Krebs, J.J.; Lubitz, P.; Chaiken, A.; Prinz, G.A. (1989). Magnetic resonance determination of the antiferromagnetic coupling of Fe layers through Cr. Phys. Reu. Lett.Vol. 63, pp.1645. Krivoruchko, V.N.; Marchenko, A.I. (2012). Spatial confinement of ferromagnetic resonances in honeycomb antidot lattices.f. Magn. Magn. Mater. Vol.324, pp.3087.

174. Krone, P.; Albrecht, M.; Schre£1, T. (2011). Micromagnetic simulation o£ ferromagnetic resonance of perpendicular granular media: Influence of the intergranular exchange on the Landau-Lifshitz-Gilbert damping constant.f. Magn. Magn. Mater. Vol.323, pp.432.

175. Kuanr, B.K.; Carnley, RE.; Celinski, Z. (2004). Relaxation in epitaxial Fe films measured by ferromagnetic resonance.f. Appl. Phys. Vol.93, pp. 6610.

176. Kuanr, B.K.; Carnley, R.E.; Celinski, Z. (2005). Extrinsic contribution to Gilbert damping in sputtered NiFe films by ferromagnetic resonance. f. Magn. Magn. Mater. Vol. 286, pp. 276.

177. Kudryavtsev, Y.V.; Oksenenko, V.A.; Kulagin, V.A.; Dubowik, J.; Lee, Y.P. (2007). Ferromagnetic resonance in Co2MnGa films with various structural ordering.f. Magn. Magn. Mater.Vol. 310, pp.2271.

178. Lacheisserie, E.T.; Gignoux, D.; Schlenker, M. (2005). Magnetism, materials and aplications. Springer science+ Business Media, Inc.Boston.

179. Landau, E.; Lifshi-tz, E. (1935). On the theory of the dispersion of magnetic permeability in ferromagnetic bodies. Physik Z. Sowjetunnion. Vol. 8, pp. 153. Layadi, A.; Lee, J.M.; Artman, J.O. (1988). Spin-wave F1v:IR in annealed NiFe/FeMn thin fihns. J. Appl. Phys. Vol. 63, pp.3808.

180. Layadi, A. (2002). Exchange anisotropy: A ferromagnetic resonance study. Phys. Reu. B. Vol. 66, pp. 184423.

181. Layadi, A. (2004). Theoretical study o£ resonance modes of coupled thin films in the rigid layer model. Phys. Reu. B. Vol. 69, pp. 144431.

182. Layadi, A.; Artman, J.O. (1990(a)). Ferromagnetic resonance in a coupled two-layer System. f. Magn. Magn. Mater. Vol.92, pp. 143.

183. Layadi A.; Artman, J.O. (1990(b)). Study of antiferromagnetic coupling by ferromagnetic resonance (F1v:IR).J. Magn. Magn. Mater. Vol.92, pp.143.

184. Lee, J.; Hong, Y.K.; Lee,W.; Abo, G.S.; Park, J.; Syslo, R.;Seong, W.M.; Park,S.H.; Ahn, W.K (2012). High ferromagnetic resonance and thermal stability spinel Nio.1Mno.a.x CoxFe2C¢ ferrite for ultra high frequency devices.f. Appl. Phys.Vol. 111, pp.07A516.

185. Limmer, W.; Glunk, M.; Daeubler, J.; Hummel, T.; Schoch, W.; Bihler, C.; Huebl, H.; Brandt, M.S.; Goennenwein, S.T.B.; Sauer, R. (2006). Magnetic anisotropy in (Ga, Mn)As on GaAs(1 1 3)As studied by magnetotransport and ferromagnetic resonance. Microelectron.f. Vol.37, pp. 1490.

186. Lindner, J.; Baberschke, K (2003). In situ ferromagnetic resonance: an ultimate tool to investigate the coupling in ultrathin magnetic films. f. Phys.: Condens. Matter. Vol. 15, pp.R193.

187. Lindner, J.; Tolinski, T.; Lenz, K.; Kosubek, E.; Wende, H.; Baberschke, K.; Ney, A.; Hesjedal, T.; Pampuch, C.; Koch, R.; Daweritz, L.; Ploog, K.H. (2004). Magnetic anisotropy of MnAs-films on GaAs(O 0 1) studied with ferromagnetic resonance. f. Magn. Magn. Mater. Vol. 277, pp.159.

188. Li, N.; Schafer, S.; Datta, R.; Mewes, T.; Klein, T. M.; Gupta, A. (2012). Microstructural and ferromagnetic resonance properties of epitaxial nickel ferrite films grown by chemical vapor deposition. Appl. Phys. Lett. Vol. 101, pp. 132409.

189. Liua, B.; Yang, Y.; Tang, D.; Zhang, B.; Lu, M.; Lu, H. (2012). The contributions o£ intrinsic damping and two magnon scattering on the ferromagnetic resonance linewidth in [Fe66C025/Si02]nmultilayer films.J. Alloy. compd. Vol.524, pp. 69.

190. Macdonald, J.R. (1956). Spin exchange effects in ferromagnetic resonance. Phys. Reu. Vol. 103, pp.280.

191. Macia, F.; Warnicke, P.; Bedau, D.; Im, M.Y.; Fischer, P.; Arena, D.A.; Kent, A.D. (2012). Perpendicular magnetic anisotropy in ultrathin Co9Ni multilayer films studied with ferromagnetic resonance and magnetic x-ray microspectroscopy. f. Magn. Magn. Mater. Vol. 324, pp.3629.

192. Maklakov,S.S.; Maklakov,S.A.; Ryzhikov, I.A.; Rozanov, K.N.;Osipov, A.V. (2012).Thin Co films with tunable ferromagnetic resonance frequency. f. Magn. Magn. Mater. Vol. 324, pp.2108.

193. Maksymowich, L.J.; Sendorek, D. (1983). Surface modes in magnetic thin amorphous films of GdCoMo alloys.f. Magn. Magn. Mater. Vol.37, pp. 177.

194. Maksymowicz, L.J.; Sendorek, D.; Zuberek, R (1985). Surface anisotropy energy o£ thin amorphous magnetic £hns of (Gd1-xCox)1-yMoy alloys; experimental results. f. Magn. Magn. Mater. Vol. 46, pp. 295. Maksymowicz, L.J.; Jankowski, H. (1992). FJv:IR experiment in multilayer structure of FeBSi/ Pd.f. Magn. Magn. Mater. Vol. 109, pp.341.

195. Marin, C.N. (2006). Thermal and particle size distribution effects on the ferromagnetic resonance in magnetic fluids.f. Magn. Magn. Mater.Vol. 300, pp.397.

196. Mastrogiacomo, G.; Fischer H.; Garcia-Rubio, I.; Gehring, A.U. (2010). Ferromagnetic resonance spectroscopic response of magnetite chains in a biological matrix. f. Magn. Magn. Mater. Vol. 322, pp. 661.

197. Mazur, P.; Mills, D.L. (1982). Inelastic scattering o£ neutrons by surface spin waves on ferromagnets. Phys. Reu. B. Vol. 26, pp. 5175. McMichael, R.D.; Wigen, P.E. (1990). High power FJv:IR without a degenerate spin wave manifold. Phys. Reu. Lett. Vol. 64, pp. 64.

198. Meiklejohn, WH.; Bean, C.P. (1956). New magnetic anisotropy. Phys. Reu. Vol. 102, pp. 1413.

199. Meiklejohn, WH.; Bean, C.P. (1957). New magnetic anisotropy. Phys. Reu.

Vol. 105, pp.904. Mercereau, J.E.; Feynman, RP. (1956). Physical conditions for ferromagnetic resonance. Phys. Rev.Vol.104, pp. 63.

200. Mibu, K.; Nagahama, T.; Ono, T.; Shinjo, T. (1997). Magnetoresistance of quasi-Bloch-wall induced in NiFe/CaSm exchange.f. Magn. Magn. Mater. Vol. 177-181, pp. 1267.

201. Mibu, K.; Nagahama, T.; Shinjo, T.; Ono, T. (1998). Magnetoresistance of Bloch-wall--type magnetic structures induced in NiFe/CoSm exchange-spring bilayers. Phys. Rev. B. Vol. 58, pp. 6442.

202. Mills, D.L.; Bland, J.A.C. (2006). Nanomagnetism ultrathin £hns, multilayers and nanostructures. Elseuier B.V. Mills, D.L. (2003). Ferromagnetic resonance relaxation in ultrathin metal £lms: The role of the conduction electrons. Phys. Rev. B. Vol. 68, pp. 014419.

203. Min, J.H.; Cho, J.U.; Kim, Y.K; Wu, J.H.; Ko, Y.D.; Chung, J.S. (2006). Substrate effects on microstructure and magnetic properties of electrodeposited Co nanowire arrays.f. Appl. Phys. Vol. 99, pp.08Q510.

204. Miyake, K.; Noh, S.M.; Kaneko, T.; Imamura, H.; Sahashi, M. (2012). Study on high frequency 3-D magnetization precession modes of circular magnetic nano-dots using coplanar wave guide vector network analyzer ferromagnetic resonance. IEEE T. Magn. Vol. 48, pp.1782.

205. Mizukami, S.; Nagashima, S.; Yakata, S.; Ando, Y.; Miyazaki, T. (2007). Enhancement o£ DC voltage generated in ferromagnetic resonance for magnetic thin £hn. f. Magn. Magn. Mater.Vol. 310, pp.2248.

206. Morrish, A. H. (1965).The physical principles of magnetism Wiley,New York.

207. Nagamine, L.C.C.M.; Geshev, J.; Menegotto, T.; Fernandes, A.A.R.; Biondo, A.; Saitovitch, E.B. (2005). Ferromagnetic resonance and magnetization studies in exchange-coupled NiFe/Cu/NiFe structures.f. Magn. Magn. Mater. Vol.288, pp.205.

208. Nakai, T.; Yamaguchi, M.; Kikuchi, H.; Iizuka, H.; Arai, K.I. (2002). Remarkable improvement of sensitivity for high-frequency carrier--type magnetic £eld sensor with ferromagnetic resonance.f. Magn. Magn. Mater. Vol.242-245, pp. 1142.

209. Neudecker, I.; Waltersdorf, G.; Heinrich, B.; Okuno, T.; Gubbiotti, G.; Back, C.H. (2006). Comparison of frequency, field, and time domain ferromagnetic resonance methods. f. Magn. Magn. Mater. Vol. 307, pp. 148.

210. Olive, E.; Lansac, Y.; Wegrowe, J.E. (2012). Beyond ferromagnetic resonance: The inertial regime of the magnetization. Appl. Phys. Lett. Vol. 100, pp. 192407.

211. Osthi:iver, C.; Grunberg, P.; Arons, RR. (1998). Magnetic properties of doped Lao.67Bao.23Mm yAy0a, A- Fe, Cr.f. Magn. Magn. Mater. Vol.177-181, pp.854.

212. Owens, F.J. (2009). Ferromagnetic resonance observation of a phase transition in magnetic field-aligned Fe:Da nanoparticles.f. Magn. Magn. Mater. Vol. 321, pp.2386.

213. Oner, Y.; Ozdemir, M.; Akta§, B.; Topacli, C.; Haris, E.A.; Senoussi, S. (1997).

The role of Pt impurities on both bulk and surface anisotropies in amorphous NiMn films. f. Magn. Magn. Mater.Vol. 170, pp. 129.

214. Ozdemir, M.; Akta§, B.;Oner, Y.;Sato, T.; Ando, T. (1996). A spin- wave resonance study on reentrant Jvfrl77lvfrna thin fihns. f. Magn. Magn. Mater. Vol. 164, pp.53.

215. Ozdemir, M.; Akta§, B.; Oner, Y.; Sato, T.; Ando, T. (1997). Anomalous anisotropy of reentrant Ni11Mruafihn.f. Phys.: Ccndens. Matter. Val.9, pp. 6433.

216. Ozdemir, M.; Oner, Y.; Akta§, B. (1998). Evidence of superparamagnetic behaviour in an amorphous Ni62Nrn2a film by ESR measurements. Phys. B. Vol.252, pp. 138.

217. Patel, R.; Owens, F.J. (2012). Ferromagnetic resonance and magnetic force microscopy evidence for above room temperature ferromagnetism in Mn doped Si made by a solid state sintering process. Solid State Commun.Vol. 152, pp. 603.

218. Patton, C.E.; Hurben, M.J. (1995). Theory of magnetostatic waves for in-plane magnetized isotropic fihns. f. Magn. Magn. Mater.Vol. 139, pp.263. Patton, C.E.; Hurben, M.J. (1996). Theory of magnetostatic waves for in-plane magnetized anisotropic films. f. Magn. Magn. Mater.Vol. 163, pp.39.

219. Park, D.G.; Kim, C.G.; Kim, W.W.; Hong, J.H. (2007). Study of GMI-valve characteristics in the Co-based amorphous ribbon by ferromagnetic resonance.f. Magn. Magn. Mater. Va!. 310, pp.2295.

220. Parkin, S.S.P.; More, N.; Roche, K.P. (1990). Oscillations in exchange coupling and magneto resistance in metallic superlattice structures: Co/Ru, Co/Cr, and Fe/Cr. Phys. Rev. Lett. Vol. 64, pp.2304.

221. Parkin, S.S.P.; Bhadra, R; Roche, KP. (1991(a)). Oscillatory magnetic exchange coupling through thin copper layers. Phys. Rev. Lett. Vol. 66, pp.2152.

222. Parkin,S.S.P. (1991(b)).Systematic variation of the strength and oscillation period of indirect magnetic exchange coupling through the 3d,4d, and Sd transition metals. Phys. Rev. Lett. Vol. 67, pp.3598.

223. Parkin, S.S.P.; Farrow, RF.C.; Marks, RF.; Cebollada, A.; Harp, G.R.; Savoy, R.J. (1994). Oscillations of interlayer exchange coupling and giant magnetoresistance in (111) oriented permalloy/Au multilayers. Phys. Rev. Lett. Vol. 72, pp.3718.

224. Parvatheeswara, R.B.; Caltun, 0.; Dumitru, I.; Spinu, L. (2006). Ferromagnetic resonance parameters of ball-milled Ni-Zn ferrite nanoparticles. f. Magn. Magn. Mater. Va!. 304, pp. 752.

225. Paul, A.; Biirgler, D.E.; Griinberg, P. (2005). Enhanced exchange bias in ferromagnet/ antiferromagnet multilayers.f. Magn. Magn. Mater. Vol. 286, pp.216.

226. Paz, E.; Cebollada, F.; Palomares, F.J.; Gonzalez, J.M.; Martins, J.S.; Santos, N.M., Sobolev, N.A. (2012). Ferromagnetic resonance and magnetooptic study of submicron epitaxial Fe(001) stripes.f. Appl. Phys.Vol. 111, pp. 123917.

227. Pires, M.J.M.; Mansanares, A.M., da Silva, E.C.; Schmidt, J.E.; Meckenstock, R.; Pelzl, J.
228. (2006). Ferromagnetic resonance studies in granular CoCu codeposited films. f. Magn. Magn. Mater. Vol. 300, pp.382.
229. Platow, W.; Anisimov, A.N.; Dunifer, G.L.; Farle, M.; Baberschke, K. (1998). Correlations between ferromagnetic-resonance linewidths and sample quality in the study of metallic ultrathin films. Phys. Rev. B. Vol.58, pp.5611.
230. Pollmann, J.; Srajer, G.; Hakel, D.; Lang, J.C.; Maser, J.; Jiang, J.S.; Bader, S.D. (2001). Magnetic imaging of a buried SmCo layer in a spring magnet.f. Appl. Phys. Vol. 89, pp. 7165.
231. Poole, C.P.;Owens J.F. (2003).Introduction to nanotechnology. john Wiley & Sons, Inc. ISBN: 0-471..()7935-9.
232. Prieto, A.G.; Fdez-Gubieda, M.L.; Lezama, L.; Orue, I. (2012). Study of surface effects on CoCu nanogranular alloys by ferromagnetic resonance. f. Appl. Phys. Vol. 111, pp. 07C105.
233. Purcell, E.M.; Torrey, H.C.; Pound, R.V. (1946). Resonance absorption by nuclear magnetic moments in a solid.Phys. Rev.Vol. 69, pp.37.
234. Puszkarski, H. (1992). Spectrum of interface coupling-affected spin-wave modes in ferromagnetic bilayer films. Phys. stat. sol.Vol. 171, pp.205.
235. Rabi, I.I.; Zacharias, J.R; Millman, S.; Kusch, P. (1938). A new method of measuring nuclear magnetic moment, Phys. Rev. Vol. 53, pp.318.
236. Rachford, F.J.; Vittoria, C. (1981). Ferromagnetic anti-resonance in non-saturated magnetic metals. J. Appl. Phys. Vol. 52, pp.2253.
237. Rado, G.T. (1958). Effect of electronic mean free path on spin-wave resonance in ferromagnetic metals. f. Appl. Phys. Vol. 29, pp. 330. Raikher, Yu.L.; Stepanov, V.I. (2003). Nonlinear dynamic response of superparamagnetic nanoparticles. Microelectron. Eng.Vol. 69, pp. 317.
238. Raita, 0.; Papa, A.; Stan, M.; Suciu, R.C.; Biris, A.; Giurgiu, L.M. (2012). Effect of Fe concentration in ZnO powders on ferromagnetic resonance spectra. Appl. Magn. Reson. Vol. 42, pp.499.
239. Rahman, F. (2008). Nanostructures in electronics and photonics. Pan Stanford Publishing Pte. Ltd. Singapore, 596224.
240. Rameev, B.Z.; Akta§, B.; Khaibullin, RI.; Zhikharev, V.A.; Osin, Yu.N.; Khaibullin, LB. (2000). Magnetic properties of iron-and cobalt-implanted silicone polymers. Vacuum. Vol. 58, pp.551.
241. Rameev, B.Z.; Y:lld1z, F.; Akta§, B.; Okay, C.; Khaibullin, R.I.; Zheglov, E.P.; Pivin, J.C.; Tagirov, L.R (2003(a)). I on synthesis and FMR studies of iron and cobalt nanoparticles in polyimides. Microelectron. Eng. Vol. 69, pp.330.
242. Rameev, B.Z.; Yilgin, R; Akta§, B.; Gupta, A.; Tagirov, L.R. (2003(b)). FMR studies of CrO epitaxial thin films. M icroelectron. Eng. Vol. 69, pp.336.
243. Rameev, B.Z.; Gupta, A.; Anguelouch, A.; Xiao, G.; Y1ld1z, F.; Tagirov, L.R.; Akta§, B. (2004 (b)). Probing magnetic anisotropies in half-metallic Cr02 epitaxial films by FMR. f. Magn. Magn. Mater.Vol. 272-276, pp. 1167.

244. Rameev, B.; Okay, C.; Y1ld1z, F.; Khaibullin, R.I.; Popok, V.N.; Aktas, B. (2004 (a)). Ferromagnetic resonance investigations of cobalt-implanted polyimides. f. Magn. Magn. Mater. Vol. 278, pp. 164.

245. Ramesh, M.; Wigen, P.E. (1988 (a)). Ferromagnetic resonance of parallel stripe domains domain wall system. f. Magn. Magn. Mater. Vol. 74, pp. 123. Ramesh, M.; Ren, E.W.; Artman, J.O.; Kryder, M.H. (1988 (b)).Domain mode ferromagnetic resonance studies in bismuth-substituted magnetic garnet films.f. Appl. Phys. Vol. 64, pp. 5483.

246. Ramprasad, R.; Zurcher P.; Petras, M.; Miller, M.; Renaud, P. (2004). Magnetic properties of metallic ferromagnetic nanoparticle composites.f. Appl. Phys. Vol.96, pp.519.

247. Ramsey, N.F. (1985).Molecular beams. Oxford University Press. New York.

248. Raposo, V.; Zazo, M.; Iiiiguez, J. (2011). Comparison of ferromagnetic resonance between amorphous wires and microwires.f. Magn. Magn. Mater. Vol.323, pp.1170.

249. Reich, KH. (1955). Ferromagnetic resonance absorption in a nickel single crystal at low temperature. Phys. Rev.Vol. 101, pp. 1647.

250. Rezende, S.M.; Moura, J.A.S.; de Aguiar, F.M. (1993). Ferromagnetic resonance in Ag coupled Ni films.f. Appl. Phys. Vol. 73, pp. 6341.

251. Riedling, S.; Knorr, N.; Mathieu, C.; Jorzick, J.; Demokritov,S.O.; Hillebrands, B.; Schreiber, R; Griinberg, P. (1999). Magnetic ordering and anisotropies of atomically layered Fe/Au(001) multilayers.J. Magn. Magn. Mater. Vol.198-199, pp.348.

252. Richard, C.; Houzet, M.; Meyer, J.S. (2012). Andreev current induced by ferromagnetic resonance. Appl. Phys. Lett.Vol. 109, pp.057002.

253. Rivkin, K; Xu, W., De Long, L.E.; Metlushko, V.V.; Tiic, B.; Ketterson, J.B. (2007). Analysis of ferromagnetic resonance response of square arrays of permalloy nanodots. f. Magn. Magn. Mater. Vol. 309, pp.317.

254. Rodbell, D.S. (1964). Ferromagnetic resonance absorption linewidth of nickel metal. Evidence for Landau- Lifshitz damping. Phys. Rev.Lett. Vol. 13, pp.471.

255. Rook, K.; Artman, J.O. (1991). Spin wave resonance in FeAlN films. IEEE Trans. Magn. Vol. 27, pp. 5450. Roy, W.V.; Boeck, J.D.; Borghs, G. (1992). Optimization of the magnetic field of perpendicular ferromagnetic thin films for device applications. Appl. Phys. Lett. Vol. 61, pp.3056.

256. Rousseau, 0.; Viret, M. (2012). Interaction between ferromagnetic resonance and spin currents in nanostructures. Phys. Rev. B. Vol. 85, pp. 144413.

257. Rfuner, F.M.; Moller, M.; Wagner, K; Gathmann, L.; Narkowicz, R.; Ziihres, H.; Salles, B.R; Torelli, P.; Meckenstock, R; Lindner, J.; Farle, M. (2012). In situ multifrequency ferromagnetic resonance and x-ray magnetic circular dichroism investigations on Fe/GaAs(110): Enhanced g-factor. J. Appl. Phys. Vol. 100, pp.092402.

258. Ruderman, M.A.; Kittel, C. (1954).Indirect exchange coupling of nuclear magnetic moments by conduction electrons. Phys. Reu. Vol. 96, pp. 99.

Rusek, P. (2004). Spin dynamics of ferromagnetic spin glass.f. Magn. Magn. Mater.Vol.272-276, pp. 1332.

259. Sarmiento, G.; Fdez-Gubieda, M.L.; Siruguri, V.; Lezama, L.; Orue, I. (2007). Ferromagnetic resonance study of Fe9lAgso granular film.f. Magn. Magn. Mater. Vol.316, pp.59.

260. Sasaki, M.; Jonsson, P.E.; Takayama, H.; Mamiya, H. (2005). Aging and memory effects in superparamagnets and superspin glasses. Reu. B. Vol. 71, pp.104405.

261. Schiifer, S.; Pachauri, N.; Mewes, C.KA.; Mewes, T.; Kaiser, C.; Leng, Q.; Pakala, M. (2012). Frequency-selective control of ferromagnetic resonance linewidth in magnetic multilayers.f. Appl. Phys.Vol. 100, pp.032402.

262. Schmool, D.S.; Barandiaran, J.M. (1998). Ferromagnetic resonance and spin wave resonance in multiphase materials:theoretical considerations.f. Phys.: Condens. Matter. Vol. 10, pp. 10679.

263. Schmool, D.S.; Schmalzl, M. (2007). Ferromagnetic resonance in magnetic nanoparticle assemblies.f. Non-Cryst. Solids. Vol.353, pp. 738.

264. Scholz, W.; Suess, D.; Schre.£1, T.; Fidler, J. (2000). Micromagnetic simulation of structure property relations in hard and soft magnets. Comp. Mater. Sci. Vol. 18, pp. 1.

265. Schreiber, F.; Frait, Z. (1996). Spinwave resonance in high conductivity films: The Fe-Co alloy system. Phys. Reu. B. Vol. 54, pp. 6473. Schre.£1, T.; Kronmiiller, H.; Fidler, J. (1993(a)). Exchange hardening in nano-structured permanent magnets. f. Magn. Magn. Mater. Vol. 127, pp.273.

266. Schre.£1, T.; Schmidts, H.F.; Fidler, J.; Kronmilller, H. (1993(b)). The role of exchange and dipolar coupling at grain boundaries in hard magnetic materials.J. Magn. Magn. Mater. Vol. 124, pp. 251. Schre.£1, T.; Fidler, J. (1998). Modelling o£ exchange-spring permanent magnets.f. Magn. Magn. Mater. Vol. 177-181, pp.970.

267. Schre.£1, T.; Forster, H.; Fidler, J.; Dittrich, R.; Suess, D.; Scholz, W. (2002). Magnetic hardening o£ exchange spring multilayers. Proof XVII Rare Earth Magnets Workshop, University o£ Delaware, ed:Hadjipanayis, G.; Bonder M.J. pp. 1006.

268. Schultz, S.; Gullikson, E.M. (1983). Measurement of static magnetization using electron spin resonance. Rev. Sci. Insrum. Vol.54, pp. 1383.

269. Seemann, K; Leiste, H.; Klever, C. (2009). On the relation between the effective ferromagnetic resonance linewidth Lif.H and damping parameter Gtell in ferromagnetic

270. Fe-Co-Hf-N nanocomposite films.f. Magn. Magn. Mater. Vol. 321, pp.3149.

271. Seemann, K.; Leiste, H.; Klever, Ch. (2010). Determination o£ intrinsic FMR line broadening in ferromagnetic (Fe44C056)77Hft2Nu nanocomposite films. f. Magn. Magn. Mater. Vol. 322, pp.2979.

272. Seib, J.;Steiauf, D.; Fiil:mle, M. (2009).Linewidth o£ ferromagnetic resonance for systems with anisotropic damping. Phys. Reu. B. Vol. 79, pp.092418.

273. Sellmyer, D.; Skomski, R (2006). Advanced magnetic nanostructures. Springer Science+Business Media, Inc.

274. Sihues, M.D.; Durante-Rincon, C.A.; Fermin, J.R. (2007). A ferromagnetic resonance study of NiFe alloy thin films.f. Magn. Magn. Mater.Vol. 316, pp.462.

275. Singh, A.; Cllowdhury, P.; Padma, N.; Aswal, D.K.; Kadam, R.M.; Babu, Y.; Kumar, M.L.J.; Viswanadham, C.S.; Goswami, G.L.; Gupta, S.K.; Yakluni, J.V. (2006). Magneto transport and ferromagnetic resonance studies of polycrystalline Lao.6Pbo.4NOa thin fihns. Solid State Commun. Vol. 137, pp.456.

276. Shames, A.I.; Rozenberg, E.; Sominski, E.; Gedanken, A. (2012). Nanometer size effects on magnetic order in La1-x CaxMn03 (x = 50.5 and 0.6) manganites, probed by ferromagnetic resonance.f. Appl. Phys.Vol. 111, pp. 07D701.

277. Sklyuyev, A.; Ciureanu, M.; Akyel, C.; Ciureanu, P.; Yelon, A. (2009). Microwave studies of magnetic anisotropy of Co nanowire arrays.f. App. Phys.Vol. 105, pp.023914.

278. Skomski, R.; Coey, J.M.D. (1993). Giant energy product in nanostructured two-phase magnets. Phys. Rev. B. Vol.48, pp.15812.

279. Skomski, R (2008) .Simple models of magnetism. Oxford University Press. ISBN 978--0-19- 857075-2.

280. Slichter, C.P. (1963). Principle of Magnetic Resonance. Harper & Row, New York. Smit, J.; Beljers, H. G. (1955). Ferromagnetic resonance absorption in BaFet20t9, a high anisotropy crystal. Philips Res. Rep.Vol. 10, pp. 113.

281. Spaldin, N.A. (2010). Magnetic materials: Fundamentals and applications. Cambridge Universzty Press. ISBN 13 978 0 52188669 7.

282. Sparks, M. (1964).Ferromagnetic Relaxation Theory. McGraw-Hill,New York.

283. Sparks, M. (1969).Theory of surface spin pinning in ferromagnetic resonance. Phys. Rev. Lett Vol. 22, pp. 1111.

284. Sparks, M. (1970(a)). Ferromagnetic resonance in thin films. III. Theory of mode Intensities. Phys. Rev. B. Vol. 1, pp.3869.

285. Sparks, M. (1970(b)).Ferromagnetic resonance in thin fihns I and II. Phys. Rev. B. Vol. 1, pp. 3831. Sparks, M. (1970(c)). Ferromagnetic resonance in thin films. I. Theory of normal mode frequencies. Phys. Rev.B. Vol. 1, pp.3831.

286. Sparks, M. (1970(d)). Ferromagnetic resonance in thin fihns. I. Theory of normal-mode intensities. Phys. Rev.B. Vol. 1, pp.3869.

287. Speriosu, V.S.; Parkin,S.S.P. (1987).Standing spin waves in FeMn/NiFe/ FeN:rn exchange bias structures. IEEE Trans. magn. Vol. 23, pp.2999.

288. Spinu, L.; Dumitru, I.; Stancu, A.; Cimpoesu, D. (2006). Transverse susceptibility as the low frequency limit of ferromagnetic resonance.f. Magn. Magn. Mater. Vol. 296, pp.1. Spoddig, D.; Meckenstock, R.; Bucher, J.P.; Pelzl, J. (2005). Studies of ferromagnetic resonance line width during electrochemical deposition of Co films on Au(1 1 1). f. Magn. Magn. Mater. Vol. 286, pp.286.

289. Song, Y.Y.; Kalarickal, S.; Patton, C.E. (2003). Optimized pulsed laser deposited barium ferrite thin films with narrow ferromagnetic resonance linewidths.f. Appl. Phys. Vol. 94, pp.5103.

290. Song, H.; Mulley, S.; Coussens, N.; Dhagat, P.; Jander, A.; Yokochi, A. (2012). Effect of packing fraction on ferromagnetic resonance in NiFe2C¢ nanocomposites. f. Appl. Phys. Vol. 111, pp.07E348.

291. Stashkevich, A.A.; Roussigne, Y.; Djemia, P.; Cllerif, S.M.; Evans, P.R.; Murphy, A.P.; Hendren, W.R; Atkinson, R.; Pollard, R.J.; Zayats, A.V.; Cllaboussant, G.;Ott, F. (2009).

292. Spin-wave modes in Ni nanorod arrays studied by Brillouin light scattering. Phys. Reu. B. Vol.80, pp.144406.

293. Stoner, E.C.; Wohlfarth, E. P. (1948). A mechanism of magnetic hysteresis in heterogeneous alloys. Phil. Trans. R. Soc. A. Vol.240, pp.599.

294. Sun, Y.; Duan, L.; Guo, Z.; DuanMu, Y.; Ma, M.; Xu, L.; Zhang, Y.; Gu, N. (2005). An improved way to prepare superparamagnetic magnetite-silica core-shell nanoparticles for possible biological application.f. Magn. Magn. Mater.Vol. 285, pp. 65.

295. Sun, Y.; Song, Y.Y.; O,ang, H.; Kabatek, M.; Jantz, M.; Schneider, W.; Wu, M.; Schultheiss, H.; Hoffmann, A. (2012(a)). Growth and ferromagnetic resonance properties of nanometer--thick yttrium iron garnet films. Appl. Phys. Lett. Vol. 101, pp. 152405.

296. Sun, Y.; Song Y.Y.; Wu, M. (2012(b)). Growth and ferromagnetic resonance of yttrium iron garnet thin films on metals. Appl. Phys. Lett. Vol. 101, pp.082405.

297. Sun, L.; Wang, Y.; Yang, M.; Huang, Z.; Zhai, Y.; Xu, Y.; Du, J.; Zhai, H. (2012(c)). Ferromagnetic resonance studies of Fe thin films with dilute heavy rare-earth impurities.f. Appl. Phys.Vol. 111, pp.07A328.

298. Suzuki, Y.; Katayama, T.; Takanashi, K;Schreiber, R.; Gtinberg, P.; Tanaka, K. (1997) . The magneto-optical effect of Cr(001) wedged ultrathin films grown on Fe(001). f. Magn. Magn. Mater. Vol. 165, pp. 134.

299. Szlaferek, A. (2004). Model exchange-spring nanocomposite. Status Solidi B, Vol. 241, pp. 1312.

300. Tannenwald, P.E.;Seawey, M.H. (1957).Ferromagnetic resonance in thin films of permalloy. Phys. Rev.Vol.105, pp.377.

301. Tartaj, P.; Gonzalez-Carrefio, T.; Bomati-Miguel, 0.;Serna, C. J. (2004). Magnetic behavior of superparamagnetic Fe nanocrystals confined inside submicron-sized spherical silica particles. Phys. Rev. B. Vol. 69, pp.094401.

302. Teale, R.W.; Pelegrini, F. (1986). Magnetic surface anisotropy and ferromagnetic resonance in the single crystal GdAh.f. Phys. F:Met. Phys. Vol. 16, pp. 621.

303. Terry, E.M. (1917). The magnetic properties of iron, nickel and cobalt above the curie point, and Keeson's quantum theory of magnetism.Phys. Reu. Vol.9, pp.394.

304. Topkaya, R.; Erkovan, M.; Oztiirk, A.; Oztiirk, 0.; Akta§, B.; Ozdemir, M.

(2010). Ferromagnetic resonance studies of exchange coupled ultrathin Py/ Cr/Py trilayers. f. Appl. Phys. Vol. 108, pp.023910.

305. Tsai, C.C.; O,oi, J.; O,o, S.; Lee, S.J.; Sarma, B.K.; Thompson, C.; O,ernyashevskyy, 0.; Nevirkovets, I.; Metlushko, V.; Rivkin, K; Ketterson, J.B. (2009). Vortex phase boundaries from ferromagnetic resonance measurements in a patterned disc array. Phys. Rev. B. Vol. 80, pp.014423.

306. Uhlenbeck, G.E.; Goudsmit,S. (1925). Ersetzung der hypothese vom unmechanischen zwang durch eine forderung beziiglich des inneren verhaltens jedes einzelnen elektrons. Die Naturwissenschaften. Vol. 13, Issue (47), pp.953. Valenzuela, R; Zamorano, R.; Alvarez, G.; Gutierrez, M.P.; Montiel, H. (2007). Magnetoimpedance, ferromagnetic resonance, and low field microwave absorption in amorphous ferromagnets.J. Non-Cryst. Solids.Vol.353, pp. 768.

307. Valenzuela, R.; Herbst, F.; Ammar, S. (2012). Ferromagnetic resonance in Ni-Zn ferrite nanoparticles in different aggregation states.f. Magn. Magn. Mater. Vol.324, pp.3398.

308. Van Vleck, J.H. (1950). Concerning the theo:ry of ferromagnetic resonance absorption. Phys. R.eu. Vol. 78, pp.266.

309. Vargas, J.M.; Zysler, RD.; Butera, A. (2007). Order-disorder transformation in FePt nanoparticles studied by ferromagnetic resonance. Appl. Surf Sci. Vol.254, pp.274.

310. Vilasi, D. (2001).Hamiltonian dynamics.World Scientific.LSBN 981..()2-3308-6.

311. Vittoria, C. (1993). Microwave Properties in Magnetic Films. World Scientific,Singapore. pp. 87.

312. Vlasko-Vlasov, K.; Welp,U.;Jiang, J.S.; Miller, D.J.; Crabtree, G.W.; Bader,S.D. (2001).Field induced biquadratic exchange in hard/soft ferromagnetic bilayers. Phys. Reu. Lett. Vol. 86, pp.4386.

313. Voges, F.; de Gronckel, H.; Osthover, C.; Schreiber, R.; Grunberg, P. (1998). Spin valves with CoO as an exchange bias layer.f. Magn. Magn. Mater. Vol. 190, pp. 183.

314. Vounyuk, B.P.; Guslienko, K.Y.; Kozlov, V.I.; Lesnik, N.A.; Mitsek, A.I. (1991). Effect on the interaction of layers on a ferromagnetic resonance in two layer feromagnetic £lms. Sov. Phys. Solid State. Vol. 33, pp.250.

315. Walker, L.R. (1957). Magnetostatic modes in ferromagnetic resonance. Phys. Reu. Vol. 105, pp.390.

316. Wallis, T.M.; Moreland, J.; Riddle, B.; Kabos, P. (2005). Microwave power imaging with ferromagnetic calorimeter probes on bimaterial cantilevers. f. Magn. Magn. Mater. Vol. 286, pp.320.

317. Wang, X.; Deng, L.J.; Xie, J.L.; Li, D. (2011).Observations of ferromagnetic resonance modes on FeCo-based nanocrystalline alloys.f. Magn. Magn. Mater. Vol. 323, pp. 635.

318. Weil, J.A.; Bolton, J.R.; Wertz, J.E. (1994). Electron Paramagnetic Resonance:

elementary Theory and Practical Application. John Wiley-Sons. New York. Weil, J.A.; Bolton, J.R (2007). Electron paramagnetic resonance. John Wiley & Sons, Inc. Hoboken, New Jersey. LSBN 978..()471-75496-1.

319. Weiss, M.T.; Anderson, P.W. (1955). Ferromagnetic resonance in ferroxdure. Phys. Reu. Vol. 98, pp.925.

320. Wegrowe, J.E.; Kelly, D.; Franck, A.; Gilbert, S.E.; Ansermet, J.Ph. (1999). Magnetoresistance of ferromagnetic nanowires. Phys. Reu. Lett.Vol. 82, pp.3681.

321. Wegrowe, J.E.; Comment, A.; Jaccard, Y.; Ansermet, J.Ph. (2000). Spin-dependent scattering of a domain wall of controlled size. Phys. Reu. B. Vol. 61, pp. 12216.

322. White, RL.; Solt, I.H. (1956).Multiple ferromagnetic resonance in ferrite spheres. Phys. Reu. Vol. 104, pp.56.

323. Wiekhorst, F.; Shevchenko, E.; Weller, H.; Kotzler, J. (2003). Anisotropic superparamagnetism of mono dispersive cobalt-platinum nanocrystals. Phys. Reu. B. Vol. 67, pp.224416.

324. Wigen P.E.; Zhang, Z. (1992). Ferromagnetic resonance in coupled magnetic multilayer systems. Braz. f. Phys. Vol. 22, pp. 267. Wigen, P.E.; Kooi, C.F.; Shanaberger, M.R; Rosing, T.R (1962). Dynamic pinning in thin-£lm spin-wave resonance. Phys. Rev. Lett. Vol. 9, pp. 206. Wigen, P.E. (1984). Microwave properties of magnetic garnet thin fihns. Thin Solid Films. Vol. 114, pp. 135.

325. Wigen, P.E.; Zhang, Z.; Zhou, L.; Ye, M.; Cowen, J.A. (1993). The dispersion relation in antiparallel coupled ferromagnetic films. f. Appl. Phys. Vol. 73, pp. 6338.

326. Wigen, P.E. (1998). Routes to chaos in ferromagnetic resonance and the return trip: Controlling and synchronizing Chaos, in Bar'yakhtar, V.G.; Wigen, P.E.; Lesnik, N.A. (Eds.). Frontiers in magnetism of reduced dimension systems Nato AS!series (Kluwer, Dordrecht) pp.29. Wojtowicz, T. (2005). Ferromagnetic resonance study of the free-hole contribution to magnetization and magnetic anisotropy in modulation-doped Gat. xN:rnxAs/Gat.yAlyAs:Be. Phys. Rev. B. Vol. 71, pp. 035307.

327. Wolfram, T.; De Warnes, R.E. (1971). Magneto-exchange branches and spin-wave resonance in conducting and insulating films: Perpendicular resonance. Phys. Rev. B. Vol. 4, pp. 3125.

328. Woltersdorf, G.; Heinrich, B.; Woltersdorf, J.; Scholz, R (2004). Spin dynamics in ultrathin fihn structures with a network of misfit dislocations. journal of Applied Physics.Vol. 95, pp. 7007-7009.

329. Woods, S.I.; Kirtley, J.R; Sun, S.; Koch, RH. (2001). Direct investigation of superparamagnetism in Co nanoparticle films. Phys. Rev. Lett. Vol.87, pp. 137205.

330. Wiichner, S.; Toussaint, J.C.; Voiron, J. (1997). Magnetic properties of exchange-coupled trilayers of amorphous rare-earth-cobalt alloys.J. Phys. Rev. B. Vol.55, pp. 11576.

331. Xu, Y.; Zhang, D.; Zhai, Y.; Chen, J.; Long, J.G.;Sang, H.; You, B.; Du, J.; Hu, A.; Lu, M.; Zhai, H.R. (2004). FMR study on magnetic thin and ultrathin Ni-Fe films. Phys. Stat. Sol. (c). Vol. 12, pp.3698.

332. Yal m, 0.; Y1ld1z, F.; Ozdemir, M.; Akta§, B.; I<Oseoglu, Y.; Bal, M.; Touminen, M.T. (2004(a)).Ferromagnetic resonance studies of Co nanowire arrays.f. Magn. Magn. Mater. Vol. 272-276, pp. 1684.

333. Yal m, 0.; Ylid1z, F.; Ozdemir, M.; Rameev, B.; Bal, M.; Tuominen, M.T. (2004(b)). FMR Studies of Co Nanowire Arrays, Nanostructures Magnetic Materials and Their Applications. Kluwer Academic Publisher. Nato Science Series. Mathematics, Physics and Chemistry. Vol. 143, pp.347.

334. Yal m, 0. (2004(c)). PhD Thesis, investigation of phase transition in inorganic spin-Peierls CuGeOa systems by ESR techgnique. Gebze institute of technology, 2004 Gebze, Kocaeli, Turkey.

335. Yal m, 0.; Akta§, B. (2003). The Effects of Zn 2·_doping on Spin-Peierls transition in CuGeOa f. Magn. Magn. Mater. Vol.258-259, pp.137.

336. Yal m, 0.; Y1lmz, F.; Akta§, B. (2007(a)). Spin-flop and spin-Peierls transition in doped CuGeOa. Spectrochim. Acta Part A. Vol. 66, pp.307.

337. Yal m, 0. (2007(b)). Comparison effects of different doping on spin-Peierls transition in CuGeOa Spectrochim. Acta Part A. Vol. 68, pp. 1320.

338. Yal m, 0.; Kazan, S.; ahingi:iz, R.; Yildiz, F.; Yerli, Y.; Ak:ta§, B. (2008(a)). Thickness dependence of magnetic properties of Co<JOFeto nanoscale thin films.f. Nanosci. Nanotech. Vol. 8, pp.841.

339. Yal :m, 0.; Erdem, R.; Ovun, S. (2008(b)). Spin-1 model of noninteracting nanoparticles. Acta Phys. Pol. A. Vol. 114, pp. 835. Yal m, 0.; Erdem,R; Demir, Z. (2012). Magnetic properties and size effects of spin-1/2 and spin-1models of core-surface nanoparticles in different type lattices, smart nanoparticles technology, Abbass Hasi:Um (Ed.), LSBN:978- 953-51..()500-8, InTech, DO!:10.5772/34706. Available from: http://www.intechopen.com/books/smart-nanoparticles.-technology/magnetic properties-and-size-effects-of-spin-1-2-and-spin-1-models-of-core-shell-nanoparticles in-di.

340. Yeh, Y.C.; Jin, J.D.; Li, C.M.; Lue, J.T. (2009). The electric and magnetic properties of Co and Fe £lms percept from the coexistence of ferromagnetic and microstrip resonance or aT type microstrip. Measurement. Vol.42, pp.290.

341. Y1ldJZ, F.; Yal m, 0.; Ozdemir, M.; Alda§, B.; Ki:iseoglu, Y.; Jiang, J.S. (2004(a)). Magnetic properties of SmCo/Fe exchange spring magnets. f. Magn. Magn. Mater. Vol. 272-276, pp. 1941.

342. Y1ldJZ, F.; Yal m,O.; Alda§, B.; Ozdemir, M.; Jiang, J.S. (2004(b)). Ferromagnetic resonance studies on Sm-Co/Fe ti:Un £lms. MSMW'04 Symposium Proceedings. Kharkcv, Ukraine.

343. Y1ldJZ, F.; Kazan, S.; Aktas, B.; Tarapov, S.; Samofalov, V.; Ravlik, A. (2004(c)). Magnetic anisotropy studies on FeNiCo/Ta/FeNiCo three layers £1m by layer sensitive ferromagnetic resonance technique. Phys. stat. sol.(c). Vol. 12, pp.3694.

344. Yosida, K (1957). Magnetic properties of Cu-Mn alloys. Phys. Rev. Vol. 106, pp. 893.

345. Yosi:Ukiyo, M.; Namai, A.; Nakajima, M.; Suemoto, T.; Ohkosi:U, S. (2012). Anomalous behavior of !:Ugh-frequency zero-field ferromagnetic resonance in aluminum substituted i-Fe203.J. Appl. Phys.Vol. 111, pp.07A726.

346. Young, J.A.; Uehling, E.A. (1953). The tensor formulation of ferromagnetic resonance. Phys. Rev.Vol. 93, pp.544.

347. Yu, J.T.; Turk, R.A.; Wigen, P.E. (1975). Exchange dominated surface spinwaves in yttrium iron-garnet films. Phys. Rev. B. Vol. 11, pp.420.

348. Zakeri, K.; I<ebe, T.; Lindner, J.; Farle, M. (2006). Magnetic anisotropy of Fe/GaAs(OOl) ultrati:Un £lms investigated by in situ ferromagnetic resonance. J. Magn. Magn. Mater. Vol. 299, pp.Ll.

349. Zakeri, Kh.; Lindner, J.; Barsukov, I.;Meckenstock, R.; Farle, M.; von Hi:irsten, U.; Wende, H.; Keune, W. (2007). Spin dynamics in ferromagnets: Gilbert damping and two magnon scattering. Phys. Rev. B. Vol. 76, pp. 104416.

350. Zavoisky, E. (1945). Spin-magnetic resonance in paramagnetics. f. Phys. USSR. Vol. 9, pp. 211.

351. Zianni, X.; Troi:Udou, KN. (1998). Monte carlo simulations the coercive behaviour of oxide coated ferromagnetic particles. f. Phys.: Condens. Matter. Vol. 10, pp. 7475.

352. Zhai, Y.; SI:U, L.; Zhang, W.; Xu, Y.X.; Lu, M.; Zhai, H.R; Tang, W.X.; Jin, X.F.; Xu, Y.B.; Bland, J.A.C. (2003). Evolution of magnetic anisotropy in epitaxial Fe £lms by ferromagnetic resonance.f. Appl. Phys. Vol.93, pp. 7622.

353. Zhang, Z.; Zhou, L.; Wigen, P.E.; Ounadjela, K. (1994 (a)). Angular dependence of ferromagnetic resonance in exchange coupled Co/Ru/Co trilayer structures, Phys. Rev. B. Vol. 50, pp. 6094. Zhang, Z.; Zhou, L.; Wigen, P.E.; Ounadjela, K. (1994 (b)). Using ferromagnetic resonance as a sensitive method to study the temperature dependence of interlayer exchange coupling. Phys. Reu. Lett. Vol. 73, pp.336.

354. Zhang, B.; O,eng, J.;Ganga, X.; Dong, X.; Liu, X.; Ma, G.; O,ang, J. (2008). Facile fabrication of multi-colors high .£luorescent/superparamagnetic nanoparticles. f. Ccllcid Interf Sci. Vol. 322, pp.485.

355. Zhu, J.; Katine, J.A.; Rowlands, G.E.; O,en, Y.J.; Duan, Z.; Alzate, J.G.; Upadhyaya, P.; Langer, J.; Amiri, P.K.; Wang, K.L.; Krivorotov, LN. (2012). Voltage-induced ferromagnetic resonance in magnetic tunnel junctions. ArXiv: 1205-2835: Cond-mat. Mes hall. pp. 1.

Chapter 7

MICROWAVE ABSORPTION IN NANOSTRUCTURED SPINEL FERRITES

Gabriela Vázquez-Victorio, Ulises Acevedo-Salas and Raúl Valenzuela

Institute for Materials Research, National Autonomous University of México, México

INTRODUCTION

Magnetic nanoparticles (MNPs) are playing a crucial role in an extensive number of potential applications and science fields. Nanotechnology industry is rapidly growing with the promise that it will lead to significant economic and scientific impacts on a wide range of areas, such as health care, nanoelectronics, environmental remediation. MNPs are mostly ferrites, i.e, transition metal oxides with ferric ions as main constituent. Although the magnetic properties of ferrites [1] are less intense than metal's, especially saturation magnetization, ferrites possess a large chemical stability (corrosion resistance), high electrical resistivity, and extended applicability at high magnetic field frequencies. The use of MNPs for biological and clinical applications [2] is undoubtedly one of the most challenging research areas in the field of nanomaterials, involving the organized collaboration of research teams formed by physicists, chemists, biologists, physiciens. The advantages of MNPs are based on their nanoscale size, large surface area, tailoring of magnetic properties and negligible side effects in living tissues. These applications include drug delivery [3], magnetic hyperthermia [4], magnetic resonance imaging [5], biosensors [6]. A field related with microwave absorption is electromagnetic interference EMI [7], as the number of electromagnetic radiation sources has growth at an exponential rate. MNPs have found applications also in environmental fields, such as soil remediation [8] and heavy metal removal [9], as MNPs provide high surface area and specific affinity for heavy metal adsorption from aqueous systems.

The reduction in scale leads to strong changes in macroscopic properties. The main reason can be attributed to the enhanced importance of the surface atom fraction as compared with core atom fraction, as the material becomes a nanoparticle. A simple estimate reveals that for a ~100 nm nanoparticle, surface atom fraction is about 6% of the total NP atoms, while for a ~5 nm NP this fraction can attain 78% [10]. Surface layer of materials exhibits different properties simply because these atoms have a very different structure than the core's. Surface atoms, for instance, have a reduced coordination number (unsatisfied bonding), crystal defects and modified crystal planes ("broken symmetry"). In the case of magnetically ordered materials (ferro, ferri, antiferromagnetic phases), additionally, several magnetic properties critically change at the nanometric scale. These properties are, for instance, the change from multidomain to single domain magnetic structure, domain wall thickness, the decrease in anisotropy energy giving rise to superparamagnetic phenomena. MNPs can thus exhibit many property changes with the reduction in size. Last (but no least), MNPs can show important macroscopic effects of interparticle magnetic interactions, which can involve additive forces (exchange), or attraction/repulsion (dipole).

Synthesis of MNPs

The most common method to synthesize MNPs are based on coprecipitation and microemulsion [11]. The coprecipitation method produces NPs by a pH change in a solution containing the desired metals in the form of nitrates or chlorides. Average size and size distribution, as well as shape depend on the pH and the ionic strength of the precipitating solution. In the microemulsion method, an aqueous metal solution phase is dispersed (entrapped) as microdroplets in a continuous oil phase within a micellar assembly of stabilizing surfactants. The advantage is that the microdroplets provide a confined space which limits the growth and agglomeration of NPs. An emerging method for preparation of uniform NPs is the polyol technique, where metallic salts (acetates, oxalates), dissolved in an alcohol (such as diethylenglycol) are directly precipitated by high temperature decomposition [12]. This method can produce metals; by addition of a controlled amount of water, it can lead to oxide MNPs. Spray and laser pyrolysis, with great commercial scale-up potential have been reported [13]. In spray pyrolysis, a solution of a ferric salt (and a reducing agent) is sprayed through a reactor to produce evaporation of the solvent within each droplet. In laser pyrolysis, the laser energy is used to heat a flowing mixture of gases leading to a chemical reaction. Under the appropriate conditions, homogeneous nucleation occurs and NPs are produced. A different method utilizes high-energy ultrasound waves to create acoustic cavitations resulting in extremely hot spots. The sound waves produced by these cavities can lead to particle size reduction and hence the formation of NPs [14]. Other methods are based on electrochemical deposition of metal in a cathode [11], and also the use of magnetotactic bacteria [15].

Microwave Absorption in MNPs

In this brief review, some of the recent developments in microwave absorption in MNPs. The response of ferromagnetic resonance (FMR) of MNPs (under different conditions) is first reviewed. FMR results on consolidated materials by spark plasma sintering (SPS) techniques are included, as this method allows the preparation of nanostructured ferrites (grains under 100 nm in size) with high densities. The behavior of electron paramagnetic resonance (EPR) of some relevant magnetically disordered MNPs systems is also presented. We devote a part of this review to the emerging low field microwave absorption technique (LFMA), which is a non-resonant method providing valuable information on magnetically-ordered materials. Results on MNPs in different aggregation states, as well as SPS-sintered materials are briefly reviewed.

As FMR [16] and EPR [17] techniques are well known, so no additional treatment of them is included here, except for some references.

FERROMAGNETIC RESONANCE (FMR) IN FERRITE NANOPARTICLES

Temperature and size Dependences

One of the most studied phenomena on MNP's is the change of their FMR spectra as a function of temperature. The main parameters describing FMR signal are plotted versus temperature and eventually compared with the bulk counterpart, as an attempt to characterize the changes associated with the nanometric scale. Such parameters are generally peak-to-peak resonance linewidth, ΔHpp, resonance field, Hres, and the intensity or height of the resonant absorption signal. Often, a simple linear dependence with T is observed [18-23]. Magnetic and structural phase transitions appear as a discontinuous event on this dependence [24].

The resonance field behavior with temperature for bulk and MNP's decreases with decreasing temperature, as a consequence of the enhancement of the contributions to the internal field associated with magnetic ordering (mainly exchange and anisotropy). This effect is stronger for small particles (figure 1) [25], revealing additional contributions to the internal field at low temperatures as a consequence of the size decreasing. These contributions can be assumed as an extra unidirectional internal field arising from surface disorder, where the magnetization processes are presumably to be isotropic and causes an extra shift of the resonance field. Therefore, as the surface area increases with decreasing the particle size, the isotropic effects on the magnetic resonance behavior are more pronounced and an additional distribution of energy barriers, promoted by surface isotropic disorder, must be assumed. As observed in many works [23,25-27], the FMR spectra for MNPs at intermediate temperatures results to be a mixing of two lines: a broad component corresponding to typical anisotropic contributions and a narrow one, presumably corresponding to the surface isotropic contributions. This leads to a characteristic FMR shape for nanoparticulated systems. The general features include a broad component (becoming wider and shifting to lower fields upon cooling, see figure 2), a narrow component, and a large broadening and shifting as the particle size decreases. Figure 2 shows the FMR signal evolution with temperature variation from room temperature down to 100 K for a well diluted magnetite suspension [26]. A double component spectrum is well observed at high temperatures, while its two components seem to overlap in a single broad signal as temperature goes down. This can be attributed to an important decrease on isotropic contributions which causes the narrow component to disappear at low temperatures. In addition, decreasing temperature makes the broad component to widen, becoming more symmetric, and shifting to lower fields, thus revealing a random distribution of the anisotropy axis enhanced at low temperatures. Inter-particle interactions are negligible and do not contribute to this broadening since the high dilution of the studied suspension promotes isolation between particles.

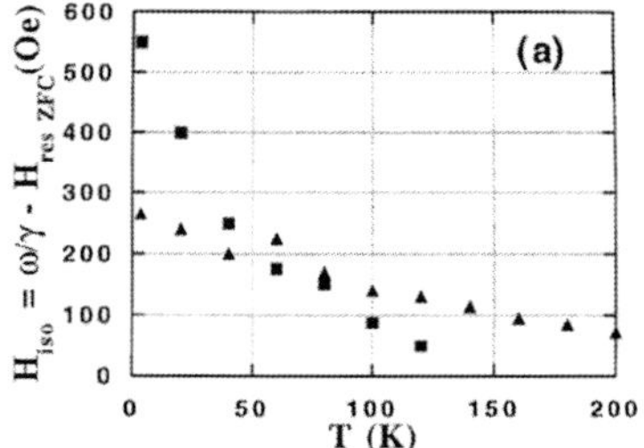

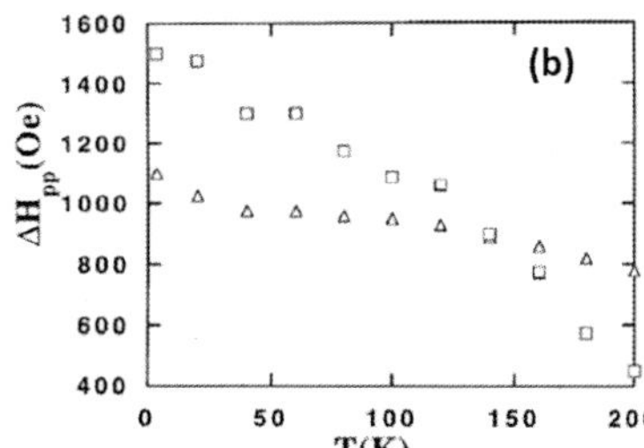

Figure 1 (a) Isotropic shift from the reference field (Hres = 3.3 kOe) as a function of temperature for ZFC non-interacting maghemite (γ-Fe2O3) nanoparticle ferrofluid samples: (Δ) d = 4.8 nm (□) d = 10 nm. (b) Peak to peak linewidth ΔHpp as a function of temperature for ZFC non-interacting maghemite (γ-Fe_2O_3) nanoparticle ferrofluid samples: (□) d = 4.8 nm (Δ) d = 10 nm [25].

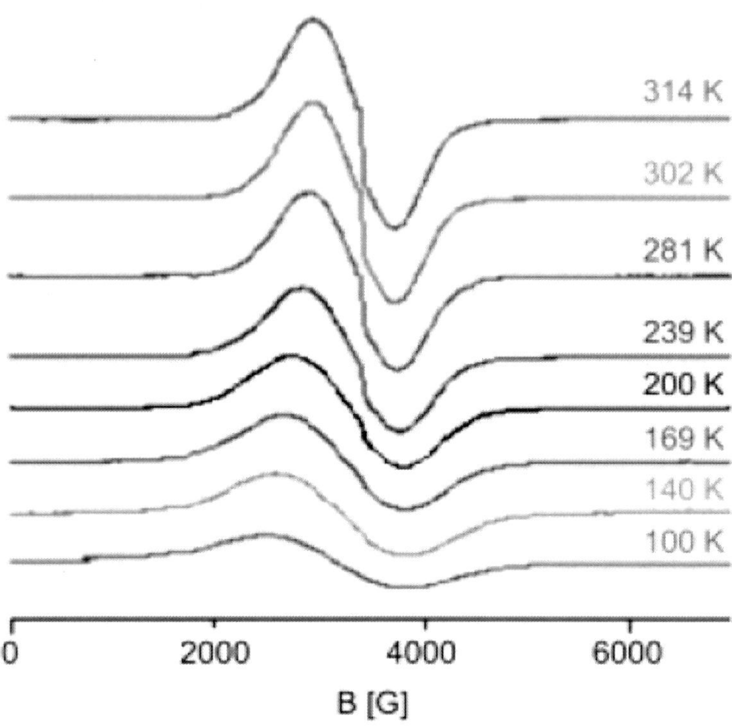

Figure 2. FMR signal evolution with temperature for a 0.1 wt% polymer Fe_3O_4 suspension [26].

Figure 3 shows the absorption signals for different particle distribution sizes of noninteracting maghemite (γ-Fe2O3) nanoparticle ferrofluid samples at room temperature [27]. The absorption signal changes drastically with changes on the particle size. Two limit cases are observed. The upper limit, for the largest size particles (sample 2, d = 10 nm), corresponds to a wide line shifted to a lower field in comparison with the reference (ω/γ = 3.3 kOe, where γ is the gyromagnetic ratio for free electrons). This line seems to be characteristic for anisotropic contributions. On the contrary, for the smaller size particles (sample 6, d = 10 nm) the spectrum shows a much narrower line, apparently characteristic for isotropic contributions and not shifted from the reference. The FMR signals for intermediate cases result in a more complex double-feature shape spectrum (samples 3, 4 and 5). Such behavior may be explained in terms of a mixing of two overlapped signals, the broad anisotropic shifted line and the narrower one, isotropic and not shifted. The coexistence of these two different contributions reveals the presence of a core-shell structure for the studied nanoparticles.

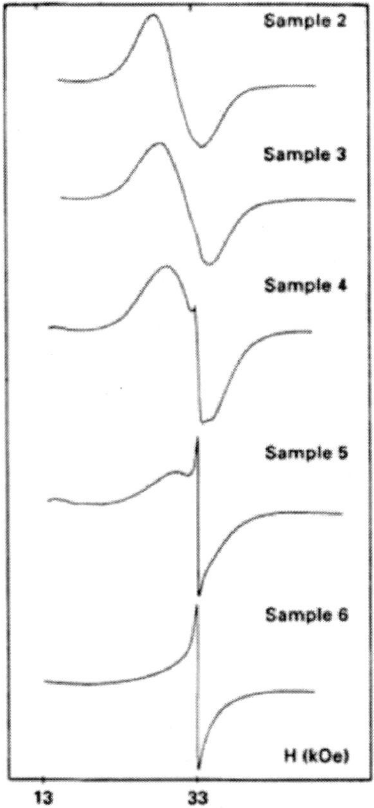

Figure 3. Influence of the particle size on X-band FMR spectra measured at room temperature for ZFC non-interacting maghemite (γ-Fe_2O_3) nanoparticle ferrofluid samples. The particle size decreases from sample 2 to 6; sample 2 has the largest particle size distribution (d = 10 nm) and sample 6 the smaller particle size distribution (d = 4.8 nm) [27].

Particle Concentration Dependence

Double component FMR spectra have also been reported for solid and liquid Fe_3O_4 suspensions [26]. Important changes were observed on the absorption signal depending on the particle concentration.

Figure 4 shows the decrease of the narrow component as the concentration increases. Both components exhibit also a broadening as the concentration increases. This important dependence of the signal linewidth with concentration is due to an increase of particle dipolar interactions at mean distances and a consequence of aggregation [28-30].

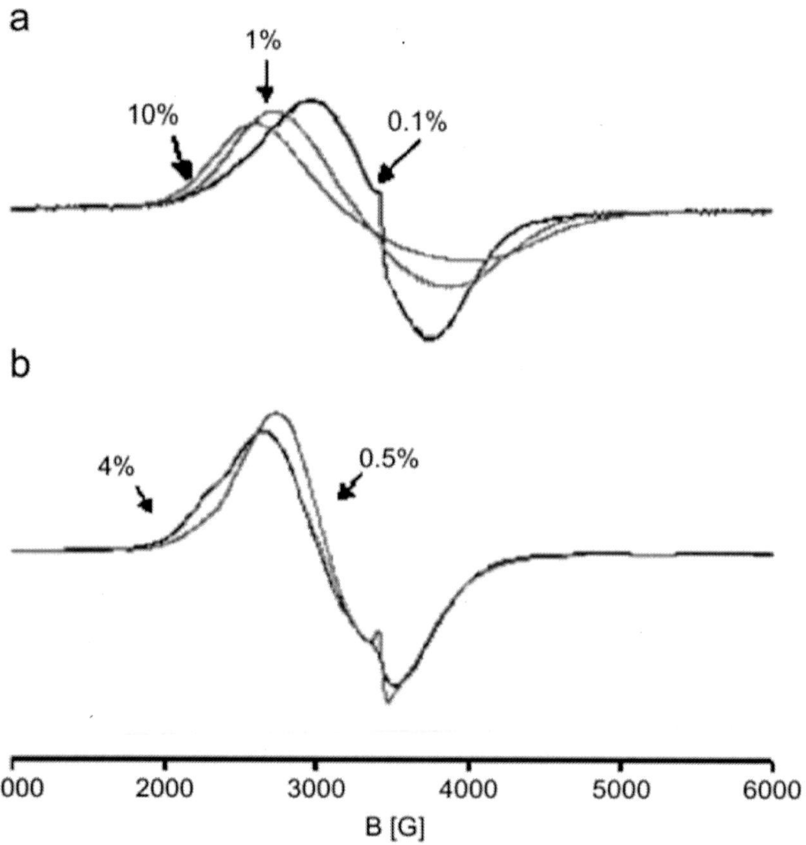

Figure 4. FMR signals for solid (a) and liquid (b) Fe_3O_4 suspensions at different concentrations [26].

Particle interactions always play an important role on the magnetic resonance absorption phenomena. FMR at different frequencies on dispersed KBr and then compressed $NiFe_2O_4$ commercial nanoparticles has recently been reported [31]. The resulting pellets showed strong shape anisotropy, as in-plane and out-of-plane analyzed measurements diverge with the increase of nanoparticle packing fraction (figure 5).

The resonance field decreases with increasing volume for in-plane measurements. In contrast, for out-of-plane measurements it increases with volume fraction. Both cases (in and out of plane), showed a resonance field shifted to a lower field in comparison with an ideal bulk as reported. As the packing fraction decreases, the in-plane and the out-of-plane curves converge to the same field value (≈ 0.04 T). This value can be identified as the average effective anisotropy field of the particles. A useful and powerful tool to estimate particle anisotropy can be based on these measurements.

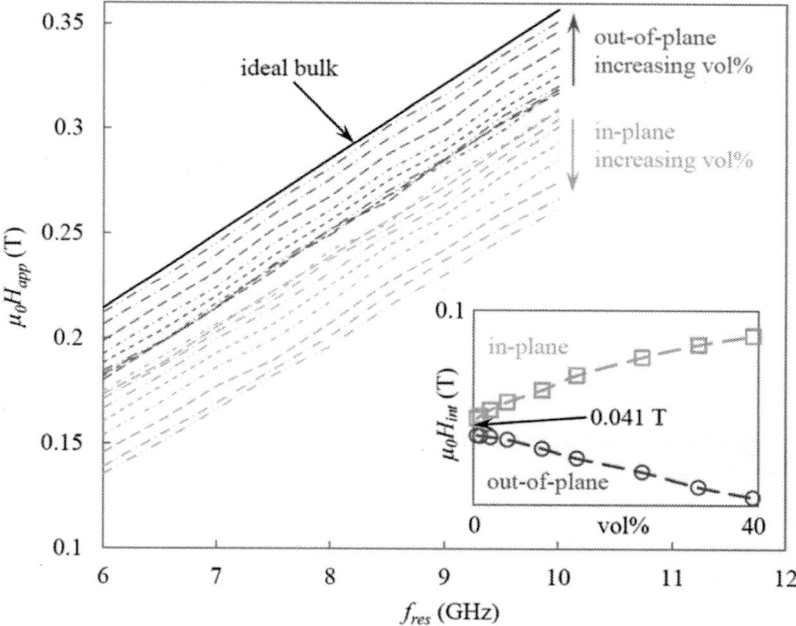

Figure 5. Frequency and shape orientation dependence of the applied field, at which FMR occurs, for commercial NiFe2O4 nanoparticles, dispersed in KBr and then compressed with different packing volume fractions. The inset shows the evolution of the resonance field as a function of the packing volume fraction [31].

Angular dependence

The anisotropy distributions and its FMR signal effects on MNP's can be carried out by measuring the angle variation between the cooling field of field-cooled (FC) samples. By comparing the obtained spectra for different orientations of FC samples, it is possible to determine the contribution of the anisotropy distribution within each particle to the FMR signal. The corresponding resonance lines often lead to a double component signal. It is then possible to separate the contributions which have a strong dependence with the angle from the weakly dependent counterpart. The narrow component can be attributed to surface anisotropy, while the the broad component should be associated with internal anisotropy of the NPs. Recent works have demonstrated the isotropic/anisotropic nature of most common MNP's, as described above. Figure 6 shows differences between FMR spectra for parallel and perpendicular configurations of FC samples. The samples consisted of maghemite nanoparticles with a particle diameter reported as 4.8 nm [27]. At the high temperature (top in figure 6), no variation as a function of θ is observed for the narrow component, while the broad component, on the contrary, shifts to higher resonance fields as the angle θ increases (bottom in figure 6). The narrow component is not affected by angular variations.

The resonance field of the anisotropic component generally is plotted against the angle θ, the dependences with θ use to be fitted, in good agreement, with sin2 θ or cos2 θ functions [24-32]. Its well established the axial symmetry nature for this kind of functions, as magnetization processes perform the higher energy absorption (energy improved by the resonance field) when the magnetizing field is perpendicular to the axial orientation, the resonance field also increases and finds a maximum at this position. The differences between the resonance field at $\theta = 0°$ and $\theta = 90°$, are related to the anisotropy field, promoted by the respective axial anisotropy within the core of the nanoparticles revealing a mono-domain configuration on this region (figure 7).

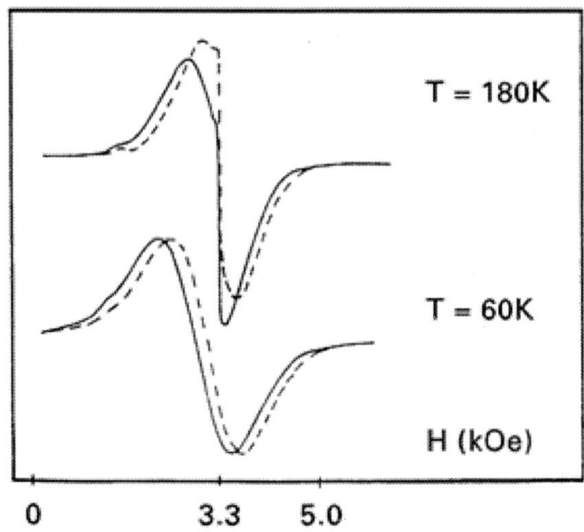

Figure 6. X-band FMR spectra of maghemite FC nanoparticles dispersed on glycerol, with a 10 kOe cooling field. Solid and dashed lines correspond respectively, to configurations with the scanning field parallel (θ = 0°) and perpendicular (θ = 90°) to the cooling field [27].

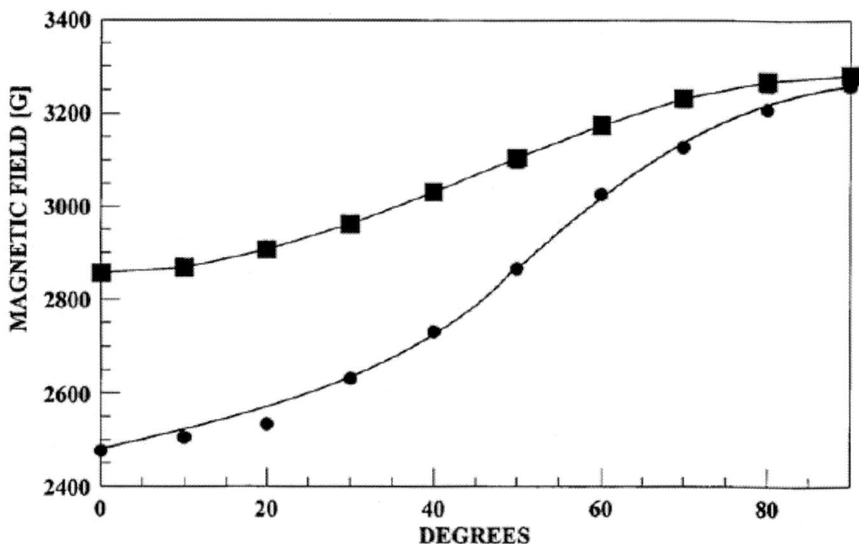

Figure 7. Dependence of the resonance field with θ for field-cooled (FC) hematite NPs at 300 and 155 K (lower curve) [24].

The fitting of the angular resonance field dependence is useful to determine anisotropy parameters. The dependence can be given by [32]:

$$H_r = H_o - H_a(1/2)(3\cos^2\theta - 1) \tag{1}$$

where Hr is the resonance field, Ha = 4|K|/M is the anisotropy field (with |K| the absolute value of the anisotropy constant and M the magnetization), and Ho measures the g value. The g value is a constant describing the relation between the energy of the microwave radiation and the dc magnetic field. For the parallel configuration the resonance field is:

$$H_r = H_o - 4|K|/M \tag{2}$$

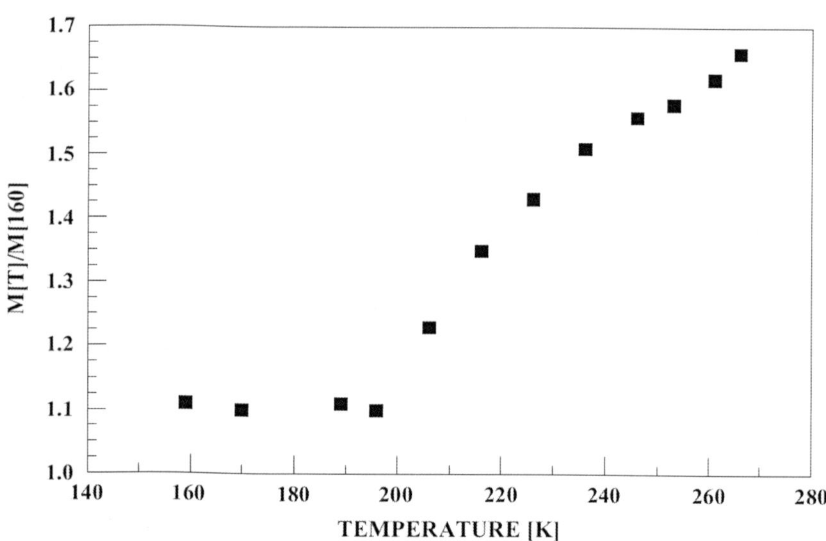

Figure 8. Reduced magnetization dependence with temperature for a FC hematite NPs [24].

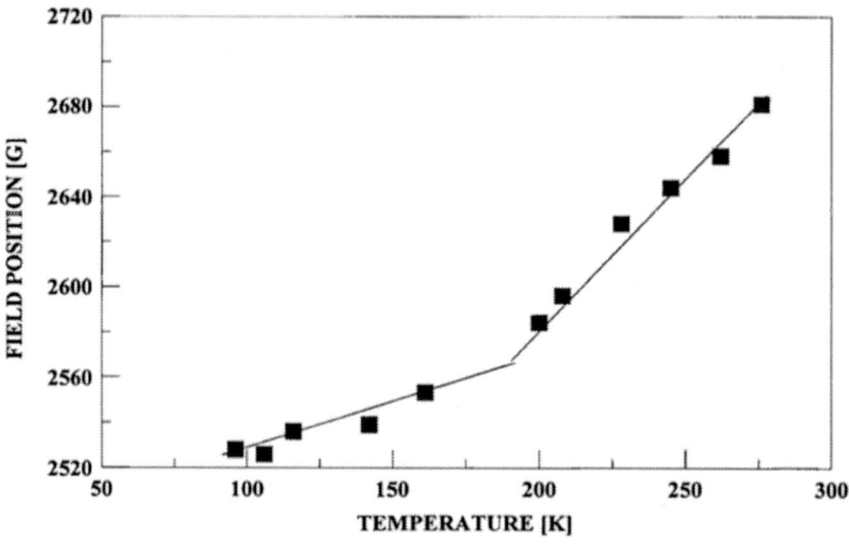

Figure 9. Resonance field dependence with temperature for hematite sample of Fig. 8 [24].

Magnetic phase transitions produce significant changes in |K|/M and hence on the resonance field. The parallel configuration can therefore be used to investigate phase transitions. Figure 8 shows the temperature dependence of magnetization of FC hematite NPs, parallel to the cooling field [24]. A change in the slope is observed in good agreement with observed transitions from a disordered phase to a magnetically ordered phase. In contrast with bulk hematite (antiferromagnetic above 260 K), these hematite NPs exhibited a superparamagnetic behavior, which can be explained in terms of a weakly ferromagnet below ~ 200 K, as shown in figure 8. The reported transition is also observed in plots of the resonance field against temperature (figure 9) and other FMR spectrum parameters, which suggest the FMR technique as a useful tool to investigate phase transitions on MNP's. Effects of the aggregation state Ferrite NPs can show the effects of two extreme aggregation states, namely monodisperse NPs, and clusters of a few hundreds of NPs. Samples of composition Zn0.5Ni0.5Fe2O4 were prepared by the polyol method, and designated as "A" for monodisperse state, and "B" for clusters. FMR spectra exhibited significant differences, as shown in

Fig. 10. The decrease in the effects of surface for sample B appear in the form of a lower resonance field and an increase in the linewidth, as compared with sample A. The former can be understood in terms of a larger internal field in the cluster sample as a consequence of the aggregation of NPs; simply, the NPs inside the cluster tend to behave as grains in a bulk material. The effects of surface (crystal defects, unsatisfied bonds, etc.) are relieved by the presence of other NPs. Magnetic exchange interactions among neighboring NPs can also be assumed, which increases the internal field and thereby decreases the applied magnetic field needed to fulfill the Larmor equation conditions. On the other hand, these interactions increase the linewidth, especially the random distribution of anisotropy axes.

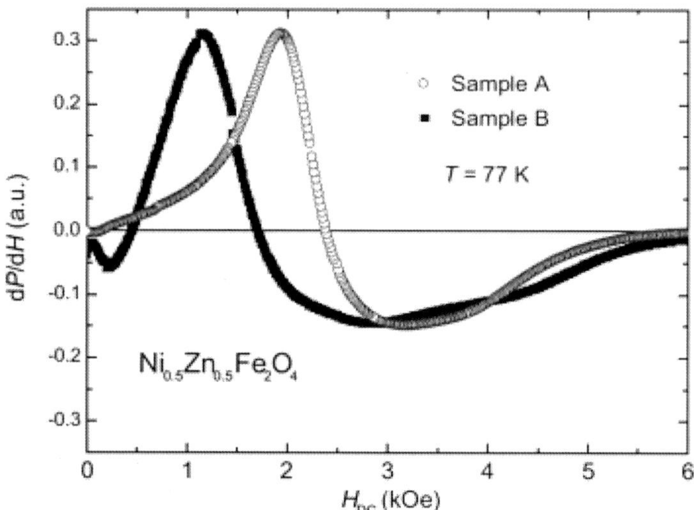

Figure 10. FMR spectra of Ni-Zn ferrites at 77 K, in monodisperse state (Sample A), and clusters (sample B) [18].

At room temperature, as shown in Fig. 11, isolated NPs show a superparamagnetic phase and again a larger resonance field with a significant reduction in linewidth as the factors just mentioned, associated with an ordered magnetic structure, are absent. The cluster sample exhibits a reduced linewidth, but always larger than the superparamagnetic state.

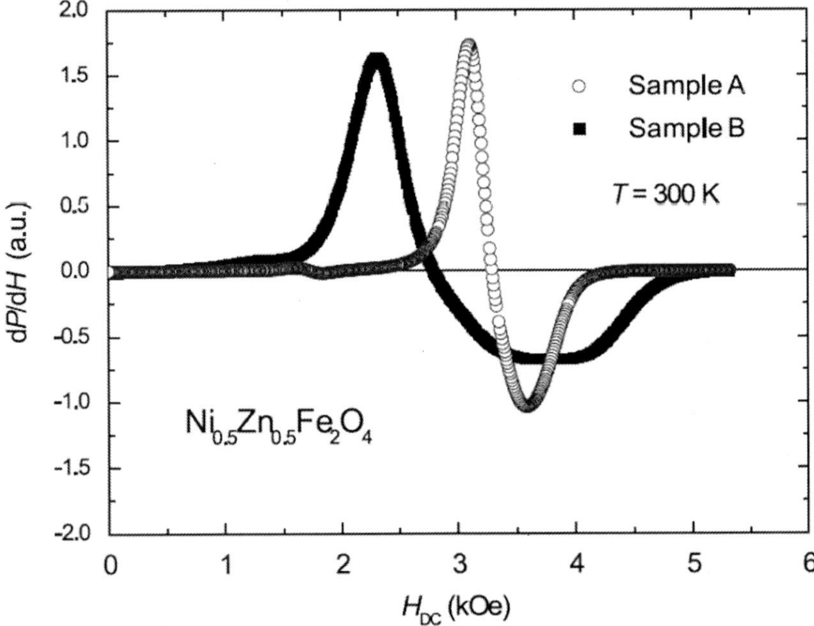

Figure 11. FMR spectra of Ni-Zn ferrites at 300 K, in monodisperse state (Sample A), and clusters (sample B) [18].

FERROMAGNETIC RESONANCE IN NANOSTRUCTURED FERRITES

Ferrite NPs have to be consolidated as a high density solid for many applications (in electronic devices, for instance), where a powder is unstable. Typical sintering processes needing high temperatures are difficult to apply, as NPs tend to grow very rapidly at temperatures above 500°C, losing the nanometric size range and thus the different properties associated with this size range. A particularly well suited method to consolidate NPs into a high density nanostructured solid preserving grains within the nanometric range is Spark Plasma Sintering (SPS for short) [33]. Also known as pulsed electric current sintering (PECS), in this technique the sample (typically in the form of a powder) is placed in a graphite die and pressed by two punches at pressures in the 200 MPa range, while a strong electric current goes through the system; the die is shown in Fig. 12. SPS therefore consolidates powders under the simultaneous action of pressure and electric current pulses (typically of a few milliseconds in duration [33]). The electric current pulses result in a very rapid heating of the sample, at rates as high as 1000°C/min. If the sample is a good conductor, current goes through it and the heating is even more efficient. A significant point is that the electric current has a significant impact on the atomic diffusion during the process [34]. The sintering process can then reach high densities at very low temperatures and extremely short times [35]. Obviously, SPS can also be utilized for reactive sintering involving a chemical reaction [36]. SPS has been used to consolidate spinel [37], garnet [38], and hexagonal ferrites [39]. In the case of spinel Ni0.5Zn0.5Fe2O4 ferrites, samples prepared in the form of 6-8 nm nanoparticles by the forced hydrolysis in a polyol method [12], were consolidated by SPS at temperatures in the 350-500°C range by times as short as 5 min. Just for comparison, the typical conditions for sintering in the classic solid state reaction are 1200°C for at least 4 hours. NPs growth was controlled, as the final grain size in the consolidated ferrite was about 60 nm, even for the highest (500°C) SPS temperature [40].The densities reached values as high as 94% of the theoretical value. FMR spectra obtained at 77 and 300 K are shown in Fig 13, together with the FMR signal corresponding to the original NPs. The resonance field exhibited a decrease as the SPS temperature increased, which can be explained in terms of the components of the total field in the Larmor expression, $\omega = \gamma H$. The main components of the total field are H = HDC + HX + Ha + …, where HDC is the applied field, HX is the exchange field responsible for magnetic ordering, and Ha is the anisotropy field. As magnetic ordering sets in, HX and Ha increase and therefore the applied field is decreased in order to fulfill the Larmor resonance conditions. As sintering progresses, the surface effects decrease and the material has the tendency to behave as a bulk ferrite.

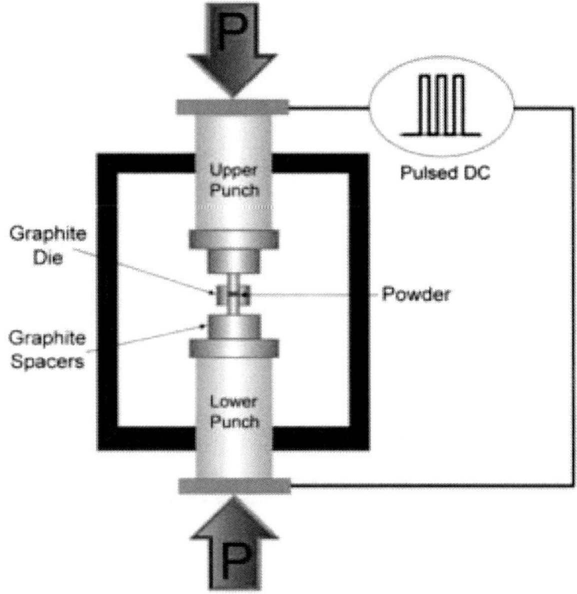

Figure 12. Schematic of spark plasma sintering apparatus [33].

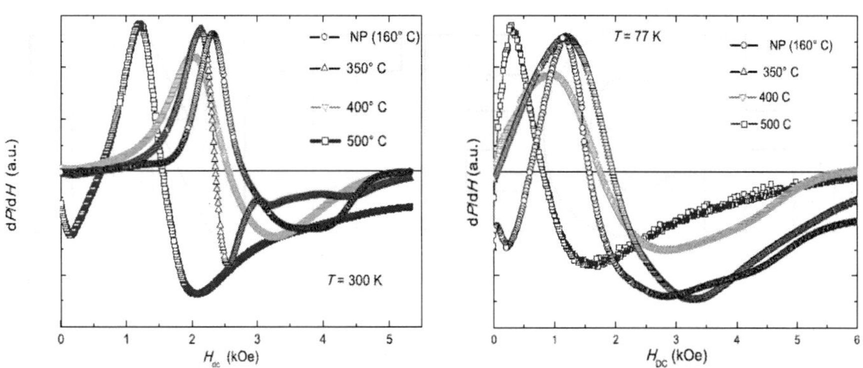

Figure 13. Ferromagnetic resonance of Ni-Zn ferrites prepared by a chimie douce method followed by SPS at temperatures in the 350-500°C range for 5 min. The FMR response of the original NPs is also shown for comparison. (Adapted from [40]).

All the samples exhibited a large broadening in the linewidth, generally interpreted by considering a random distribution of the anisotropy axis in single domain NPs [41]. The broadening decreases as consolidation increases, as the surface effects are diminished by formation of grain boundaries. At room temperature the linewidth decreases since all the samples approach the paramagnetic state where the internal field is eliminated and only the applied field is involved in the Larmor expression. The NPs signal exhibited the lowest linewidth as at room temperature these NPs are superparamagnetic; all the consolidated samples showed a ferrimagnetic behavior at 300 K as compared with the as-produced sample which has a blocking temperature about 90 K.

PARAMAGNETIC RESONANCE IN FERRITE NANOPARTICLES

Zinc ferrite is a very good material to investigate the site occupancy by cations. In spite of a relatively large cation radius, Zn+2 has a tendency to occupy tetrahedral sites, while Fe3+ fill octahedral sites [1], in other words, a "normal" spinel. This arrangement leads to B-O-B superexchange interactions between iron cations and therefore to an antiferromagnetic structure with a Néel temperature about 9 K. In the case of nanosized Zn ferrite, however, the cation distribution can be significantly different; some degree of inversion occurs [1,42], with a fraction of Zn2+ on octahedral sites, and hence Fe3+ on both sites. The spinel becomes ferromagnetic, with a Curie temperature well above 9 K. The variations in linewidth, DBPP, and g-factor, geff in these ferrites depend mainly on the interparticle magnetic dipole–dipole interactions and intraparticle super-exchange interactions. On the other hand, the interparticle superexchange between magnetic ions (through oxygen) can reduce the value of linewidth. The magnitude of this interaction is determined by the relative position of metallic and oxygen ions. When the distance between the metallic and oxygen ions is short, the metal cations have an exactly half-filled orbital, and the angle between these two bonds is close to 180°, the superexchange interaction is the strongest [43].

Several factors can produce a distribution of local field, such as unresolved hyperfine structure, g-value anisotropy, strain distribution, crystal defects. The field strength on a particular spin is then modulated by local field distribution and leads to an additional linewidth broadening [44,45]. The strain distribution produced by a small average particle size can cause a small resonance signal [1]. As the particle size is increased, such small signal disappears and only the broad signal remains. This suggests that the particle size has a threshold value above which the strains are relieved. In order to gain some insight into the relationship between internal structure and EPR spectra, it can be useful to compare the properties of several iron-based oxide NPs embedded in a polytethylene matrix, prepared by the same method: Fe_2O_3, $BaFe_2O_4$, and $BaFe_{12}O_{19}$ [46]. The experimental EPR spectra of the samples are presented in Figs. (14-20). At room temperature the EPR spectra of all the samples show a "two-line pattern" (Fig. 14) which is typical of superparamagnetic resonance (SPR) spectra. The relative intensity of these lines depends on the particle size and shape distribution function [47]. For the Fe_2O_3 and $BaFe_2O_4$ samples, the broad line predominates in the room temperature spectra; the opposite is observed for the $BaFe_{12}O_{19}$ sample, where the narrow line is more pronounced.

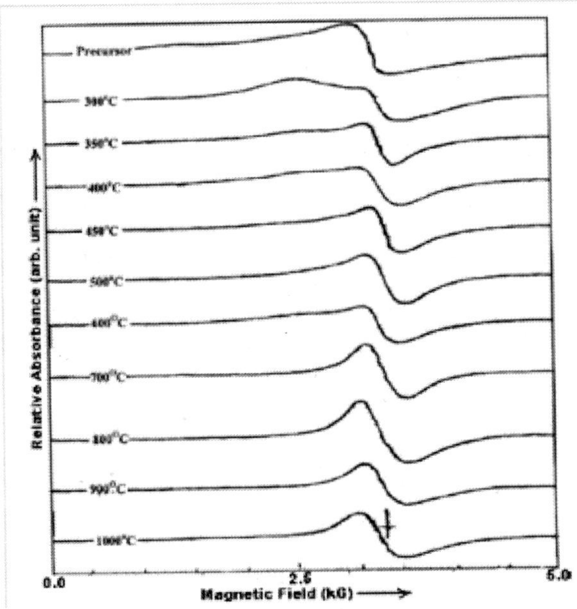

Figure 14. EPR spectra of zinc ferrite sintered at different temperatures

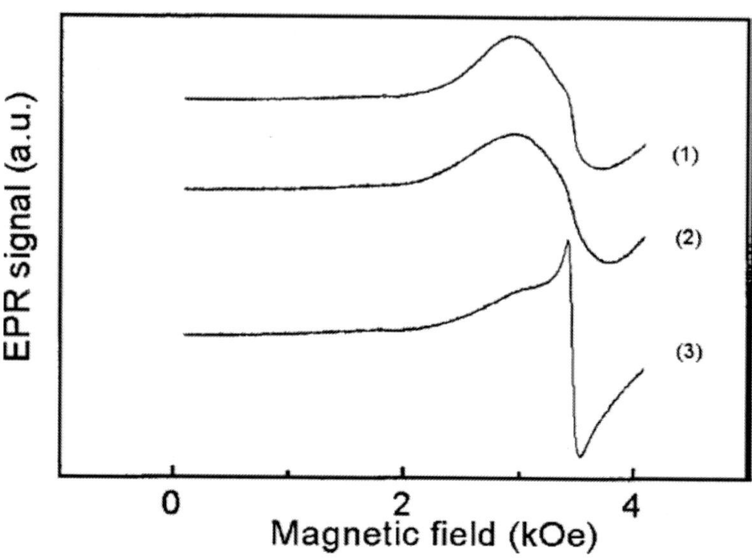

Figure 15. Room temperature EPR spectra of nanoparticles: Fe_2O_3 (curve1), $BaFe_2O_4$ (curve 2), and $BaFe_{12}O_{19}$ (curve3) [46].

At room temperature, the spectra of all samples show a "two-line pattern" (Figure 15) which is typical of superparamagnetic resonance. These spectra can be considered as a broader line superimposed on a narrow line. The relative intensity of these lines depends on the particle size and shape distribution function, as well as on the magnitude of the magnetic anisotropy. For Fe2O3 and BaFe2O4 samples, the broad line predominates in the RT spectra. This line is characterized by a peak-to-peak linewidth of $\Delta H \sim 850$ Oe and an effective gvalue of 2.07. In the RT spectrum of BaFe12O19 sample, in contrast, the narrow line is more visible, with $\Delta H \sim 120$ Oe and $g \sim 2.0$. At low temperatures, the EPR spectra of Fe2O3 change significantly (Fig.16). On cooling below 100 K, the broad line S1 shows a monotonous increase of the linewidth ΔH and a decrease of the amplitude A. However, below about 50 K new resonances, S2 and S3, appear in the spectra of Fe2O3 (Figs. 16 and 17). It is interesting that S2 behaves like a typical paramagnetic resonance signal, namely, the amplitude increases and the linewidth decreases as the temperature is diminished. The EPR spectra suggest that Fe2O3 nanoparticles contain both ferromagnetic and antiferromagnetic phases. The weak EPR signal of the rhombic symmetry (g ≈ 4.3) that appears at low temperatures may be attributable to an Δ- Fe2O3 phase that undergoes an antiferromagnetic-like transition near 6 K. [46]

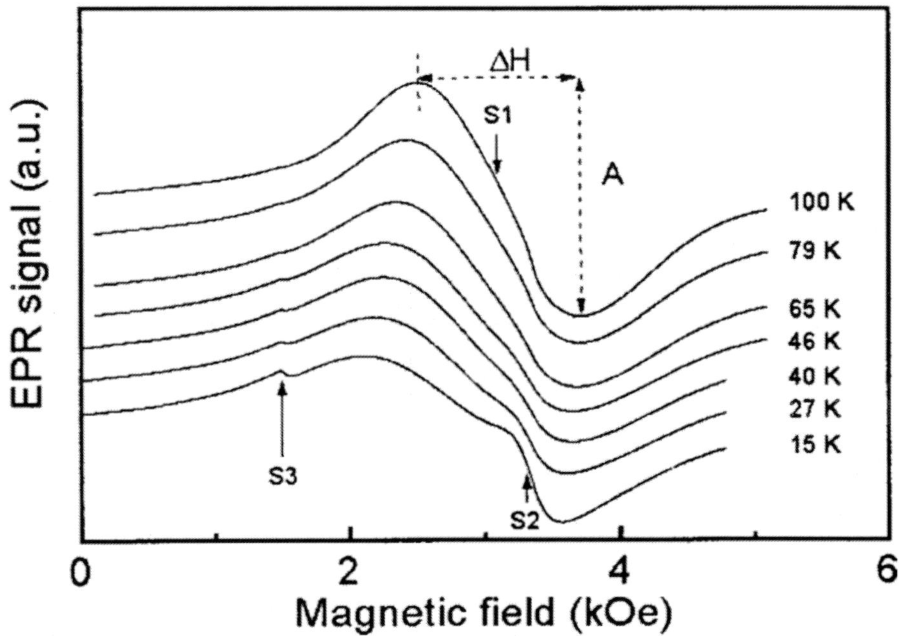

Figure 16. EPR spectra of Fe_2O_3 nanoparticles at different temperatures. [46]

EPR spectra of the $BaFe_2O_4$ nanoparticles at different temperatures are shown in Fig. 18. At all temperatures the spectra are broad and rather asymmetric. A significant shift of the line position to low magnetic fields and a marked spectrum broadening are observed at low temperatures.

On the other hand, the thermal variations of EPR spectra of the $BaFe_{12}O_{19}$ sample is typical of superparamagnetic resonance. The relatively narrow line that dominates at room temperature disappears as temperature decreases (Figs. 19 and 20). The $BaFe_{12}O_{19}$ nanoparticles reveal an EPR signal that is significantly narrowed at high temperatures by superparamagnetic fluctuations. This is evidence of the reduced magnetic anisotropy energy that may be due to the particle's nanosize (effective diameter <10 nm).

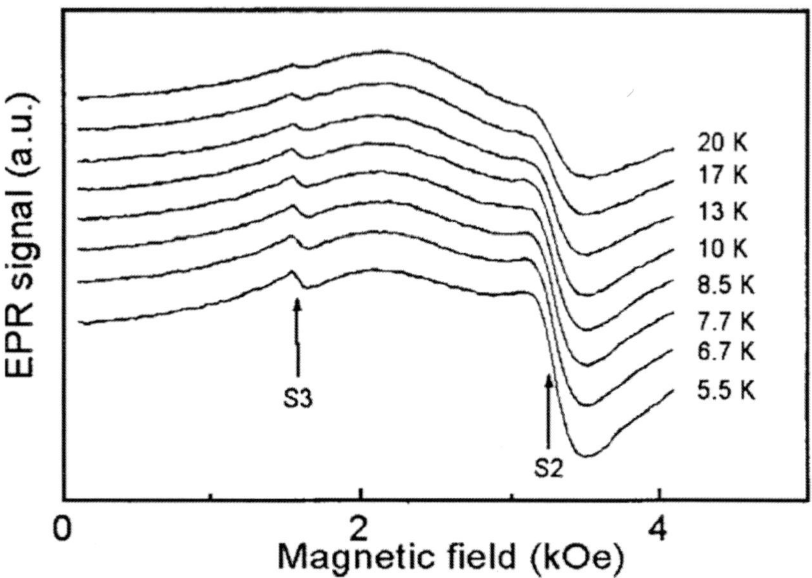

Figure 17. EPR spectra of Fe_2O_3 nanoparticles at low temperatures. [46]

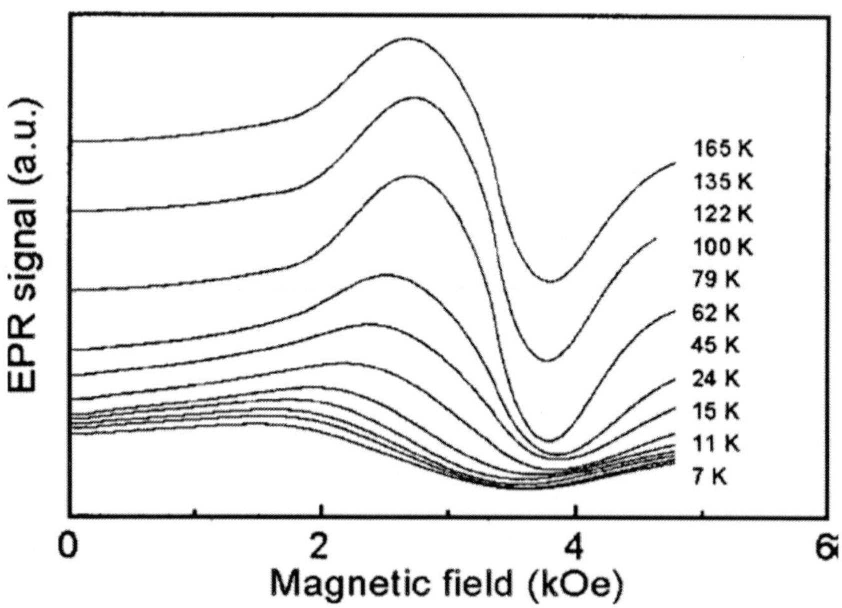

Figure 18. EPR spectra of the $BaFe_2O_4$ nanoparticles at different temperatures. [46]

These results show that EPR spectra at low temperatures are desirable for the correct identification of NPs and a comparison with high temperature experiments allows a better understanding of phenomena related with variations associated with nanosized materials.

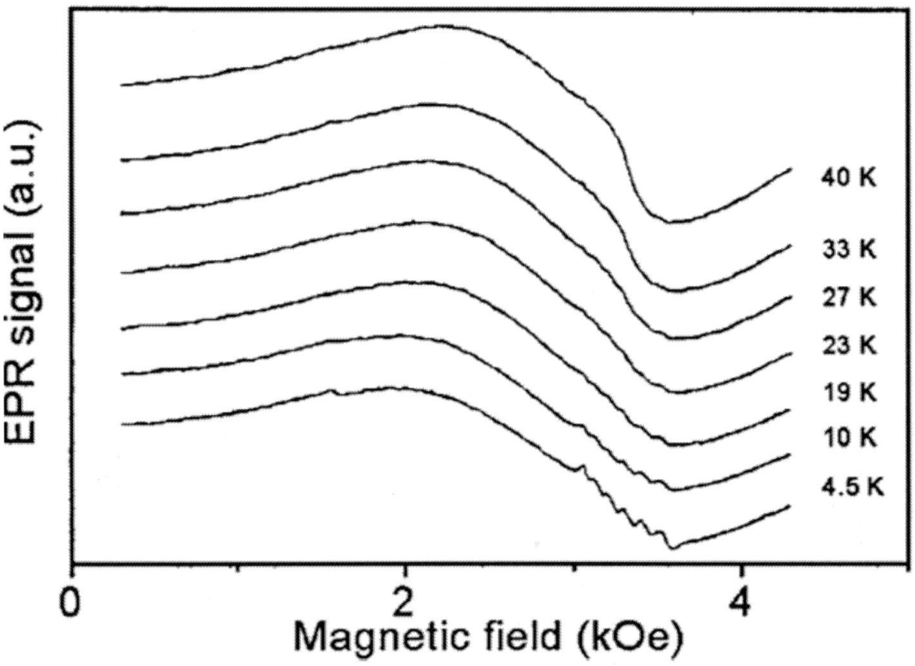

Figure 19. Low-temperature EPR spectra of $BaFe_{12}O_{19}$ nanoparticles. [46]

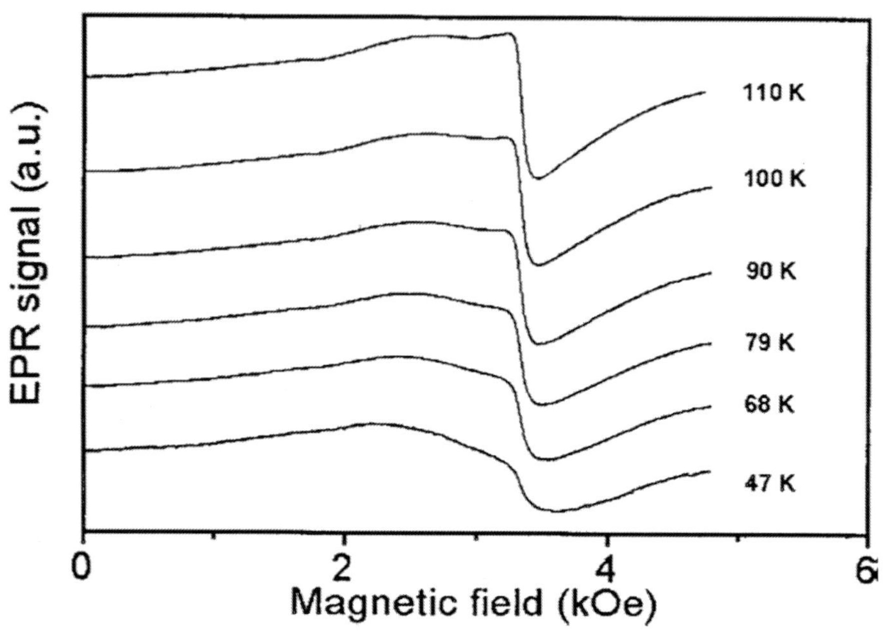

Figure 20. EPR spectra of $BaFe_{12}O_{19}$ nanoparticles at different temperatures. [46]

LOW FIELD MICROWAVE ABSORPTION (LFMA)

Low field microwave absorption (LFMA for short) refers to the non-resonant, hysteretical losses of a material subjected to a high frequency electromagnetic field. Recently, it has become a useful method to investigate magnetization processes [48], magnetoelastic effects [49], phase transitions [50], non-aligned ferromagnetic resonance [51,52], spin arrangements [53,54]. LFMA is similar to giant magnetoimpedance (GMI) [55], but physically different to ferromagnetic resonance (FMR) [56]). GMI, generally defined as the variations of impedance of a magnetic conductor carrying an alternate electrical current when subjected to an external magnetic field [57], extends into a very wide frequency range. Clearly, GMI is a non-resonant phenomenon as confirmed by two facts: GMI does not fulfill the resonant Larmor conditions, and exhibits magnetic hysteresis.

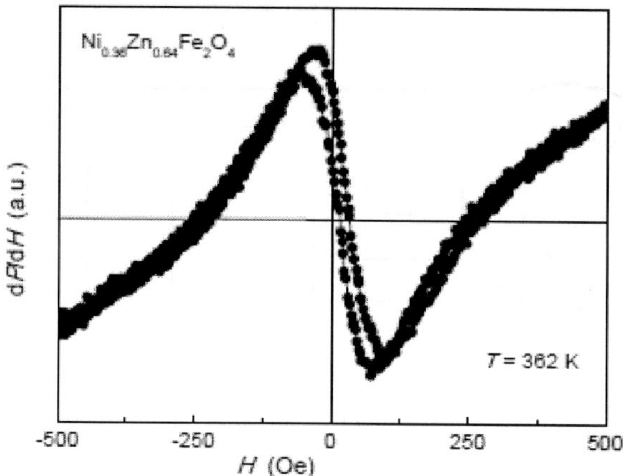

Figure 21. Typical "positive" LFMA signal from a Ni-Zn bulk ferrite (Adapted from [59]).

LFMA is associated with magnetization processes in magnetically ordered materials, in the process from the unmagnetized state to the magnetic saturation. In bulk ferro and ferrimagnetic materials, LFMA exhibits a flat response in the paramagnetic phase. To measure experimentally LFMA in a typical FMR/EPR facility, the applied field has to be cycled; usually between -1 kOe and +1kOe is enough. Also, a device to compensate for the remanent magnetization is needed in most electromagnets. An important parameter is, of course, the total anisotropy field of the particular sample. In most cases, LFMA exhibits a critical behavior at the total anisotropy field in the form of a maximum and a minimum, leading to a characteristic signal as shown in Fig. 21 [59]. In bulk Ni-Zn ferrite, a correlation exists between the magnetocrystalline anisotropy and the halfpeak-to-peak, measurement of LFMA. Figure 22 shows a comparison between the peak-topeak LFMA field (divided by 2) [60] and a calculation from magnetocrystalline constant, K1,published results [61]. The small differences can be attributed to shape anisotropy, as the calculation is based solely on K1. LFMA seems to be associated with spin structure. By convention, this signal can be assigned as "positive", simply because it has the same shape than the FMR/EPR signal, i.e., a maximum and a minimum when observed from left to right. A positive LFMA sign has been observed in most insulator and semiconductor materials, while a "negative" signal appears for most metallic conductors, as shown in Fig. 23, for a Co-rich CoFeBSi amorphous microwire [59]. An interesting result was found in bulk Ni-Zn ferrites showing the Yafet- Kittel triangular arrangement [62], by measuring the LFMA as a function of temperature in the 150 K-240K temperature range [63]. The LFMA sign changed from negative at ~ 154 K to positive at T ≥ 240 K.

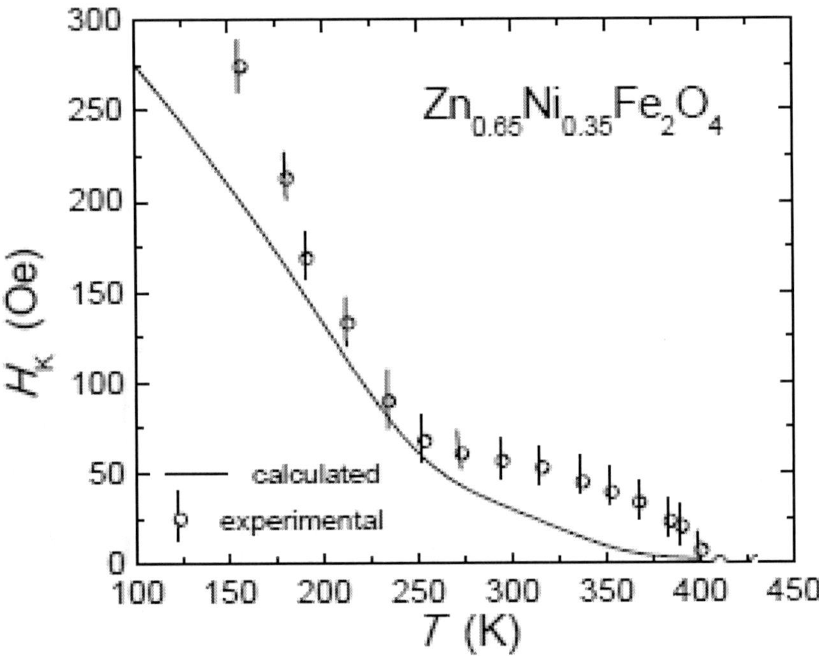

Figure 22. Fig. 21. Correlation between the field on LFMA critical points (divided by 2) and a direct calculation of magnetocrystalline anisotropy in Zn0.65Ni0.35Fe$_2$O$_4$ bulk ferrites [60]. K1 anisotropy constant data for the calculation was obtained from ref [61].

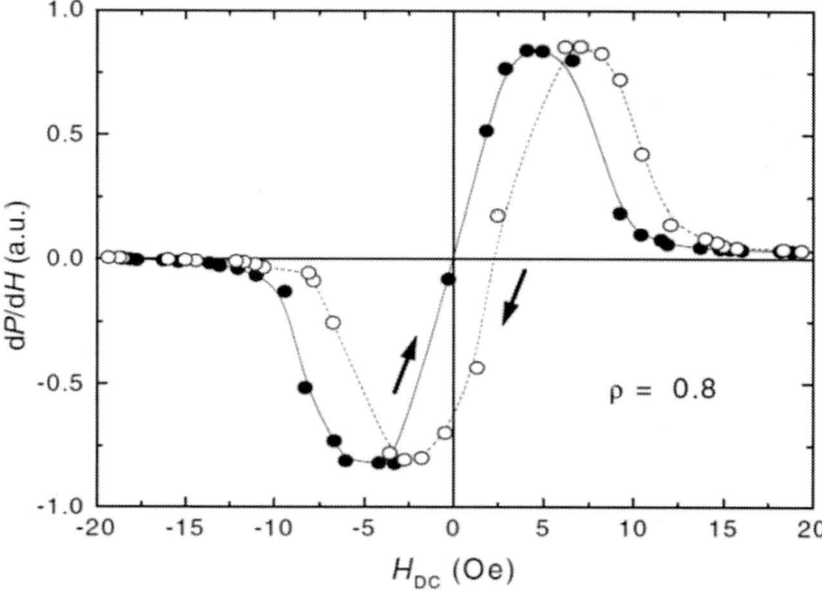

Figure 23. LFMA from a Co-rich CoFeBSi glass-covered amorphous microwire [59]. This is an example of a "negative" LFMA signal, mostly observed in conductor materials (compare with Fig. W above).

LFMA in NPs

LFMA can provide useful insights into the structure of NPs. When two different aggregation states are compared, i.e., monodisperse and clustered NPs with the same composition and NP diameter, clearly different spectra are obtained. By varying the synthesis conditions in the forced hydrolysis in a polyol method [12], Ni-Zn ferrites can be obtained as monodisperse, well crystallized ~ 6 nm NPs on one hand, and labeled as sample "A"; on the other, clusters about ~ 100 NPs constituted by NPs with the same composition and diameter [58], labeled as sample "B". It is interesting to mention that high resolution transmission electron microscopy (HRTEM) showed some epitaxial arrangements within the clusters of sample B. Sample A showed a superparamagnetic behavior at room temperature, and a blocking temperature about 50 K [58]. Sample B, in contrast, exhibited a ferromagnetic behavior up to 300 K (blocking temperature > 300

K), in spite of being constituted by NPs of the same composition and NP diameter. Clusters effectively decrease the effects of surface producing samples with a general behavior between that of NPs and bulk materials. Figure 24 shows the LFMA signal from sample A at 77 and 300 K. AT the low temperature, a positive LFMA behavior is observed, corresponding to a non-conductor material. At 300 K, however, a flat response appears, associated with the superparamagnetic phase of these monodisperse NPs. Clearly, the superparamagnetic phase behaves like a paramagnetic signal. This flat signal with a small slope is associated with the microwave absorption of non-interacting dipoles [64].

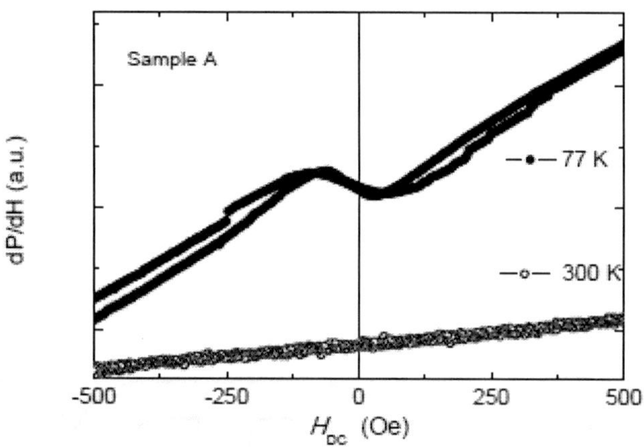

Figure 24. LFMA signal from monodisperse, ~ 6 nm diameter, Zn0.5Ni0.5Fe_2O_4 NPs, at 77 and 300 K (adapted from [58]).

Clusters of the same NPs, in contrast, showed a clear positive LFMA signal at room temperature, see Fig. 254. This is consistent, as this sample B behaved as a ferromagnetic phase close to the bulk properties. For low temperatures, however, sample B exhibited a very different signal as shown in Fig. 26, with an important hysteresis and no similarity to either positive or negative character. As mentioned above, bulk Ni-Zn ferrites with Zn content x > 0.5 (in ZnxNi1-xFe_2O_4) show a triangular Yafet-Kittel spin arrangement, which appears as an evolution from negative LFMA signal at low temperatures towards a positive signal at temperatures close to room temperature [63]. Inspection of Figs. 24-26 reveals that in the case of sample A at 77 K (monodisperse NPs), no YK arrangement is manifested. Some evolution is found in sample B (clusters) at low T, but no full inversion of LFMA signal is observed. A preliminary explanation of these findings is that the surface effects inhibit the formation of the YK structure. Another possibility is based on the fact that cation distribution in NPs as synthesized by the polyol method is often different [42] as compared with bulk ferrites (prepared by the solid state reaction at high temperatures). A different cation distribution (in the present case a non-negligible concentration of Zn on B sites, for instance) could hence lead to different conditions for the formation of the YK structure.

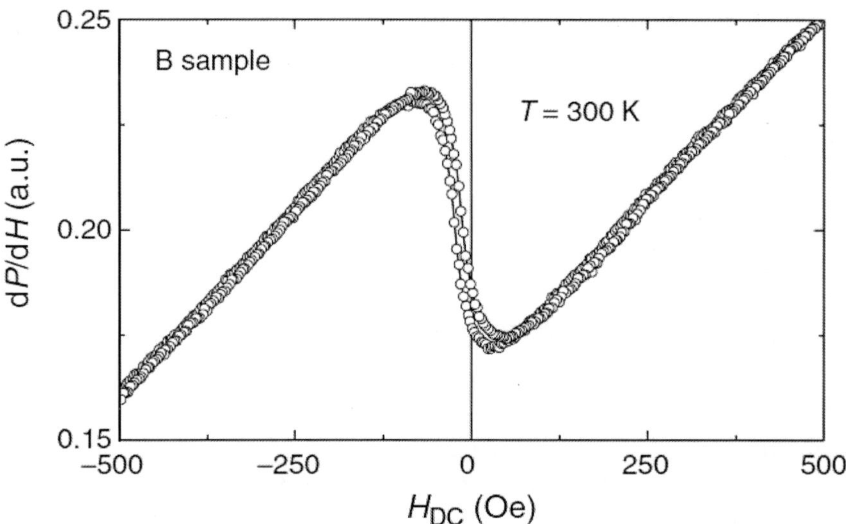

Figure 25. LFMA signal from epitaxial clusters of Zn0.5Ni0.5Fe_2O_4 ferrites, formed by a few hundreds of ~ 6 nm NPs, at 300 K.

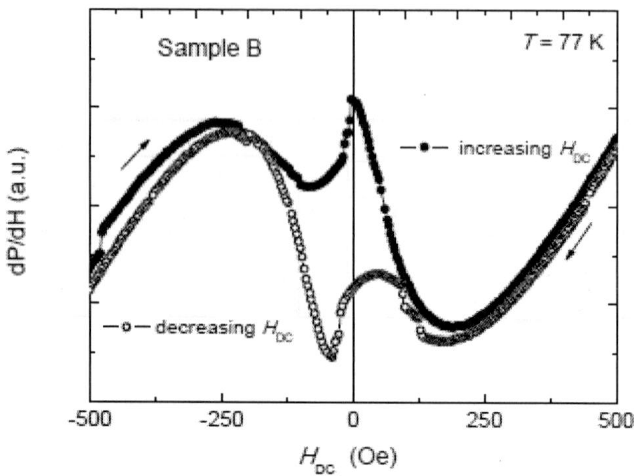

Figure 26. LFMA signal of epitaxial clusters of Zn0.5Ni0.5Fe_2O_4 ferrites, formed by a few hundreds of ~ 6 nm NPs, at 77 K.

CONCLUSIONS

Microwave absorption (MA) is a very sensitive phenomenon and has become an extremely powerful characterization tool. MA accurately depends on all the factors surrounding unpaired electrons; it can play a significant role in the characterization of the complex and fascinating development of magnetic nanoparticles. In this brief review, recent results on the characterization of magnetic nanoparticles and consolidated spinel ferrites by means of ferromagnetic resonance, paramagnetic resonance, and low field microwave absorption have been presented.

Author Details

Gabriela Vázquez-Victorio, Ulises Acevedo-Salas and Raúl Valenzuela

Institute for Materials Research, National Autonomous University of México, México

ACKNOWLEDGEMENT

Authors acknowledge partial support for this work from ANR-CONACyT grant 139292, as well as PAPIIT-UNAM grant IN141012.

REFERENCES

1. Valenzuela R. Magnetic Ceramics. Cambridge University Press (2005) (ISBN: 0-521- 01843-9).
2. Amiri S, Shokrollahi H. The role of cobalt ferrite magnetic nanoparticles in medical science. Materials Science and engineering C 2013: 33(1) 1-8.
3. Mahmoudi M, Sant S, Wang B, Laurent S, Sen T. Superparamagnetic iron oxide nanoparticles (SPIONs): development, surface modification and applications. Advanced Drug Delivery Reviews 2011: 63(1) 24-46-
4. Kashevsky BE, Agabekov BE, Kashevsky SB, Kekalo KA, Manina EY, Prokhorov IV, Ulashchik VS. Study of cobalt ferrite nanosuspensions for low frequency ferromagnetic hyperthermia. Particuology 2008; 6(5) 322-333.
5. Kliza E, Strijkers GJ, Nicolay K. Multifunctional magnetic resonance imaging probes. Recent Results in Cancer Research 2013; 187 151-190.
6. Xu Y, Wang E. Electrochemical biosensors based on magnetic micro/ nanoparticles. Electrochimica Acta 2012; 84(1) 62-73.
7. Dahawan SK, Singh K, Bakhshi AK, Ohlan A. Conducting polymer embedded with nanoferrite and titanium dioxide nanoparticles for microwave absorption. Synthetic Metals 2009, 159(21-22) 2259-2262.
8. Liu R, Lal R. Nanoenhanced materials for reclamation of mine lands and other degrade soils: a review. Journal of Nanotechnology 2012; 461468-1-461468-18. doi: 10.1155/2012/461468.
9. Hua M, Zhang S, Pan B, Zang W, Lv L, Zhang Q. Heavy metal removal from water/wastewater by nanosized metal oxides. Journal of Hazardous Materials 2012; 211- 212 317-331.
10. Hosokawa M, Nogi K, Naito M, Yokoyama T. Nanoparticle Technology Handbook (Elsevier, Amsterdam) 2007.
11. Tartaj P, Morales MP, Gonzalez-Carreño V, Veintemillas-Verdaguer S, Serna CJ. Advances in magnetic nanoparticles for biotechnology applications. Journal of Magnetism and Magnetic Materials 2005; 290(1) 28-34.
12. Beji Z., Ben Chaabane T., Smiri L.S., Ammar S., Fiévet F., Jouini N., and Grenèche J.M. Synthesis of nickel-zinc ferrite nanoparticles in polyol: morphological, structural and magnetic properties. physica status solidi (a) 2006; 203(3) 504-512.
13. Tartaj P, Morales MP, Gonzalez-Carreño V, Veintemillas-Verdaguer S, Serna CJ. The preparation of magnetic nanoparticles for applications in biomedicine. Journal of Physics D: Applied Physics 2003; 36 R182.

14. Dang F, Enomoto N, Hojo J, Enpuku K. A novel method to synthesize monodispersed magnetite nanoparticles. Chemical Letters 2008; 37(5) 530-531.

15. Matsunaga T, Okamura Y, Tanaka T. Biotechnological application of nano-scale engineered bacterial magnetic particles. Journal of Materials Chemistry 2004; 14(14) 2099-2105.

16. Guimaraes AP. Magnetism and magnetic resonance in solids. Wiley VCH (1998). (ISBN: 0471197742).

17. Well JA, Bolton JR. Electron paramagnetic resonance: elementary theory and practical applications 2nd edition. John Wiley and Sons, 2007. (ISBN: 047175496X); Brustolon MR. Electron paramagnetic resonance: a practitioner toolkit, 1st edition. John Wiley and Sons (2009) (ISBN: 0470258829).

18. Valenzuela R, Herbst F, Ammar S. Ferromagnetic resonance in Ni–Zn ferrite nanoparticles in different aggregation states. Journal of Magnetism and Magnetic Materials 2012; 324(21) 3398-3401.

19. Guskos N, Zolnierkiewicz G ,Typek J, Guskos A, Czech Z. FMR study of γ-Fe2O3 agglomerated nanoparticles dispersed in glues. Reviews on Advanced Materials Science 2007;14 57-60.

20. Thirupathi G, Singh R. Magnetic Properties of Zinc Ferrite Nanoparticles. Institute of Electrical and Electronics Engineers Transactions on Magnetics 2012; 48 3630-3633.

21. De Biasi E, Lima E, Ramos C A, Butera A, Zysler R D. Effect of thermal fluctuations in FMR experiments in uniaxial magnetic nanoparticles: Blocked vs superparamagnetic regimes . Journal of Magnetism and Magnetic Materials 2013; 326(1) 138-146.

22. Sobón M, Lipinski I E, Typek J, Guskos A. FMR Study of Carbon Coated Cobalt Nanoparticles Dispersed in a Paraffin Matrix. Solid State Phenomena 2007; 128 193-198.

23. Shames A I, Rozenberg E, Sominski E, Gedanken A. Nanometer size effects on magnetic order in La12xCaxMnO3 (x = 0.5 and 0.6) manganites, probed by ferromagnetic resonance. Journal of Applied Physics 2012; 111 07D701-1-07D701-2.

24. Owens F. Ferromagnetic resonance observation of a phase transition in magnetic field aligned Fe2O3 nanoparticles. Journal of Magnetism and Magnetic Materials 2009; 321(15) 2386-2391.

25. Gazeau F, Bacri J C, Gendron F, Perzynski R, Raikher Yu L, Stepanov V I, Dubois E. Magnetic resonance of ferrite nanoparticles: Evidence of surface effects. Journal of Magnetism and Magnetic Materials 1998; 186(2) 175-187.

26. Noginov M, Noginova N, Amponsah O, Bah R, Rakhimov R, Atsarkin V A. Magnetic resonance in iron oxide nanoparticles: Quantum features and effect of size. Journal of Magnetism and Magnetic Materials 2008; 320(18) 2228-2232.

27. Gazeau F, Bacri J C, Gendron F, Perzynski R, Raikher Yu L, Stepanov V I, Dubois E. Magnetic resonance of nanoparticles in a ferrofluid: Evidence of

thermofluctuational effects. Journal of Magnetism and Magnetic Materials 1999; 202(2-3) 535-546.

28. Edelman I, Petrakovskaja E, Petrov D, Zharkov S, Khaibullin R, Nuzhdin V, Stepanov A. FMR and TEM studies of Co and Ni nanoparticles implanted in the SiO2 matrix. Applied Magnetic Resonance 2011; 40 363-375.

29. Martinez B, Obradors X, Balcells Ll, Rouanet A, Monty C. Low temperature surface spin-glass transition in gamma-Fe2O3 nanoparticles. Physical Review Letters 1998; 80 181-184.

30. Winkler E, Zysler R D, Fiorani D. Surface and magnetic interaction effects in Mn3O4 nanoparticles. Physical Review B 2004; 70 174406-1-174406-5.

31. Song H, Mulley S, Coussens N, Dhagat P, Jander V. Effect of packing fraction on ferromagnetic resonance in NiFe2O4 nanocomposites. Journal of Applied Physics 2012; 111 (07E348) 07E348_1 - 07E348_3.

32. De Biasi R, Devezas T. Anisotropy field of small magnetic particles as measured by resonance. Journal of Applied Physics 1978; 49 2466-2470.

33. Munir Z.A., Quach D.V. and Ohyanagi M. Electric current activation of sintering: a review of the pulsed electric current sintering process. Journal of the American Ceramic Society 2011; 94(1) 1-19.

34. Anselmi-Tamburini U, Garay JE and Munir ZA. Fundamental investigation on the spark plasma sintering/synthesis process III. Current effect on reactivity. Materilas Science and Engineering A 2005; 407 24-30.

35. Mizuguchi T, Guo S and Kagawa Y. Transmission electron microscopy characterization of spark plasma sintered ZrB2 ceramicc. Ceramics International 2010; 36(3) 943-946.

36. Regaieg Y, Delaizir G, Herbst F, Sicard L, Monnier J, Montero D, Villeroy B, Ammar- Merah S, Cheikhrouhou A, Godart C, Koubaa M. Rapid solid state synthesis by spark plasma sintering and magnetic properties of LaMnO3 perovskite manganite. Materials Letters 2012; 80(1) 195-198.

37. Valenzuela R, Ammar S, Nowak S, and Vázquez G. Low field microwave absorption in nanostructured ferrite ceramics consolidated by spark plasma sintering. Journal of Superconductivity and Novel Magnetism 2012; 25(7) 2389-2393.

38. Gaudisson T, Acevedo U, Nowak S, Yaacoub N, Greneche JM, Ammar S, Valenzuela R. Combining soft chemistry and Spark Plasma Sintering to produce highly dense and finely grained soft ferrimagnetic Y3Fe5O12 (YIG) ceramics. (to be published).

39. Nakamura T, Okano Y, Tabuchi M, Takeuchi T. Synthesis of hexagonal ferrite via spark plasma sintering technique. Journal of the Japan Society of Powder and Powder Metallurgy 2001; 48(2) 166-169.

40. Valenzuela R, Beji Z, Herbst F, and Ammar S. Ferromagnetic resonance behavior of spark plasma sintered Ni-Zn ferrite nanoparticles produced by a chemical route. Journal of Applied Physics 2011; 109(*) 07A329-1-07A329-3.

41. Sukhov A, Usadel KD, and Nowak U. Ferromagnetic resonance in an ensemble of nanoparticles with randomly distributed anisotropy axes.

Journal of Magnetism and Magnetic Materials 2008; 320(1) 31-33.

42. Ammar S, Jouini N, Fièvet F, Beji Z, Smiri L, Moliné P, Danot M, Grenèche JM. Magnetic properties of zin-ferrite nanoparticles synthesized by hydrolysis in a polyol medium. Journal of Physics: Condensed Matter 2006; 18(39), 9055-9069.

43. Sui Y, Xu DP, Zheng FL and Su WH. Electrons spin resonance study of NiFe2O4 nanosolids compacted under high pressure. Journal of Applied Physics 1996; 80 719- 723.

44. A. Abragam and B. Bleaney, Electron Paramagnetic Resonance of Transition Ions (Clarendon Press, Oxford,1970).

45. C. P. Poole and H. A. Farach, Relaxation in Magnetic Resonance (Academic Press, London, 1971).

46. Koksharov Yu. A, Pankratov D A., Gubin SP, Kosobudsky ID, Beltran M, Khodorkovsky Y and Tishin AM. Electron Paramagnetic Resonance of Ferrite Nanoparticles. Journal of Applied Physics 2001;89 2293-2298.

47. Kliava J and Berger R. Size and Shape Distribution of Magnetic Nanoparticles in Disordered Systems: Computer Simulations of Superparamagnetic Resonance Spectra. Journal of Magnetism and Magnetic Materials 1999; 205 328-342.

48. Valenzuela R, Alvarez G, Montiel H, Gutiérrez MP, Mata-Zamora ME, Barrón F., Sanchez AY, Betancourt I, Zamorano R. Characterization of magnetic materials by lowfield microwave absorption techniques. Journal of Magnetism and Magnetic Materials 2008; 320(14) 1961-1965.

49. Montiel H, Alvarez G, Gutiérrez MP, Zamorano R, and Valenzuela R. The effect of metal-to-glass ratio on the low field microwave absorption at 9.4 GHz of glass coated CoFeBSi microwires. IEEE Transactions on Magnetics 2006; 42(10) 3380-3382.

50. Montiel H., Alvarez G., Gutiérrez M.P., Zamorano R., and Valenzuela R. Microwave absorption in Ni-Zn ferrites through the Curie transition. Journal of Alloys and Compounds 2004; 369(1) 141-143.

51. Prinz G.A., Rado G.T. and Krebs J.J. Magnetic properties of single-crystal {110} iron films grown on GaAs by molecular beam epitaxy. Journal of Applied Physics 1982; 53 (3) 2087-2091.

52. Gerhardter F, Li Yi, and Baberschke K. Temperature-dependent ferromagneticresonance study in ultrahigh vacuum: magnetic anisotropies of thin iron films. Physical Review B 1993; 47(17) 11204-11210.

53. Valenzuela R. The temperature behavior of resonant and non-resonant microwave absorption in Ni-n ferrites. In Electromagnetic Waves / Book 1, InTech Open Access Publisher, edited by Vitaliy Zhurbenko, pp. 387 - 402 (2011), ISBN: 978 - 953 - 307 - 304 - 0. Online June 24, 2011 at: http://www.intechopen.com/articles/show/title/thetemperature- behavior-of-resonant-and-non-resonant-microwave-absorption-in-ni-znferrites.

54. Alvarez G., Montiel H., Barron J.F., Gutiérrez M.P., Zamorano R. Yafet-Kittel-type magnetic ordering in Ni0.35Zn0.65Fe2O4 ferrite detected

by magnetosensitive microwave absorption measurements. Journal of Magnetism and Magnetic Materials 2009; 322(3) 348-352.

55. Montiel H, Alvarez G, Betancourt I, Zamorano R. and Valenzuela R. Correlation between low-field microwave absorption and magnetoimpedance in Co-based amorphous ribbons. Applied Physics Letters 2005 ; 86(7) 072503-1-072503-3.

56. Barandiarán M., García-Arribas, A., and de Cos, D. Transition from quasistatic to ferromagnetic resonance regime in giant magnetoimpedance. Journal of Applied Physics 2006; 99 103904-1-103904-4.

57. Knobel M, Kraus L, and Vázquez M. Magnetoimpedance. Handbook of Magnetic Materials, edited by K.H.J. Buschow, Elsevier Amsterdam 2003; 15 497-563.

58. Valenzuela R., Ammar S., Herbst F., Ortega-Zempoalteca R. Low field microwave absorption in Ni-Zn ferrite nanoparticles in different aggregation states. Nanoscience and Nanotechnology Letters 2011; 3(4) 598-602.

59. Valenzuela R., Montiel H., Alvarez G., and Zamorano R. Low-field non-resonant microwave absorption in glass-coated Co-rich microwires. Physica Status Solidi A 2009; 4, 652-655.

60. Valenzuela R. (Unpublished results)

61. Broese van Groenou A., Schulkes J.A., and Annis D.A. Magnetic anisotropy in some nickel zinc ferrites. Journal of Applied Physics 1967; 38(3), 1133 (1967).

62. Yafet Y. and Kittel, C. Antiferromagnetic arrangements in ferrites. Physical Review 1952; 87(2) 290-294.

63. Alvarez G., Montiel H., Barrón J.F., Gutiérrez M.P. Zamorano R. Yafet-Kittel-type magnetic ordering on NiZnFe2O4 ferrite detected by magnetosensitive microwave absorption measurements. Journal of Magnetism and Magnetic Materials 2010; 322(3) 348-352.

64. Alvarez G., Font R., Portelles J., Zamorano R., and Valenzuela R. Microwave power absorption as a function of temperature and magnetic field in the ferroelectromagnet Pb(Fe1/2Nb)O3. Journal of the Physics and Chemistry of Solids 2007; 68(7) 1436-1442.

Chapter 8

CARBON AND SILICON FLUORESCENT NANOMATERIALS

Joaquim G. G. Esteves da Silva

Centro de Investigação em Química da Universidade do Porto (CIQ-UP), Department of Chemistry and Biochemistry, Faculty of Sciences, University of Porto, Porto, Portugal

INTRODUCTION

The most known fluorescence nanoparticles are the semiconductor nanocrystals usuallyb called quantum dots (QD) (Smith & Nie, 2010). These nanoparticles, with typical sizesb between 1 to 12 nm, are being used in many advanced technological applications, forb example in bioimaging (Gerion et al., 2001; Murcia et al., 2008; Williams et al., 2009).b However, some of the materials that compose the QD are rare in the earth crust and highlyb toxic (Lovric et al., 2005).b Recently, the elements carbon and silicon, which are among the most abundant elements inb the earth crust and are intrinsically non-toxic, become quite important elements inb nanochemistry because they originate fluorescent nanostructures with relatively highb quantum yield. Bulk carbon and silicon materials are not fluorescent but the correspondingb nanomaterials are strongly fluorescent, allow emission colour tuning and are non-blinkingb nanoparticles with high scientific and technological potential.b

Carbon nanomaterials are already well known, like for example the fullerenes, carbon nanotubes (CNT), either single-wall (SWNT) or multiple-wall (MWNT), carbon nanofibers and graphene (Liu et al., 2010). Highly fluorescent carbon nanomaterials, here called carbon dots (CD), were only accidently discovered in 2004 during the electrophoretic purification of SWNT derived from arc-discharge soot (Xu et al., 2004). However, in the years 2000 and 2007, studies observed strongly fluorescence shortened MWNT and shortened SWNT (Luo et al., 2007; Riggs et al., 2000). Indeed, CD are carbon based nanomaterials that possesses similar size and surface functionality they constitute different families of nanomaterials and are constituted mainly by carbon with sp2 hybridization characteristic of monocristaline graphite with relatively high oxygen contents (Baker & Baker, 2010; Esteves da Silva & Gonçalves, 2011; Fan & Chu, 2010; Xu et al., 2004). CD are different from nanodiamonds because these last nanoparticles are constituted by about 98% carbon with a sp3 hybridization with small amounts of graphitic carbon on the surface that are synthesized from milling microdiamonds, chemical vapour deposition, shockwave or detonation processes (Baker & Baker, 2010). New simplified synthetic pathways are being proposed and the number of potential technological applications of CDs is increasing in the last years (Baker & Baker, 2010; Esteves da Silva & Gonçalves, 2011; Fan & Chu, 2010).

Another class of potentially non-toxic and biocompatible fluorescent nanoparticles are those of silicon (silicon dots - SD). One fluorescent silicon nanoparticle (silicon porous nanoparticles) were first prepared in the 1950s by electrochemical etching in hydrofluoric acid and ultraviolet irradiation (Parkhutik & Timashev, 2000) but its red luminescence was only discovered in 1990 (Canham, 1990). However, porous silicon is fragile and highly reactive limiting its straightforward applications. SD with typical sizes in the order of 1 to 10 nm are mechanically more rough and show high quantum yields. Also, SD has one common property to QD, i.e. the emitted fluorescence is red shift with increasing size particles (Canham, 1999). These properties confer SD a quite interesting role for optoelectronic devices. The research and modulation of SD is more advanced than

CD and the application of computational theoretical methods tom SD is a quit interesting research field (Pudzer et al., 2003; Belomoin et al., 2002; Trabi & Barone, 2011). Indeed, SD have been studied by density functional theory (DFT) and quantum Monte Carlo (QMC) methods to understand and simulate their structural and optical properties. This theoretical approach together with experimental confirmation will allow further developments in the understanding of the fluorescent properties tuning.

This chapter will focus on the description of the fluorescent nanoparticles carbon dots (CD) and silicon dots (SD) with a brief description of the synthesis methodologies, the current state of the art about the comprehension of the fluorescence mechanisms, their multiphoton excitation properties and its relevance to new bioimaging methodologies, and their most recent scientific and technological applications.

SYNTHESIS OF CARBON AND SILICON NANOMATERIALS

The synthesis of stable carbon and silicon based high fluorescent nanoparticles usually requires a two step methodology: the synthesis of the raw nanoparticle; and, reaction of the surface atoms with other substances to passivation and/or functionalization. Raw nanomaterials can usually be obtained by top-down or bottom-up approaches. Topdown approaches usually include physical techniques, like for example laser ablation, or chemical reactions, like for example chemical catalyzed etching. Bottom-up approaches are usually chemical synthesis methods that use chemical precursors to synthesise the nanoparticles. Figures 1 and 2 show the main synthetic routes of CD and SD, respectively. (Baker & Baker, 2010; Esteves da Silva & Gonçalves, 2011; Fan & Chu, 2010; Gonçalves and Esteves da Silva, 2011; Kang et al., 2011)

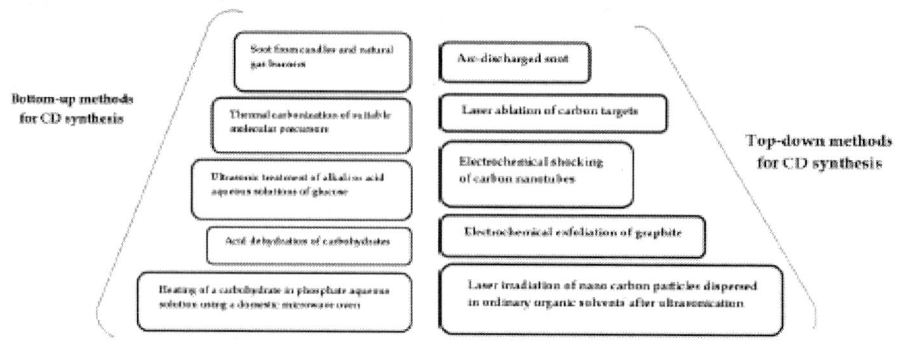

Figure 1. Synthetic methodologies for CD.

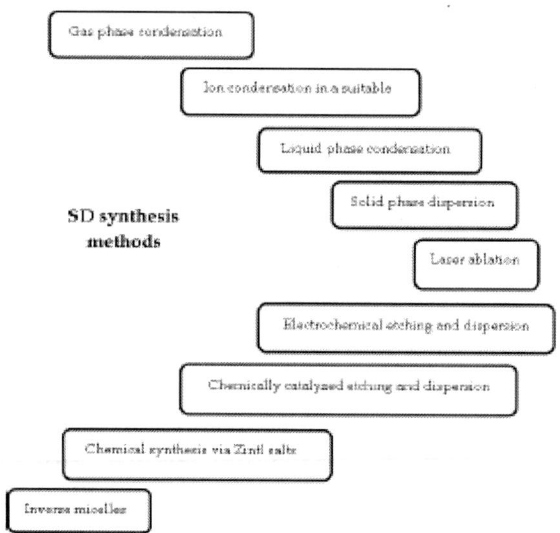

Figure 2 Synthetic methodologies for SD.

FLUORESCENCE MECHANISM OF SILICON AND CARBON NANOMATERIALS

QD are usually constituted by elements from groups 12 to 16, for example III-V materials (GaN, InP, etc.), II-VI materials (ZnO, CdS,

CdTe, etc.) and IV-VI materials (PbS, PbTe, etc.), (Smith & Nie, 2010). Two of the most spectacular properties of QD, which is not observed in the bulk material that is not fluorescent, are their quite strong luminescence and the dependence of their fluorescence properties with the size of the nanocrystal. The dependence of the luminescence properties of QDs with their size is well established. QD show an electronic energy states distribution between that of a discrete molecule and of the bulk semiconductor, characterized by a bandgap energy (Eg) corresponding to the energy difference between the valence and conduction band (Fig. 3). The absorption of energy higher than Eg by an electron in the valence band provokes its excitation into the conduction band originating a negative charge and a positive hole in the valence band (the pair negative charge and positive hole is called the exciton). In a nanoparticle the annihilation of the exciton corresponds to the emission of fluorescence. However, the exciton has a finite nanometric size defined by the Bohr exciton diameter that, for relatively small nanocrystals, may be greater than the size of the nanoparticle (Esteves da Silva and Gonçalves, 2011). When this condition is observed the nanoparticle is under a quantum confinement regime and their electronic properties are size dependent - the nanocrystal growth provokes a redshift in the emission of the fluorescence (Fig. 3) (Smith & Nie, 2010).

Another important characteristic of nanoparticles is that, due to its small size, a relatively high fraction of the atoms that constitutes the nanoparticle are localized at the surface. The atoms at the surface shows reduced coordination number and there is corrosion and imperfections which may affect the electronic properties of a nanoparticle - the smaller the nanocrystal the more important become the surface defects (Esteves da Silva and Gonçalves, 2011). Consequently, the electronic properties of QD may have contribution of surface defects besides quantum confinement effects.

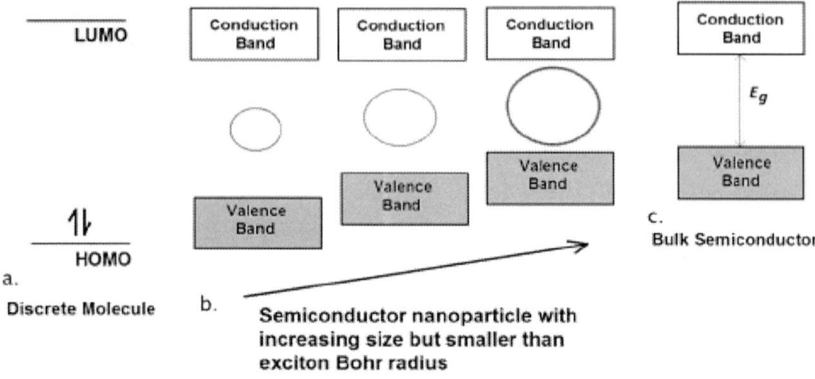

Figure 3. Schematic representation of the electronic energy states in a semiconductor where quantum confinement is observed in the nanoparticles.

Silicon is a semiconductor with a bandgap energy of 1.14 eV (Shirahata et al., 2010; Wilcoxon et al., 1999). However, Silicon has no significant optical performance due to its indirect bandgap character (Fig. 4). The exciton Bohr radius of silicon is 4 nm and, under strong quantum confinement regime, the probability of radiative recombination through the direct bandgap transitions is increased and through phonon-assisted indirect bandgap transitions is reduced (Fig. 4) (Warner et al., 2005). Nanoparticles of silicon are highly luminescent and show quantum confinement size dependent properties resulting in the widening of the bandgap from 1.14 eV of the bulk to about 3.26 eV (380 nm) (Holmes et al., 2001; Shirahata et al., 2010).

Quantum confinement size dependent properties of SD have been observed for Si nanoparticles of size between 3 and 8 nm (Ledoux et al., 2002). Indeed, the energy of the emitted photons shifts to higher values when the size of the nanoparticles is reduced. However, for smaller SD, the surface composition becomes more important than the quantum confinement effect and the tuning of the emitted light becomes more difficult (Kang et al., 2011). SD of size between 1 and 4 nm showed quantum confinement effects, namely the maximum emission wavelength of 1 nm nanoparticles particles was 450 nm, 2 nm 520 nm, 3 nm 640 nm and 4 nm 740 nm (Kang et al., 2007; Kang et al., 2011). The emission of these SD could be tuned by chemically modification of the surface of the nanoparticles showing controllable surface effects.

The fluorescence of SD is still not complete understood but both quantum confinement effects and/or surface defects are involved in the fluorescence emission mechanism.

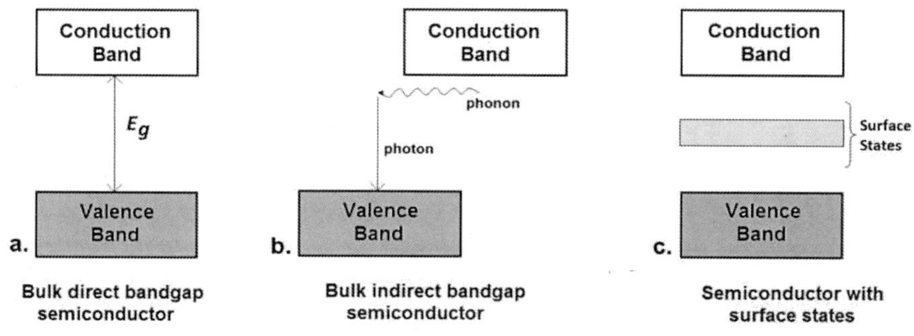

Figure 4. Schematic representation of the electronic energy states in a direct (a.), indirect (b.) and with surface states semiconductor.

Similarly to QD and SD, CD are highly fluorescent nanoparticles. However, the fluorescence properties of CD do not show size dependency suggesting different fluorescence mechanism. Nevertheless, CD also shows the particular property of multicolour emission when different excitation wavelengths are used (Li et al., 2010; Liu et al., 2007; Liu et al., 2009; Mao et al., 2010; Peng & Travas-Sejdic, 2009; Zhu et al., 2009). Indeed, carbon is an insulator and quantum confinement effects are not expected for CD. Although the fluorescent mechanism of CD is still not clearly understood the fluorescence properties of these nanomaterials should be due to surface defects that originates energy levels that justifies the light emission (Fig. 4.c). These electronic levels distribution allow radiative recombination of excitons (Peng & Travas-Sejdic, 2009; Zhu et al., 2009). Also, and supporting the main effect of surface defects on the fluorescence properties, CD become fluorescent and the quantum yield increases when the nanoparticles are subjected polymer passivation and nitric acid treatment (Li et al., 2010; Liu et al., 2007; Liu et al., 2009; Mao et al., 2010; Peng & Travas-Sejdic, 2009;

Zhu et al., 2009). Different starting materials and different fractions of similar size carbon nanoparticles also show different fluorescence emission properties.

MULTIPHOTON EXCITATION PHENOMENON

When two low-energy photons arrive 'simultaneously' at a fluorophore and, through interaction with it, provokes the excitation of an electron that normally is excited by one higher energy photon, a two-photon excitation phenomenon (TP) is observed (Fig. 5) (Kim & Cho, 2009; Xu et al., 1996; Williams et al., 2001). TP based fluorescence microscopy using lower energy radiations as excitation sources (red or NIR photons) is becoming a quite popular technique for in vivo analysis (bioimaging) because of the following: (i) deeper penetration depth (>500 mm); (ii) lower tissue auto-fluorescence and self-absorption; (iii) reduced photodamage and photobleaching; and, (iv) localized excitation. One characteristic of the TP probes is the excitation cross sections (δTP) usually expressed in GM (Göppert-Mayer) units (1 GM = 10-50 cm4 s/photon). Larger δTP the better because the TP probe is more fluorescent and higher sensitivities are achieved. Apparently, the progress in this area is limited for the lack of appropriate probes (Kim & Cho, 2009).

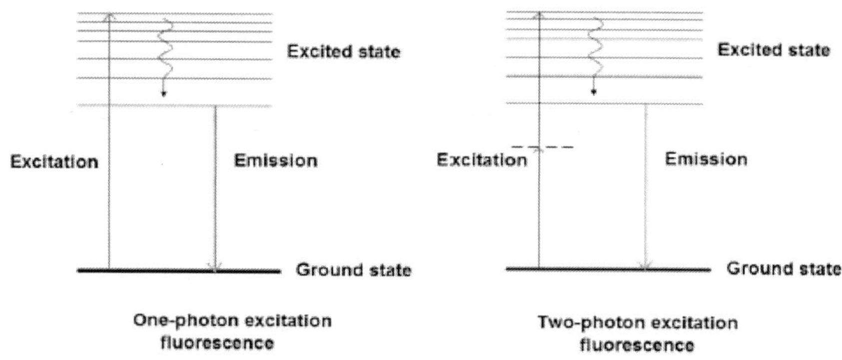

Figure 5 Schematic representation of the electronic energy states of a fluorescent substance when one and two-photon excitation is observed.

Carbon and silicon nanoparticles can be excited by single (UV or near UV) or TP (red or NIR). TP and the low cytotoxicity and biocompatibility of carbon and silicon nanomaterials make them excellent sensors for bioimaging applications with a simple and straightforward application methodology (Fig. 6). CD and SD show luminescence with TP excitation in the NIR (800 nm fentosecond pulsed Ti:sapphire laser) (Cao et al., 2007) or long-wavelength light (from 500 to 1000 nm) (Akcakir et al., 2000; Li et al., 2010b). The maximum of the emission of fluorescence is in the 400 to 500 nm wavelength range as usually observed when one-photon excitation is used. The TP excitation is confirmed because a quadratic relationship between excitation laser power and the measured luminescence intensity is observed (Cao et al., 2007; Li et al., 2010b).

APPLICATIONS OF CARBON AND SILICON NANOMATERIALS

In this section some selected applications of fluorescent CD and SD are described. The selection criteria were their recent publication and potential for the future development of scientific and technological real applications. Indeed, taking into consideration the unlimited resources of both carbon and silicon in the earth crust, and their no toxicity and biocompatibility potential, major nanotechnological advances and future applications will have a strong contribution from CD and SD. These fluorescent nanoparticles will have an active role in the near future in bioimaging in biomedicine and bioanalytical applications, as nanocatalysts in technology and biochemistry, in sustainable energy devices and advanced analytical chemistry applications in general.

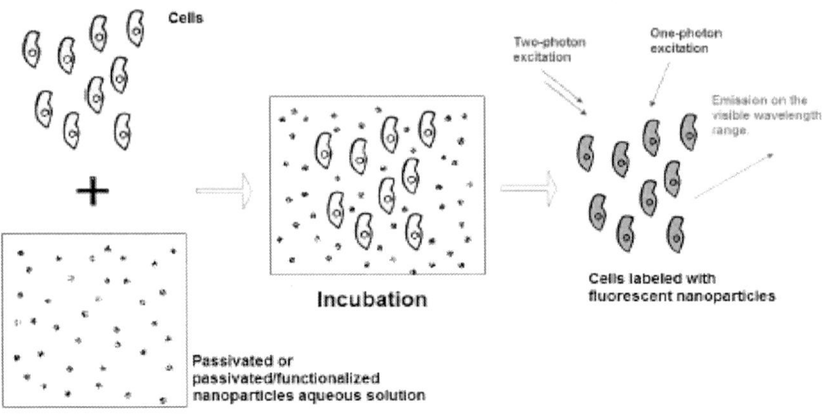

Figure 6 Steps necessary to label cells with fluorescent nanoparticles before imaging with (either one-photon or two-photon excitation).

Carbon and Silicon Dots in Bioimaging

Bioimaging sensors for in vivo diagnostics must be non-toxic and biocompatible. QDs based bioimaging methodologies have appeared in the last years together with toxicity concerns because they are constituted by intrinsically toxic elements like cadmium (Hardman, 2006; Lovric et al., 2005). CD and SD are constituted by intrinsically non-toxic elements which make them a particularly useful and promising bioanalytical tools as their performance approaches that of QDs. CD and SD have been demonstrated to have a very low cytotoxicity and were shown to internalize cells, probably by endocytosis mechanism, which, taking into consideration the MPE properties discussed above, makes these nanoparticles suitable for bioimaging purposes (Akcakir, 2000; Erogbogbo, 2011; Li et al., 2010; Li & Ruckenstein, 2004; Manhat, 2011; Ray et al., 2009; Sun et al., 2006; Veinot, 2006; Yang et al., 2009a; Yang et al., 2009b).

The development of bioimaging agents that are selectively uptake by cancer cells is a particularly active research field. Usually CD or SD are functionalized with an aminoacid (lysine or glutaric acid), folic acid, antimesothelin or transferring to be selectively uptake by cancer cells (Erogbogbo, 2011; Li et al., 2010a; Manhat, 2011).

Carbon Dots as Peroxidase Mimetics

Surprisingly CDs have shown to possess enzymatic properties because they have a peroxidase like activity (Shi et al., 2011). Indeed, CDs catalyses the oxidation of peroxidase substrates by hydrogen peroxide. For example, 3,3′,5,5′-tetramethylbenzidine (TMB) is oxidised by hydrogen peroxide in the presence of CDs and detected by the production of a blue coloured compound (oxidised TMB). This new function of CDs opens new perspectives in their bioanalytical potential as a nanosensor for hydrogen peroxide and glucose (Shi et al., 2011).

The comparison of CDs with horseradish peroxidase (HRP) showed that CDs required a hydrogen peroxide concentration about two orders of magnitude higher than HRP to reach a maximum level of peroxidase activity (Shi et al., 2011). This result suggested that CDs are more stable than HRP at high hydrogen peroxide concentration. Also, CDs are much more stable than HRP.

Because the colour development (absorbance at 652 nm) was found proportional to the hydrogen peroxide concentration when TMB was used as substrate a quantitative method was obtained with a limit of detection of 0.2 mM (linear plot in the hydrogen peroxide concentration between 0.0010-0.10 mM) (Shi et al., 2011). The coupling of CDs with other enzymes can generate new bioanalytical methodologies, like for example for the quantification of glucose. Indeed, by combining CDs and TMB with glucose oxidase a straightforward colorimetric method for glucose is obtained with a limit of detection of 0.4 mM (linear plot in the glucose concentration between 0.0010-0.50 mM) (Shi et al., 2011). This method is sensible enough for serum glucose quantification because the concentration in healthy and diabetic individuals range from 3 to 8 mM and 9 to 40 mM. 5.3 Silicon dots as nanocatalyst and photodynamic therapy SD shows great potential as nanocatalyst with strong and tunable chemical activity, specificity and selectivity (Kang et al., 2011). These properties of SD are attributed to their tunable band gap energy and photoconductivity properties.

Due to the larger exciton (electron/hole pair) energy of SD with 1 to 2 nm size, these nanoparticles induces the photochemical reduction of carbon dioxide to carbonate anion and degradation of methyl red. Also, SD with 3 to 4 nm size photocatalyses the hydroxylation of aromatic hydrocarbons - benzene is transformed to phenol with a 100% yield with high selectivity. (Kang et al., 2011)

These catalyst properties of SD are probably due to their efficient photosensitizers of singlet oxygen (Timoshenko et al., 2006). This property has been used to suppress the division of cancer cells as consequence of the oxidation of cell material by singlet oxygen (Steller, 1995).

Also, the formation of superoxide ions was observed when SD are present (Fujii et al., 2005). The photochemical synthesis of reactive oxygen species appears promising for use in photodynamic therapy of cancer.

Metallic Plasmons and Carbon dots Luminescence Interaction

An important recent scientific result was the successful coupling of plasmonic metal nanoparticles with luminescent CD (Zhang et al., 2011; Li et al., 2011). Indeed, the coupling of CD with plasmon metals enhances the CD brightness (metal enhanced fluorescence - MEF) and photostability resulting into better detectability in bioanalytical applications. Two different coupling strategies have been successfully tested that show MEF: CDs immobilized in glass substrates containing silver islands films (SiFs) (Zhang et al., 2011); and, deposition of CDs on the surface of silver nanoparticles mediated with a silica layer (Li et al., 2011).

MEF results from the interaction, within the wavelength of light (near-field conditions), of the luminescent material with the metallic surface plasmons. This interaction provokes an enhancement of the fluorescence of the nanoparticle and reduces the corresponding excited decay times leading to enhanced photostability (Zhang et al., 2011). When the fluorescent nanoparticle is more than one wavelength of light way (far-field conditions) the corresponding quantum yield (Qo) is given by (Zhang et al., 2011):

$$Q_o = \Gamma / (\Gamma + K_{nr})$$

where Γ isthefluorophores' radiativedecayrateandKnrarethenonradiative decay rates for excited state relaxation. Under near-field conditions the quantum yield (Qm) is given by:

$$Q_m = (\Gamma + \Gamma_m)/ (\Gamma + \Gamma_m + K_{nr})$$

Where Γ m is the system modified radiative rate. The corresponding far (τ) and near-field (τ_m) lifetimes are given by (Zhang et al., 2011):

$$\tau = 1 / (\Gamma + K_{nr})$$
$$\tau_m = 1 / (\Gamma + \Gamma_m + K_{nr})$$

These equations show that when Γ m increases, the quantum yield increases (MEF) and the lifetime is reduced (enhanced photostability).

Silicon dots Coupled with multi-polar Plasmonic hot Spots

The metal enhanced fluorescence (MEF) has been investigated for SD (Nychyporuk et al., 2011). Indeed, as discussed above for CD, the photo-stimulated emission of fluorescence of fluorophores suffers a significant enhancement when they are in the vicinity of metal nanoparticles. Stronger fluorescence enhancement is achieved when SD are localized inregions where the photo-induced electric fields from several metal nanoparticles are superimposed (hot spots) (Nychyporuk et al., 2011).

SD were embedded in a silicon nitride matrix (SiNx) and deposited on a quartz substrate using a chemical vapour deposition technique followed by the fabrication of a monolayer of silver nanoislands on the surface of the SiNx dielectric film and covered by a thin silicon nitride film (SiNy) containing SD dispersed inside it (Nychyporuk et al., 2011). This procedure resulted in the preparation of 208 nm films (SiNx - 170 nm; SiNy - 38 nm) containing SD with mean diameters between 2 and 4 nm, silver nanoclusters with a mean diameter of 18±3 nm and two adjacent silver islands are separated by about 20 nm distance. Strong MEF was observed in the mean distance between two adjacent silver islands because these places correspond to the photoexcited plasmon hot spots (Nychyporuk et al., 2011). The local plasmon resonance of the silver islands could be tuned in the visible spectral range (400 to 600 nm) by adjusting the SiNy dielectric constant allowing an optimization of the MEF for the SD.

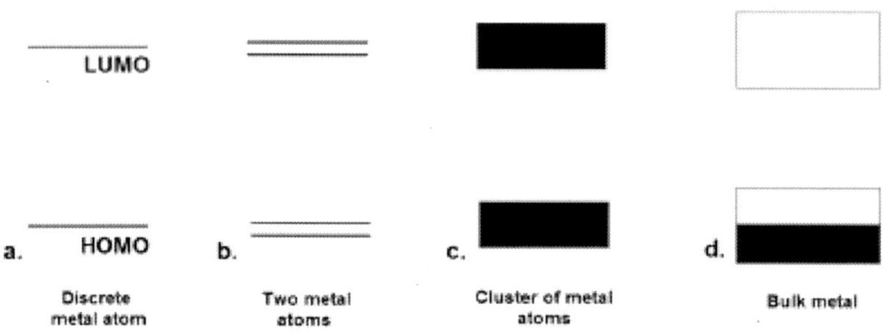

Figure 7 Schematic representation of the electronic energy level diagram of discrete metal atoms (a.), two metal atoms (b.), clusters of metal atoms (c.) and bulk metallic element (d.).

Carbon dots Immobilized in Fiber Optics

The use of optical sensors in chemical and biological analytical measurements is an expanding area of research with growing importance, especially in biomedical and environmental applications. Indeed, optical sensors are intrinsically immune to electromagnetic interferences and, if properly designed, can show an increased sensitivity, fast response times and suitability to remote monitoring when used with optical fibers (Jorge et al., 2005). The combination of optical fiber technologies with fluorescence spectroscopy has greatly contributed to the progress of optical chemical sensors (Gonçalves et al., 2010; Jorge et al., 2005; Maule et al., 2010). However, the design of sensing heads based on fiber optics requires high quantum yield fluorescent sensors and fluorescent nanomaterials are strong candidates to be coupled to fiber optics (Gonçalves et al., 2010; Jorge et al., 2005; Maule et al., 2010).

CD synthesised by laser ablation and and functionalized with N-acetyl-l-cysteine were successful immobilized (thin of about 750 nm) in the tip of a fiber optic using a sol-gel technology (Gonçalves et al., 2010). Due to the CD functionalization the sensing head selectively and reversibly responds to mercury(II) ions by a quenching mechanism. The response of the nanosensor is fast (less than10s) and stable. Also, submicronmolar concentrations of mercury(II) could be detected and quantified - Stern-Volmer constant (pH=6.8) of 5.3×105 M-1.

Carbon dots for Photocatalyst Design

Photoexcited CD have been shown to have redox active properties namely as electron-donor capabilities in reduction reactions (Wang et al., 2009). This property was observed when silver ion was reduced to silver metal upon irradiating with a Xe arc lamp (450 and 600 nm) an aqueous solution of CD silver ions (Wang et al., 2009).

The upconverted photoluminescent and redox properties of CD opened interesting perspectives in its application as energy-transfer agents in photocatalyst design (Li et al., 2010). CD were dispersed on the surface of TiO_2 and SiO_2 and their potential in the photocatalytic degradation of methyl blue was assessed using a 50 mg/L solution – after 25 minutes the reduction of methyl blue was almost completed (Li et al., 2010). When undoped TiO_2 and SiO_2 or CD alone were used the reduction of methyl blue was nearly inexistent.

The observed difference in the reduction yield is due to the relatively large intrinsic band gap of TiO_2 which has as consequence of only less than 5% of the sunlight is used. When TiO_2 is doped with CD the full spectrum of sunlight can be used because the nanoparticles absorb visible light and emit at shorter wavelength (325 to 425 nm) which can be used by TiO_2 and SiO_2 originating the exciton. This excitation species is responsible for the formation of reactive oxygen species that provokes the degradation of the dyes (Li et al., 2010). Moreover, the presence of CD attached to the oxides surfaces permits the transfer of electrons allowing charge separation, stabilization and hindered recombination with an overall higher activity of the photocatalyst (Li et al., 2010).

Carbon dots for White LED

White light-emitting devices (WLED) are currently an alternative as low energy consumption light sources and liquid-crystal displays among many other technological applications with significant economical advantages (Jang et al., 2010; Wang et al., 2011). CD have been successfully used in the fabrication of an electroluminescent (EL) device that emits a spectrum that approximates natural sun light (Wang et al., 2011).

WLED based CD were assembled using a tri-layer nanocrystal EL device (Wang et al., 2011): a 40 nm thick poly(3,4-ethylenedioxyt hiophene):poly(stryrenesulfonate) (PEDOT:PSS) layer on the anode; CD with 5 nm diameter with a quantum yield of more than 60%, dispersed in toluene, were spun-casted onto the PEDOT:PSS forming a 20 nm thick film; a 40 nm thick 1,3,4-tris(N-phenylbenzimidazol-2-yl) benzene (TPBI) was used as the electron transporting layer; finally, a 1 nm thick LiF and 120 nm thick Al electrodes were deposited by thermal evaporation. The EL spectrum of this device covers the entire visible zone with a turn-on voltage of about 6 V and a maximum brightness output of 35 cd m 2.

Silicon dots for Photovoltaic Applications

The use of solar cells for the sustainable production of electric energy is a well known strategy that is characterized by its relatively high cost due to the cost of the cell material and relatively low efficiency. SD may contribute significantly for efficiency improving and cost reducing of solar cells (Luo et al., 2011).

One of the most important properties of SD useful for solar cells is their tunability because different bandgap layers, with chemical and thermal compatibility (the same material), can be assembled in multi-junction solar cells. Indeed, a large number of different bandgap layers that match the solar spectrum and raising the efficiency can be assembled using the same growth process because the same SD are being used. Another property of SD, that is common to QDs, is the multi-exciton generation (MEG) and consists in the generation of multiple electron-hole pairs from the absorption of a single photon. MEG may increases considerably the power conversion of solar cells. Also, the possibility of MPE of SD constitutes another factor for increasing the efficiency of solar cells.

PERSPECTIVES

The progresses that are being observed in the synthesis of high quantum yield fluorescent carbon and silicon nanoparticles are opening new technological applications for these intrinsically no toxic and highly abundant materials. Indeed, in the very near future CD and SD will have an important role in energy conversion systems, new illumination apparatus and catalytic processes. In the human health sector, namely in medical diagnosis and imaging, CD and SD are becoming an important area of scientific and technological development. Indeed, these nanomaterials with a suitable functionalization can be used in bioimaging methodologies replacing the already established protocols including classical organic dyes and the potentially toxic cadmium based QD. Moreover, CD and SD show multiphoton excitation properties which markedly increase their bioimaging potential.

Consequently, fluorescent carbon and silicon nanomaterials will play an important role in a sustainable development of the earth.

ACKNOWLEDGMENTS

Financial support from Fundação para a Ciência e a Tecnologia (FCT, Lisbon) (Programa Operacional Temático Factores de Competitividade (COMPETE) e comparticipado pelo Fundo Comunitário Europeu FEDER) (Project PTDC/QUI/71001/2006) is acknowledged.

REFERENCES

1. Akcakir, O.; Therrien, J.; Belomoin, G.; Barry, N.; Muller, J.D.; Gratton, E. & Nayfeh, M. (2000) Detection of luminescent single ultrasmall silicon nanoparticles using fluctuation correlation spectroscopy. Applied Physics Letters, 76, 1857-1859, ISSN: 1077-3118.

2. Baker, S.N. & Baker, G.A. (2010) Luminescent Carbon Nanodots: Emergent Nanolights. Angewandte Chemie International Edition, 49, 6726 - 6744, ISSN: 1521-3773.

3. Belomoin, G.; Rogozhina, E.; Therrien, J.; Braun, P.V.; Abuhassan, L. & Nayfeh, M.H. (2002) Effects of surface termination on the band gap of ultrabright Si29 nanoparticles: Experiments and computational models, Physical Review B, 65, 193406, ISSN: 1550- 235X.

4. Canham, L.T. (1990) Silicon quantum wire array fabrication by electrochemical and chemical dissolution of wafers, Applied Physics Letters, 57, 1046-1048, ISSN: 1077-3118.

5. Canham, L.T. (1999) Lewis Acid Mediated Hydrosilylation on Porous Silicon Surfaces. Journal of the American Chemical Society, 121, 11491-11492, ISSN: 1520-5126.

6. Cao, L.; Wang, X.; Meziani, M.J.; Lu, F.; Wang, H.; Luo, P.G.; Lin, Y.; Harruff, B.A.; Veca, L.M.; Murray, D.; Xie, S.Y. & Sun, Y.P. (2007) Carbon dots for multiphoton bioimaging. Journal of the American Chemical Society, 129, 11318-11319, ISSN: 1520- 5126.

7. Erogbogbo, F.; Tien, C.-A.; Chang, C.-W.; Yong, K.-T.; Law, W.-C.; Ding, H.; Roy, I.; Swihart, M.T. & Prasad, P.N. (2011) Bioconjugation of luminescent silicon quantum dots for selective uptake by cancer cells. Bioconjugate Chemistry, 22, 1081-1088, ISSN: 1043- 1802.

8. Esteves da Silva, J.C.G. & Gonçalves, H. (2011) Analytical and bioanalytical applications of carbon dots. Trends in Analytical Chemistry, doi:10.1016/j.trac.2011.04.009, ISSN: 0167-2940.

9. Fan, J & Chu, P.K. (2010) Group IV Nanoparticles: Synthesis, Properties, and Biological Applications. Small, 6, 2080–2098, ISSN: 1613-6829.

10. Fujii, M.; Kovalev, D.; Goller, B.; Minobe, S.; Hayashi, S. & Timoshenko, V.Y. (2005) Timeresolved photoluminescence studies of the energy transfer from excitons confined in Si nanocrystals to oxygen molecules. Physical Review B, 72, 165321-165329, ISSN: 1550-235X.

11. Gerion, D.; Pinaud, F.; Williams, S.C.; Parak, W.J.; Zanchet, D.; Weiss, S. & Alivisatos, A.P. (2001) Synthesis and Properties of Biocompatible Water-Soluble Silica-Coated CdSe/ZnS Semiconductor Quantum Dots. Journal of Physical Chemistry B, 105, 8861- 8871, ISSN: 1520-5207.

12. Gonçalves, H. & Esteves da Silva, J.C.G. (2011) A New Insight on Silicon dots. Current Analytical Chemistry, In revision, ISSN: 1573-4110.

13. Gonçalves, H.; Duarte, A. & Esteves da Silva, J.C.G. (2010) Carbon dots based optical fiber sensor for Hg(II). Biosensors and Bioelectronics, 26, 1302-1306, ISSN: 0956-5663.

14. Hardman, R. (2006) A Toxicological Review of Quantum Dots: Toxicity Depends on Physicochemical and Environmental Factors. Environmental Heath Perspectives, 114, 165-172, ISSN: 0091-6765.

15. Holmes, J.D.; Ziegler, K.J.; Doty, R.C.; Pell, L.E.; Johnston K.P. & Korgel, B.A. (2001) Highly Luminescent Silicon Nanocrystals with Discrete Optical

Transitions. Journal of the American Chemical Society, 123, 3743-3748, ISSN: 1520-5126.

16. Jang, E.; Jun, S.; Jang, H.; Lim, J.; Kim, B. & Kim, Y. (2010) White-Light-Emitting Diodes with Quantum Dot Color Converters for Display Backlights. Advanced Materials, 22, 3076–3080, ISSN: 1521-4095.

17. Jorge, P.A.S.; Caldas, P.; Esteves da Silva, J.C.G.; Rosa, C.C.; Oliva, A.G.; Farahi, F. & Santos, J.L. (2005) Luminescence-based optical fiber chemical sensors. Fiber and Integrated Optics, 24, 201-225, ISSN: 1096-4681.

18. Kang, Z.; Liu, Y. & Lee, S.T. (2011) Small-sized silicon nanoparticles: new nanolights and nanocatalysts. Nanoscale, 3, 777-791, ISSN: 0306-0012.

19. Kang, Z.; Tsang,C.; Wong, N.; Zhang, Z. & Lee, S. (2007) Silicon Quantum Dots: A General Photocatalyst for Reduction, Decomposition, and Selective Oxidation Reactions. Journal of the American Chemical Society, 129, 12090-12091, ISSN: 1520-5126.

20. Kim, H.M. & Cho, A.B.R. (2009). Two-photon probes for intracellular free metal ions, acidic vesicles, and lipid rafts in live tissues. Accounts of Chemical Research, 42, 863-872, ISSN: 1520-4898.

21. Ledoux, G.; Gong, J.; Huisken, F.; Guillois, O. & Reynaud, C. (2002) Photoluminescence of size-separated silicon nanocrystals: Confirmation of quantum confinement. Applied Physics Letters, 80, 4834-4836, ISSN: 1077-3118. Li, Z.F. & Ruckenstein, E. (2004) Water-Soluble Poly(acrylic acid) Grafted Luminescent Silicon Nanoparticles and Their Use as Fluorescent Biological Staining Labels. Nano Letters, 4, 1463-1467, ISSN: 1530-6984.

22. Li, Q.; Ohulchanskyy, T.Y.; Liu, R.; Koynov, K.; Wu, D.; Best, A.; Kumar, R.; Bonoiu, A. & Prasad, P.N. (2010a) Photoluminescent carbon dots as biocompatible nanoprobes for targeting cancer cells in vitro. Journal of the Physical Chemistry C, 114, 12062- 12068, ISSN: 1932-7447.

23. Li, H.; He, X.; Kang, Z.; Huang, H.; Liu, Y.; Liu, J.; Lian, S.; Tsang, C.H.A.; Xiaobao, X. & Lee, S.T. (2010b) Water-soluble fluorescent carbon quantum dots and photocatalyst design. Angewandte Chemie International Edition, 49, 4430-4434, ISSN: 1521-3773.

24. Li, J.; Zhang, B.; Wang, F. & Liu, C. (2011) Silver/carbon-quantum-dot plasmonic luminescence nanoparticles. New Journal of Chemistry, 35, 554-557, ISSN: 1369-9261.

25. Liu, H.; Ye, T. & Mao, C. (2007) Fluorescent Carbon Nanoparticles Derived from Candle Soot. Angewandte Chemie International Edition, 46, 6473-6475, ISSN: 1521-3773.

26. Liu, R.; Wu, D.; Liu, S.; Koynov, K.; Knoll, W. & Li, Q. (2009) An Aqueous Route to Multicolor Photoluminescent Carbon Dots Using Silica Spheres as Carriers. Angewandte Chemie International Edition, 48, 4598-4601, ISSN: 1521-3773.

27. Liu, Z.; Zhou X. & Qian, Y. (2010) Synthetic Methodologies for Carbon Nanomaterials. Advanced Materials, 22, 1963-1966, ISSN: 1521-4095.

28. Lovric, J.; Bazzi, H.S.; Cuie, Y.; Fortin, G.R.A.; Winnik, F.M. & Maysinger, D.

(2005) Differences in subcellular distribution and toxicity of green and red emitting CdTe quantum dots. Journal of Molecular Medicine, 83, 377-385, ISSN: 1432-1440.

29. Luo, Y.; X. Xia, X.; Liang, Y.; Zhang, Y.; Ren, Q.; Li, J.; Jia, Z. & Tang, Y. (2007) Highly visible-light luminescence properties of the carboxyl-functionalized short and ultrashort MWNTs. Journal of Solid State Chemistry, 180, 1928-1933, ISSN: 0022-4596.

30. Luo, J.-W.; Stradins, P. & Zunger, A. (2011) Matrix-embedded silicon dots for photovoltaic applications: a theorectical study of critical factors. Energy & Environmental Science, doi:10.1039/CIEE01026C, ISSN: 1754-5706.

31. Manhat, B.A.; Brown, A.L.; Black, L.A.; Ross, J.B.A.; Fichter, K.; Vu, T.; Richman, E. & Goforth, A. (2011) One-step melt synthesis of water-soluble, photoluminescent, surface-oxidized nanoparticles for cellular imaging applications. Chemistry of Materials, 23, 2407-2418, ISSN: 1520-5002.

32. Mao, X.J.; Zheng, H.Z.; Long, Y.J.; Du, J.; Hao, J.Y.; Wang, L.L. & Zhou, D.B. (2010) Study on the fluorescence characteristics of carbon dots. Spectrochimica Acta Part A, 75, 553– 557, ISSN: 1386-1425.

33. Maule, C.D.; Gonçalves, H.M.R.; Mendonça, C.; Sampaio, P.; Jorge, P.A.S. & Esteves da Silva, J.C.G. (2010) Wavelength encoded analytical imaging and fiber optic sensing with pH sensitive CdTe quantum dots. Talanta, 80 1932–1938, ISSN: 0039-9140.

34. Murcia, M.J.; Shaw, D.L.; Long, E.C. & Naumann, C.A. (2008) Fluorescence correlation spectroscopy of CdSe/ZnS quantum dot optical bioimaging probes with ultra-thin biocompatible coatings. Optics Communications, 28, 1771-1780, ISSN: 0030-4018.

35. Nychyporuk, T.; Zakharko, Y.; Serdiuk, T.; Marty, O.; Lemiti, M. & Lysenko, V. (2011) Strong photoluminescence enhancement of silicon quantum dots by their rearresonant coupling with multi-polar hot spots. Nanoscale, 3, 2472-2475, ISSN: 0306- 0012.

36. Parkhutik, V. & Timashev, S. (2000) Kinetics of porous silicon growth studied using flickernoise spectroscopy. Journal of Applied Physics, 87, 7558-7560, ISSN: 1089-7550.

37. Peng, H. & Travas-Sejdic, J. (2009) Simple Aqueous Solution Route to Luminescent Carbogenic Dots from Carbohydrates. Chemistry of Materials, 21, 5563–5565, ISSN: 1520-5002.

38. Pudzer, A.; Williamson, A.J.; Grossman, J.C. & Galli, G. (2003) Computational studies of the optical emission of silicon nanocrystals. Journal of the American Chemical Society, 125, 2786-2791, ISSN: 1520-5126.

39. Ray, S.C.; Saha, A.; Jana, N.R. & Sarkar, R. (2009) Fluorescent carbon nanoparticles: synthesis, characterization, and bioimaging application. Journal of the Physical Chemistry C, 113, 18546-18551, ISSN: 1932-7447.

40. Riggs, J.E.; Guo, Z.; Carroll, D.L. & Sun, Y.P. (2000) Strong Luminescence of Solubilized Carbon Nanotubes. Journal of the American Chemical Society, 122, 5879-5880, ISSN: 1520-5126.

41. Shi, W.; Wang, Q.; Long, Y.; Cheng, Z.; Chen, S.; Zheng, H. & Huang, Y. (2011) Carbon nanodots as peroxidase mimetics and their applications to glucose detection. Chemical Communications, 47, 6695-6697, ISSN: 1364-548X.

42. Shirahata, N.; Hasegawa, T.; Sakka, Y. & Tsuruoka, T. (2010) Size-Tunable UV-Luminescent

43. Silicon Nanocrystals. Small, 6, 915-921, ISSN: 1613-6829.

44. Smith, A.M. & Nie, S. (2010) Semiconductor nanocrystals: structure, properties, and band gap engineering. Accounts of Chemical Research, 43, 190-200, ISSN: 1520-4898.

45. Steller, H. (1995) Mechanisms and genes of cellular suicide. Science, 267, 1445-1449, ISSN: 0036-8075.

46. Sun, Y.P.; Zhou, B.; Lin, Y.; Wang, W.; Fernando, K.A.S.; Pathak, P.; Meziani, M.J.; Harruff, B.A.; Wang, X.; Wang, H.F.; Luo, P.J.G.; Yang, H.; Kose, M.E.; Chen, B.L.; Veca, L.M. & Xie, S.Y. (2006) Quantum-sized carbon dots for bright and colourful photoluminescence. Journal of the American Chemical Society, 128, 7756-7757, ISSN: 1520-5126.

47. Timoshenko, V.Y.; Kudryavtsev, A. A.; Osminkina, L. A.; Vorontsov, A. S.; Ryabchikov, Y.V.; Belogorokhov, I.A.; Kovalev, D. & Kashkarov, P.K. (2006) Silicon nanocrystals as photosensitizers of active oxygen for biomedical applications. JETP Letters, 83, 423-426, ISSN: 0370-274X.

48. Trani, F. & Barone, V. (2011) Silicon Nanocrystal Functionalization: Analytic Fitting of DFTB Parameters, Journal of Chemical Theory and Computation, DOI: dx.doi.org/ 10.1021/ct1006086, ISSN: 1549-9626.

49. Veinot, J.G.C. (2006) Synthesis, surface functionalization, and properties of freestanding silicon nanocrystals. Chemical Communications, 4160-4168, ISSN: 1364-548X.

50. Wang, X.; Cao, L.; Lu, F.S.; Meziani, M.J.; Li, H.; Qi, G.; Zhou, B.; Harruff, B.A.; Kermarrec, F. & Sun, Y.P. (2009) Photoinduced electron transfers with carbon dots Chemical Communications, 3774-3776, ISSN: 1364-548X.

51. Wang, F.; Chen, Y.H.; Liu, C.Y. & Ma, D.G. (2011) White light-emitting devices based on carbon dots' electroluminescence. Chemical Communications, 3502-3504, ISSN: 1364- 548X.

52. Warner, J.H.; Dunlop, H.R. & Tilley, R.D. (2005) Highly Luminescent Silicon Nanocrystals with Discrete Optical Transitions. Journal of the Physical Chemistry B, 109, 19064- 19067, ISSN: 1520-5207.

53. Wilcoxon, J.P.; Samara, G.A. & Provencio, P.N. (1999) Optical and electronic properties of Si nanoclusters synthesized in inverse micelles. Physical Reviews B, 60, 2704-2714, ISSN: 1550-235X.

54. Williams, Y.; Sukhanova, A.; Nowostawska, M.; Davies, A.M.; Mitchell, S.; Oleinikov, V.; Gun'ko, Y.; Nabiev, I.; Kelleher, D. & Volkov, Y. (2009) Probing cell-type-specific intracellular nanoscale barriers using size-tuned quantum dots. Small, 5, 2581-2588, ISSN: 1613-6829.

55. Williams, R.; Zipfel, W.R & Webb, W.W. (2001). Multiphoton microscopy in biological research. Current Opinion in Chemical Biology, 5, 603-608, ISSN:

1367-5931.

56. Xu, X.Y.; Ray, R.; Gu, Y.L.; Ploehn, H.J.; Gearheart, L.; Raker, K. & Scrivens, W.A. (2004) Electrophoretic Analysis and Purification of Fluorescent Single-Walled Carbon Nanotube Fragments. Journal of the American Chemical Society, 126, 12736-12737, ISSN: 1520-5126.

57. Xu, C.; Zipfel, W.; Shear, J.B.; Williams, R.M. & Webb, W.W. (1996). Multiphoton fluorescence excitation: new spectral windows for biological nonlinear microscopy. Proceedings of the National Academy of Sciences, 93, 10763-10768, ISSN: 1091-6490.

58. Yang, S.T.; Wang, X.; Wang, H.; Lu, F.; Luo, P.G.; Cao, L.; Meziani, M.J.; Liu, J.H.; Liu, Y.; Chen, M.; Huang, Y. & Sun, Y.P. (2009a) Carbon dots as nontoxic and highperformance fluorescence imaging agents. Journal of the Physical Chemistry C, 113, 18110-18114, ISSN: 1932-7447.

59. Yang, S.T.; Cao, L.; Luo, P.G.; Lu, F.S.; Wang, X.; Wang, H.F.; Meziani, M.J.; Liu, Y.F.; Qi, G. & Sun, Y.P. (2009b) Carbon dots for optical imaging in vivo. Journal of the American Chemical Society, 131, 11308-11309, ISSN: 1520-5126.

60. Zhang, Y.; Gonçalves, H.; Esteves da Silva, J.C.G. & Geddes, C.D. (2011) Metal-enhanced photoluminescence from carbon nanodots. Chemical Communications, 47, 5313-5315, ISSN: 1364-548X

61. Zhu, H.; Wang, X.; Li, Y.; Wang, Z.; Yanga, F. & Yang, X. (2009) Microwave synthesis of fluorescent carbon nanoparticles with electrochemiluminescence properties. Chemical Communications, 5118–5120, ISSN: 1364-548X.

Citations

CHAPTER 1

FMR Measurements of Magnetic Nanostructures, Ferromagnetic Resonance - Theory and Applications, Dr. Orhan Yalçın (Ed.), ISBN: 978-953-51-1186-3, InTech, DOI: 10.5772/56615.

CHAPTER 2

FMR Studies of [SnO2/Cu-Zn Ferrite] Multilayers, Ferromagnetic Resonance - Theory and Applications, Dr. Orhan Yalçın (Ed.), ISBN: 978-953-51-1186-3, InTech, DOI: 10.5772/56330.

CHAPTER 3

New Methods and New Types of Functionalised Nanocomposites Intended for the Ecological Depollution of Waters, Nanomaterials, Prof. Mohammed Rahman (Ed.), ISBN: 978-953-307-913-4, InTech, DOI: 10.5772/34078.

CHAPTER 4

FMR Measurements of Magnetic Nanostructures, Ferromagnetic Resonance - Theory and Applications, Dr. Orhan Yalçın (Ed.), ISBN: 978-953-51-1186-3, InTech, DOI: 10.5772/56615.

CHAPTER 5

Instrumentation for Ferromagnetic Resonance Spectrometer, Ferromagnetic Resonance - Theory and Applications, Dr. Orhan Yalçın (Ed.), ISBN: 978-953-51-1186-3, InTech, DOI: 10.5772/56069.

CHAPTER 6

Ferromagnetic Resonance, Ferromagnetic Resonance - Theory and Applications, Dr. Orhan Yalçın (Ed.), ISBN: 978-953-51-1186-3, InTech, DOI: 10.5772/56134.

CHAPTER 7

Microwave Absorption in Nanostructured Spinel Ferrites, Ferromagnetic Resonance - Theory and Applications, Dr. Orhan Yalçın (Ed.), ISBN: 978-953-51-1186-3, InTech, DOI: 10.5772/56058.

CHAPTER 8

Carbon and Silicon Fluorescent Nanomaterials, Nanomaterials, Prof. Mohammed Rahman (Ed.), ISBN: 978-953-307-913-4, InTech, DOI: 10.5772/25809.

INDEX